Also by Michael Selzer from KeepAhead Press

Theomonarchism in the Biblical Text (2014).

Footnotes from an Antiquarian Bookseller's Life (2015).

Renewing the Fear: A Jew Goes to Berlin (2nd ed 2015).

Rochefoucauld's Maxims French text and English translation by M.S. (2018).

The Natural Garden in 18th century England: Literary Texts. Edited and with an introduction by M.S. (2018).

My Florence: Forty-four photographs and introductory essay by M.S. (2019).

Byzantine Aesthetics and the Concept of Symmetry (2021)

KeepAhead Press Architectural Theory Series

The series offers texts that are significant for an understanding of the history of architecture but that have been out of print, or not available in English translation.

Heinrich Wölfflin: *Prolegomenon to a Psychology of Architecture* (1886). German text, and English translation by M.S.

August Thiersch: *Proportion in Architecture* (1883) AND Heinrich Wölfflin: *A Theory of Proportion* (1889). English translations by M.S.

Roland Freart, sieur de Chambray: *A Parallel of the Ancient Architecture with the Modern in a collection of ten principal authors who have written upon the five orders;* WITH, John Evelyn: *An Account of Architects and Architecture in an Historical and Etymological Explanation of certain terms affected by architects;* AND WITH, Leon Battista Alberti: *Treatise of Statues* (translations by John Evelyn).

Symmetry Fallacies

THE CATHEDRAL OF NOTRE DAME, ST. LO, NORMANDY The early-medieval façade of the cathedral was almost completely destroyed during World War II. This 19th. cent. stereoscopic photograph by S. Thompson shows that the structure was asymmetric in virtually every respect. The flanking doors and windows do not mirror each other; and the towers, with their spires, have different dimensions and designs. Even the main portal into the cathedral is off-center.

SYMMETRY FALLACIES

Michael Selzer

"There is no center about which everything turns in the universe." – *Leon Lederman*

Second Edition

KeepAhead Press
Colorado Springs
2021

Earlier versions of this book
were published under the title of
The Symmetry Norm & the Asymmetric Universe

The quotation on the titlepage is from
Lederman 2008, p. 65

ISBN- 9798677549885

10 9 8 7 6 5 4 3 2

This book is dedicated
with abiding affection and gratitude
to the memory of
Lilly and Morris Fleming

CONTENTS

	Table of Illustrations	I
	Addenda & Corrigenda	VI
	Preface	1
	Foreword: The Concept of Symmetry	3
1	The Idea of Symmetry; Asymmetry	7
2	The Asymmetric Universe	17
3	Symmetry in the Classical World	54
	Appendix 1: Is the Parthenon Symmetric?	90
	Appendix 2: Restoring the Parthenon	107
4	The Question of Medieval Symmetry	110
	Appendix 1: Corbusier's Notre Dame	129
	Appendix 2: Ruskin on Gothic Architecture	131
	Illustrations	146
5	The Asymmetry of Primitive Art	259
6	The History of the Concept of Symmetry	279
	Appendix: Alberti's *Collocatio*	331
7	The Natural Garden in England	340
	Appendix: Symmetry and Political Power	363
8	Wittkower and the S. Maria Novella Façade	368
	Works Cited or Consulted	391
	Index	413

I

ILLUSTRATIONS

Cathedral of Notre Dame, St.Lo   *Frontispiece*

0.1   Ancient Greek *hydria*   1
0.2   Nebula Hen-2-437   5
1.1   First Christian Church, Columbus IN   7
1.2   North Christian Church. Columbus, IN   7
1.3   Leonardo da Vinci, head of "Vitruvian Man"   147
1.4   "Vitruvian Man", left side doubled   147
1.5   "Vitruvian Man", right side doubled   147
2.1   Arizona landscape   17
2.2   *Licmophora juegensii*   148
2.3   Microscopic photograph of human tear   148
2.4   Spider web and spider   149
2.5   Cosmic background radiation from Big Bang   149
2.6   Series of NBE molecules   150
2.7   Eagle Nebula, "Pillars of Creation"   151
2.8   Starburst galaxy   152
2.10 Olaus Magnus' frost and snowflakes   153
2.11 Honeycomb drawn by Wyman   153
2.12 Bees' cells   154
2.13 Weyl's purportedly symmetric snowflake   155
2.14 Libbrecht's "especially precise sixfold symmetry"   156
2.15 Water molecule   157
3.1   10th century BCE cinerary urn   54
3.2   Minoan vessel c.2100-1700 BCE   158
3.3   Vase from Thera c. 9th-7th century BCE   159
3.4   Prosthesis vase 8th century BCE (detail)   159
3.5   11th.century BCE Gorgon from Rhodes   160
3.6   Euoboean amphora c.570-560 BCE (detail)   161
3.7   Attic amphora c.540 BCE attributed to Exekias   161
3.8   Krater, Apuleia c.330-320 BCE   162
3.9   Cretan Mitra 6th.cent. BCE   163

3.10	Argive Vase 5th.cent BCE	164
3.11	Female statuete, Cyclades 3rd millennium BCE	165
3.12	Kouros 6th. Cent. BCE	166
3.13	Head of Apollo (?) from Delphia, 6th cent. BCE	167
3.14	Head of Poseidon (?) of Artemision	168
3.15	Engraved gems 6th-5th cent. BCE	169
3.16	Gold libation bowl from Olympia, 7th cent. BCE	170
3.17	Capital from Gamla, Israel,4th-5th cent.BCE	171
3.18	Votive tablet from Pergamon, c.200-250 CE	171
3.19	Plan of unidentified Roman structure, England	172
3.20	Greco-Roman votive relief, Syria	173
3.21	Murals from Pompei villas c.20-10 BCE	174
3.22	Rome - relief from Pantheon lintel (?)	175
3.23	Portland vase	176
3.24	Rome – Senate (curia Julia)	177
3.25	Endymion sarcophagus	178
3.26	Rome – Trajan's Forum marketplace plan	179
3.27	Rome – *Palatina Domus* wall decorations	179
3.28	Rome – *Ara Pacis* sacrificial table	180
3.29	Rome – *Domus Aurea* mural	181
3.30	Rome – *Domus Aurea* apse	182
3.31	Rome – Arch of Constantinople (east side)	183
3.32	Rome – catacomb of S.Domotilla fresco	183
3.34	Dura Europos – *Mithraeum*	184
3.35	Crete – Minoan palace complex	185
3.36	Parthenon stylobate elevations, east and west	186
3.37	Parthenon metope widths	186
3.38	Temple of Apollo Bassitas column offsets	187
3.39	Temple of Apollo Bassitas column heights	187
4.1	Chartres Cathedral	110
4.2	Chartres Cathedral rose window	188
4.3	Paris – Notre Dame	189
4.4	Rouen Cathedral	190
4.5	St. Denis, West front	191

III

4.6	St. Denis West front detail	191
4.7	Lübeck, Marienkirche towers	192
4.8	Regensburg Cathedral	193
4.9	Norwich Cathedral	194
4.10	Exeter Cathedral	195
4.11	Gloucester Cathedral	196
4.12	Oviedo Cathedral	197
4.13	Leon Cathedral	198
4.14	Fiesole – Badia	199
4.15	Padua – S. Maria Assunta	200
4.16	Padua – S. Maria Assunta baptistery	201
4.17	Otranto cathedral	202
4.18	Ferrara Cathedral	203
4.19	Venice – Piazza San Marco	204
4.20	Lucca – S. Giusta	205
4.21	Florence – Santa Croce medieval façade	206
4.22	Florence – Santa Croce today	207
423	Florence – Santa Maria del Fiore (Poccetti)	208
4.24	Siena cathedral, nave	209
4.25	Siena Baptistery, detail of mosaic panel	210
4.26	Siena Baptistery, carving	211
4.27	Rome – San Clemente apse	212
4.28	Verona, San Zeno door panel	213
4.29	Verona, San Zeno door	214
4.30	Arezzo – S Maria del Pieve doorway and arches	215
4.31	Arezzo- S.Maria del Pieve doorway and window	216
4.32	Hebrew prayer book – 13th cent. Germany	217
4.33	Leaf from 15th century Book of Hours, Paris	218
4.34	Ashburnham Gospels, rear cover	219
4.35	Chateau de Chambord	220
4.36	Florence – Palazzo Vecchio	221
4.37	Giotto's symmetric architecture	222
5.1	Giant stone structures, Azraq Oasis, Iraq	259
5.2	Fuegan facial painting	222

5.3	Andaman islander with body decoration	223
5.4	Haida dish	224
5.5	Wooden mask, Urua, Congo	225
5.6	Two halves of a Tlingit blanket	226
5.7	Chilkat pattern board and blanket	227
5.8	Haida dog-fish painting	227
5.9	Bowl, 6th or 7th century BCE. Dieteldorf	228
5.10	Venus of Vestonice	229
5.11	Ivory idols, Dolni	230
5.12	Skavberg, rock drawings	231
5.13	Female idol, Cucuteni-Baiceni	232
5.14	Bronze belt hook from Hoelzelsau	233
5.15	Bronze mirror, Desborough	234
5.16	Togoland: tribal house	235
5.17	King and attendants, Benin	236
5.18	Dance cap, Cameroons	237
5.19	Dance mask, Congo	238
5.20	Royal grave, Borneo	238
5.21	Tomb reliefs, Croizard	239
5.22	Silver buckle, Fornass	239
6.1	Serlio, the "miser's" house renovated	279
6.2	Late medieval garden	240
6.3	Palladio villas – Saraceno and Godi	241
6.4	Rome – Palazzo Farnese	242
6.5	Windsor Castle gardens, c. 1725	242
6.6	Villa d'Este	243
6.7	Villa di Castel Pulci	243
6.8	Botticelli's "dense and living forest"	244
6.9	Uccello, "The Hunt"	245-6
6.10	Piero di Cosmo, "The Hunt"	247
6.11	Pollaiuolo, "Combat"	248
6.12	Giovanni di Paolo, "Death on horseback"	248
6.13	Giovanni di Paolo, "Florence in the Plague"	249
6.14	Baccio del Biondo, "Plague in Florence"	250

V

6.14	Pierart dou Tielt, "Burying Black Death victims"	250
6.15	Pienza – cathedral and papal palace	251
6.16	Pienza – cathedra piazza and buildings	251
6.17	Serlio – asymmetric façade "corrected"	252
6.18	Serlio – new symmetric façade for two buildings	253
6.19	Le Muet – "symmetric house façade"	254
6.20	Downing's "symmetric irregularity"	254
7.1	High Wycombe park by Humphrey Repton	340
7.2	Stourhead	255
7.3	Plan of Pope's garden in Twickenham	255
7.4	London – Chiswick House	256
7.5	Paris – Rue Castiglione	257
7.6	London – MI6 Intelligence Agency headquarters	257
8.1	S. Maria Novella façade with Wittkower drawing superimposed	366
8.2	Wölfflin's drawing of S. Maria Novella Facade	258

ADDENDA & CORRIGENDA

On Pasteur's *"L'univers est un ensemble dissymétrique"*:
(add after p. 53)

On the titlepage of previous editions of this book I had placed as a motto Pasteur's extraordinary pronouncement that, (in my translation), "the universe is asymmetric, and I am persuaded that life, as it is known to us, is a direct result of the asymmetry of the universe, or of its direct consequences". I regarded this quotation as a tour de force, *as confirmation, by one of the great figures of Western science, of the basic thesis of my book. The term used by Pasteur in this statement was "dissymetrique", and I had assumed that he used it as a synonym of "asymmetric". A specialist in crystallography, however, contacted me to say that during the early 19th century that had ceased to be the term's meaning, and that Pasteur had used it instead to refer to certain crystals whose mirror images cannot be superimposed upon each other. (Think of two gloves, for example, whose relationship is a symmetric one, but which cannot be superimposed one upon the other.) I am not qualified to debate with a specialist what Pasteur had meant by the term. It was clear to me, however, that "dissymetrique" had been used as a synonym of "asymmetry" well after the time that my correspondent said it had lost this meaning. My evidence for this is the title that the architect and architectural historian Choisy had given to a paper that he published in 1865: "Note sur la courbure dissymétrique des degrés qui limitent au couchant la plate-forme du Parthenon".[1]*

Although I saw no reason to yield to my correspondent's views, my encounter with him forced me to recognize that I should not have used Pasteur's statement – least of all as a motto for my book – if I did not understand what it meant. I still do not understand Pasteur's meaning. And so, I no longer use his words on this book's titlepage.

---

[1] *Academie des Inscriptions et Belles-Lettres*, NS v. I (1865). The emphasis is mine.

PREFACE

Shortly before I began this revision of *Symmetry Fallacies*, I finished work on another book, *Byzantine Aesthetics and the Concept of Symmetry*. One of my objectives for that book was to see whether the 15th-century Italians, whom I identified in *Symmetry Fallacies* as the first people in the West to know the concept of symmetry, had learned it from Byzantine sources. This in turn naturally led me to investigate whether the concept of symmetry had been known in Byzantium long before it appeared in the Latin world.

These were not possibilities that I dreamed up out of thin air, as it were, but ones I encountered in something written by Goethe, the German poet. Goethe claimed that the Byzantines had the concept of symmetry from an early period; that their art was "always" (*immerfort*) symmetric; and that it was from them that the concept of symmetry to the West.[1] The evidence I presented in my book showed that Goethe's notions were devoid of merit. (Goethe, as it turned out, knew very little about Byzantine art, and cared for it even less; his notion of symmetry was ... idiosyncratic, and not useful.)

The Byzantine inquiry led me to modify some of my views, but for practical reasons it would have taken me too far off the course I follow in *Symmetry Fallacies* to incorporate them in a revised text. What I have done here, instead, is to indicate, in footnotes, and with brief explanations, that certain statements I made in the text have been superceded in *Byzantine Aesthetics and the Concept of Symmetry*. I have however borrowed from the latter to revise my remarks on the distinction between the idea and the concept of symmetry; and also some comments on Goethe.

Acknowledgments

I would like to acknowledge the assistance of librarians at the Alumni Library of Simon's Rock College in Great Barrington, Mass; the Clark Art Institute Library in Williamstown, Mass; the Architecture and Environmental Design Library of Arizona State University in Tempe; the Special

---

[1] Goethe 1963, section on Heidelberg. My discussion of Goethe's views is in Selzer 2021, chapter 3. See also pp. 322-323, *below*.

Collections Library of the University of Arizona in Tucson; the Tutt Library of Colorado College in Colorado Springs; and at the Archivio dell'Opera di Santa Maria del Fiore and the Biblioteca Nazionale Centrale, both in Florence.

I would also like to express my thanks to audiences, both academic and lay, on three continents, as well as to friends, family members, and even the occasional stranger, for the opportunity to talk about and hear their views on many of the ideas in this book.

Earlier editions of this book were published under the title of *The Symmetry Norm and the Asymmetric Universe*, which seems to have confused some people. Under its new (and I hope less obscure) title, this book appears in my 81st year, and is likely to be the last revision of an undertaking that has added much value and pleasure to my life over the course of two decades. As an avowed amateur, I have been able to enjoy many of the rewards of a scholar's life while yet shielding myself from the pernicious forces that sadly now vitiate almost every part of the academic world.[2]

> *Michael Selzer*
> Colorado Springs
> July, 2021

---

[2] As I write this, Princeton University has announced that, "to address systemic racism on campus", Classics majors will no longer be required to demonstrate proficiency in Latin or Greek. The director of undergraduate studies in Princeton's Classics Department is quoted as saying, "Having people who come in who might not have studied classics in high school and might not have had a previous exposure to Greek and Latin, we think that having those students in the department will make it a more vibrant intellectual community" Not even at Princeton, I am confident, can ignorance make an intellectual community "more vibrant". And how does a Classics department dropping the requirement for proficiency in Classical languages "address systemic racism? What kind of madness is this?

FOREWORD:
THE CONCEPT OF SYMMETRY

0.1: ancient Greek *hydria,* or water vessel

Fallacy. A delusive notion
- *OED*

There is a baffling phenomenon that has persisted for more than five centuries and has shaped notable aspects of Western culture: but that has gone largely unnoticed by scientists, scholars, and others who usually make it their business to notice this sort of thing.

Their ignorance of it is the more surprising because the phenomenon itself is not at all ambiguous or elusive. On the contrary, it is readily apparent, and there for all to see; and the typical response of people when I call their attention to it is not to deny the point but to express surprise that they had not noticed it earlier.

Two illustrations, with extracts from descriptions that originally accompanied them, will make clear what this phenomenon is.

Fig. 0.1 (previous page) shows an ancient Greek water vessel, or *hydria*; and *fig.*0.2 shows the planetary nebula Hen 2-437 in the constellation Vulpecula. The two objects, completely dissimilar in every other respect, have in common the fact that each has right and left halves that do not mirror one another.

That is to say, the appearance of each is asymmetric.

The asymmetry of the *hydria*'s decoration is immediately evident. All the elements on one side of the notional tree trunk which also appear, in some form or the other, on the other side, differ from the latter in size and relative location, and often in shape, too. Moreover, not all the elements on one side of the tree trunk are, in any form, repeated on the other.[1] (We might note that it is indeed the irregularity – the asymmetry – of the decorative scheme that imparts to the jar a feeling of vivacity, and almost *compels* one's eye to continue roaming over the unpredictable surface.)

I have superimposed a grid on the image of the nebula (*fig.* 02, *next page*) to make the differences between its two halves more apparent. With this aid, it can readily be seen that, for instance, the upper contour in the two sections to the right of the central axis describes a graceful curve, while the upper contour on the left is made up of angular lines with a peak, in the first section,

---

[1] The handles, too, do not mirror each other in size or location.

and a trough in the second. In fact, with the possible exception of the lower section on either side of the central axis, the contour in every section of the grid on one side of the axis is different from that in its equivalent section on the other side of it.

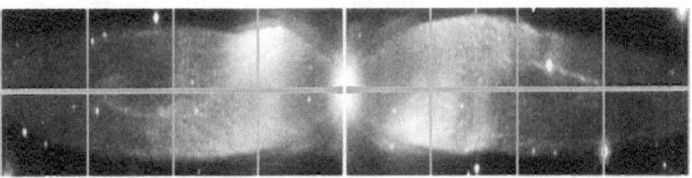

0.2: nebula Hen 2-437 with superimposed grid

In dwelling on differences between the left and right halves of these objects, it may seem that I have been belaboring the obvious. Yet to some people, among them highly-trained experts, those differences not only are *not* obvious but are not even apparent. They do not see them!

Thus the art historian Licia Ragghianti, in a catalog issued by the Greek National Archeological Museum in Athens, described the decorations on the *hydria* shown here as "symmetrically and rigidly ordered".[2]

Similarly, the NASA website from which I took the image of the nebula, identifies it, not merely as symmetric, but as "spectacularly symmetrical".[3]

Misperceptions like these are not at all unusual. Literally scores of further instances appear in this book. They illustrate the strong bias in Western culture that leads people to see asymmetric forms as symmetric: or perhaps more precisely, that *prevents them from recognizing* asymmetric forms for what they are.

One characteristic of these misperceptions is that they are almost never challenged. Usually, they seem to be accepted as accurate.

This bias is a principal tool of what I call *the concept* of symmetry, which I distinguish from *the idea* of symmetry. (My use of these terms is idiosyncratic.)

The idea of symmetry - I discuss it in Chapter 1 - is solely

[2] Ragghianti 1979, p.46.
[3] nasa.gov/image-feature/goddard/2016/hubble-watches-the-icy-blue-wings-of-hen-2-437 .

descriptive: it describes what a symmetric shape is. It is not prescriptive or normative: no value or "meaning" inheres in it. It is complete in itelf, it stands on it own.

The concept of symmetry, on the other hand, subsumes the idea of symmetry: it builds on it and attaches value and meaning to it. Where the idea of symmetry is descriptive, the concept of symmetry is prescriptive, or normative. It asserts that all the things we make – from small artifacts to large buildings – *must* be shaped symmetrically. The history, character and manifestations of this concept are discussed at length in Chapters 2 to 8, that is to say, in the greater part of this book.

The concept of symmetry was first formulated in the middle of the 15th century, at the outset of the Italian Renaissance, and remains firmly entrenched to this day. It has given rise to fallacies about Nature and aesthetics, culture and history – the "symmetry fallacies" of my title - that buttress each other in a pernicious and tenacious structure of error that has played, and continues to play, a more significant role in Western civilization, and in our lives today, than I believe is generally recognized.

A mainstay of these fallacies is the bias I have alluded to here, which leads to the misperceptions that ostensibly validate the symmetry concept.

[1.1] *Columbus Indiana:*
First Christian Church
Eliel Saarinen, architect

CHAPTER ONE
THE IDEA OF SYMMETRY;
ASYMMETRY

[1.2] *Columbus Indiana:*
North Christian Church
Eero Saarinen, architect

Omnis enim, ratione suscipitur de aliqua re
institutio, debet a definitione proficisci,
ut intellegatur, quid sit id, de quo disputetur
Cicero, *De Officiis* I,7

The idea of symmetry is the idea of a form whose lateral halves mirror each other. That is, in a symmetric form the reflection of one half appears on the other, so that reversed versions of the details on the left (and only those details) are present, in the equivalent locations, on the other half, thus:

}}}>> <<{{{

The middle of such a configuration may be, but is not necessarily, indicated by a vertical object or line, thus:

}}}>> | <<{{{

(Where there is a row of identical elements that are themselves symmetric - a row of identical windows, for example -

OOOO OOOO

- the left and right halves duplicate each other, but are also legitimately regarded as symmetric. However, a duplicated arrangement *per se* –

XOOO XOOO

- is not necessarily symmetric.)

The origin of the idea of symmetry is obscure. To be sure, forms such as the square or the sphere that we see as symmetric have been known for a long time, but there is no evidence that their symmetry was recognized before the 15th century (see p. 35,fn.31; and pp. 56, 334, *below* for a fuller discussion). There is some – rather ambiguous - evidence that the idea of symmetry may have been known to the Byzantines, but none that they ever connected it to the concept of symmetry, as would happen in the

West[1]. In our first explicit encounter with the idea of symmetry in the West, it was already subsumed by the concept of symmetry (see pp. 336, *below*). The Byzantine example suggests, however, that in the West, too, it may have preceded the advent of the concept of symmetry.

Although symmetry, as defined here, is a simple and unambiguous idea, it is often misunderstood. Sometimes this is because of the failure to recognize that "symmetry", our usual term for the idea, nowadays has not one, but in fact *three*, entirely different meanings. A statement that may be true when used with one meaning of the word, accordingly, is not likely to be true when either of the other two meanings is understood.

Today's conventional meaning of the term originated in Italy at the beginning of the Renaissance, and is what we mean by "symmetry" in this book.

For the ancient Greeks, however (see pp. 56*ff. below*), "*symmetria*" had a very wide range of meanings, many of which are inconsistent with each other, and not a few of which seem quite obscure. Plato, for example, as we will see in Chapter 3, used the term a number of times in his *Philebus*, but the meaning or meanings that he attached to it are not at all clear. The term was later adopted by the Romans and, although the meanings they (and notably Vitruvius) gave it are again often unclear, we find that "*symmetria*" as often as not now came to mean something like "good or harmonious or pleasing proportions". It continued to be used in more or less this sense in the Renaissance and thereafter – most famously, perhaps, by Blake with the "fearful symmetry" of his tiger. Even today the term sometimes occurs with this meaning (for an example see p. 315, fn. 96, *below*).

However, there is no evidence that either the Greeks or the Romans ever used "*symmetria*" in the conventional modern sense, for a shape whose left and right halves mirror each other. When this meaning was adopted in the Renaissance it did not displace an earlier meaning or meanings, but existed alongside it or them – sometimes (as in the *Hypnerotomachia Poliphili* of 1499) even in the same book (see Chapter 6).

A third meaning of the word was introduced by scientists in the twentieth century. It derives from the observation that symmetric shapes remain the same when revolved at 180 degrees around their central axes: which is to say that they "change

---

[1] Selzer 2021, chapters 5 and 6.

without changing"[2]; and this quality is called "symmetry" by scientists when they refer to invariances that they have discovered in Nature's operations. In this usage, of course, "symmetry" is not understood literally but as a metaphor, so that when scientists refer to the symmetry of the universe they are not claiming that the universe is divided into two mirroring halves or that it has harmonious proportions. Rather, they are stating something about the invariant character of the laws that govern Nature. There is really no compelling reason to employ the term "symmetry" for this purpose, and Wilczek, for one, acknowledges that "it seems rather distant from the everyday meaning of the word". He writes: "Some people find this use of the word 'symmetry' jarring … If you have that difficulty, you might want to keep 'invariance' in mind as a supplement or substitute. After some deliberation I decided to stick with 'symmetry' because it is deeply embedded in the [scientific] literature, and not without resonance. Whatever you call it, the big idea remains Change without Change".[3]

The adoption of "symmetry" for bilaterally-mirrored shapes meant that, from the middle of the 15th century, a single word has been used for two distinct ideas in the field of aesthetics. Inevitably, this became – and remains - a source of befuddlement. Modern scientists' own idiosyncratic usage has added another layer of confusion.

Not all that infrequently, scientists themselves fall victim to it. Forgetting that their's is a metaphorical use of the term, they sometimes seem to assume that because natural forces are symmetric in *their* – scientific - sense, almost everything that Nature creates is also symmetric in the *conventional* sense. Wilczek, for example, asserts that bees' cells are symmetrically shaped. He is by no means alone among scientists in succumbing to this and similar fallacies, as we will see in detail in Chapter Two.

---

[2] Wilczek 2015, p. 73. As an example, Wilczek (*ibid*, p. 167) cites "Galilean symmetry", according to which "we [can] change the world … by moving everything at a common velocity, without change to the way things behave". According to Wilczek (*ibid*, p. 165) symmetry in this sense "has come to dominate our best understanding of the fundamental laws of Nature. So say the masters". Comp. the statement of P. W. Anderson (1972), like Wilczek a Nobel prize-winning physicist, that it is only a slight overstatement to say that "physics is the study of symmetry".

[3] Wilczek, *op. cit.*, p. 238.

There are other reasons, too, why it is important to clarify at the outset what "symmetry" means – and what it does not mean. Symmetry, in the words of the noted mathematician and physicist Hermann Weyl (himself author of an influential study of the subject), is an "absolutely precise" idea.[4] By that token, of course, asymmetry ("without symmetry") too is an absolutely precise idea. The terms moreover are not only unambiguous but mutually exclusive, which is to say that symmetry and asymmetry are opposites and not points on a continuum like "hot" and "cold", for example, are. Thus, if a shape is not symmetric it must be asymmetric, and if it is not asymmetric it must be symmetric: there is no such thing as a shape that partakes of both qualities.[5] Nor is there a shape that is neither the one nor the other. Careless usage sometimes obscures the distinction between the two with solecisms such as "almost symmetric" or "broken symmetry", and so on. Terms like these refer to shapes that are asymmetric, and should be called that. It is curious that we never encounter their opposites, which would be terms like "almost asymmetric" or "broken asymmetry". This, I suspect, reflects the widespread but false assumption that symmetry is the norm, and that asymmetry is a damaged or defective version of it – see my discussions of this point on pages 15-16 and 329-30, below.[6]

---

[4] Weyl, 1952, p. 4. The insistence that symmetry requires the two halves of a form repeat each other precisely goes back to the earliest days of the idea, when Alberti (1966, IX.7; see also the appendix, "Alberti's *Collocatio*", to Chapter Six) insisted that the right must always accord with the left (*dextra sinistris convenirent*) even in the most minute detail (*in minutissimis*). For Alberti's use of *convenire* see Chap. 6, fn. 147. Alberti never specified that the two halves *mirror* each other, but it is clear from his architectural designs (eg the façade of S. Sebastiano, Mantua) that that is in fact what he had in mind. The explicit requirement that the two halves *mirror* or reflect each other appears quite late and, as far as I am aware, has not yet been studied except in the limited context of chirality.

[5] I use "symmetric" and "asymmetric" throughout this book. Their longer forms add sound but not meaning to the words, and should not be encouraged. See the entry "-ic(al)" in Fowler's *Dictionary of Modern English Usage*.

[6] See Prigogine and Stengers (1984, p.163) for an instance of this fallacy. They claim, "In the world around us some basic symmetries seem to be broken". Examples they cite include the "preferential chirality" of some shells, and the right-handed helix of DNA; they marvel that Pasteur (see

That said however it is also the case that asymmetry (and sometimes symmetry, too) can be more apparent in some shapes than in others. Indeed, on occasion the asymmetry (or symmetry) of a shape can only be established after careful scrutiny. But no matter how discrete the asymmetry is, it must always be recognized as such, and not be thought of as a weakened or defective form of symmetry.[7]

(It should be noted, too, that while many shapes have a pronounced *bilateral* quality, they must not be considered symmetric unless their two halves accurately mirror each other. The two lateral halves of our bodies, for example, in their outward appearance, resemble each other but are not their mirror images. Our bodies accordingly are not symmetric, let alone

---

p. 53, *below*) went so far as to see "in the breaking of symmetry" the very characteristic of life. They go on to ask, "how did this dissymmetry arise?", and confess, "we have not yet found a satisfactory answer." However, I suspect that part of their inability to answer the question may stem from their insistence that asymmetry represents "the breaking of symmetry": in other words, that it is a defective, damaged, form of symmetry; and with the implication that, at some previous stage, what is now broken had been whole, or intact: an oblique but possibly unintended allusion to the fallacious notion that everything "in the world" was originally symmetric!

[7] The fact that some asymmetries are not recognized consciously need not mean that they do not register themselves on the unconscious mind. White for example (1967, p. 158) plausibly suggests that people whose attention is held by the appearance of the façade of Santa Maria Novella in Florence are unwittingly responding to its "hidden asymmetry, a half-felt difference on the fringes of consciousness". The same point is made more generally by Ruskin, who noted (*Stones of Venice*, 2.5.12) that the eye "is continually influenced by what it cannot detect". But whether such "discrete" asymmetries were created intentionally – whether, for instance, the builders of Santa Maria Novella's façade made a point of not flaunting but of "hiding" its asymmetry, is another question. Goodyear claims that the asymmetry and other "refinements" of Greek temples were not intended to be noticed consciously; and in this he appears to echo Ruskin's claim in *The Lamp of Life* that medieval builders sometimes created a "pretended symmetry" to conceal asymmetric arrangements. One doubts, however, whether it is possible to know the undeclared intentions of craftsmen who lived centuries or even millennia ago. The evidence considered in this book indicates, moreover, that the idea of symmetry was almost certainly unknown before the beginning of the Italian Renaissance, and therefore could not have informed the work of craftsmen in earlier times.

almost symmetric. They are, rather, asymmetric. And that of course is also so of our *internal* structure.)

There is nevertheless a legitimate question about the degree of precision to which one must resort when determining if a shape is symmetric or asymmetric. Clearly, in all but the very rarest instances it would be fatuous to demand for this purpose that the two sides mirror each other down to a sub-atomic level! Yet sometimes extremely fine measurements, whose precision far surpasses the capabilities of the naked eye, may reveal unsuspected symmetries. Such is possibly the case with certain portions of the Parthenon, whose symmetry is claimed to be accurate to the astonishing degree of 1:10,000 or even more.[8] Precision of this order can only be detected using modern optical or, better yet, electronic instruments. The fact that it is not apparent to the naked eye does not mean that we should ignore it or assume that it was achieved unwittingly, for except in the rarest instances, forms are symmetric only because their makers intended them to be. But it would be unreasonable, surely, to insist that the alleged >1:10,000 accuracy of parts of the Parthenon sets the standard for determining whether a form is symmetric.

What, then, does Weyl's "absolutely precise" criterion mean? Is it, indeed, an *absolute* criterion: or is it one that is subject to the contexts in which forms are created? If parts of the Parthenon come as close as one could expect to Weyl's *absolute* precision, what of the majority of extant and measurable portions of the same structure that do *not* meet that standard? Are we to consider them asymmetric, even if the degree to which they mirror each other could be accepted as symmetric on another structure? And what about the possibility that, in cultures where standards of accuracy are far lower than in our own, an object that *we* recognize as asymmetric was both intended and seen by its creator to be symmetric, instead? I am not sure that there are conclusive answers to such questions.

Sometimes, however, we can be entirely confident that a bilateral design was *not* intended to be symmetric. The point can be made with regard to a 10th-century cinerary urn amphora now in the Kerameikos Museum in Athens (*fig. 3.1*). The three

---

[8] I would stress that the reliability of these measurements cannot be taken for granted, and still awaits confirmation. The matter is discussed in some detail in the appendix to Chapter Three.

principal vertical strips of the center section are asymmetrically related to each other. So are the vertical strips on the flanks, that on the left having a triangular design in each of its four panels, while that on the right has six-and-a-half pairs of a different motif. We can be absolutely confident that this not only *is* not, but was not *intended to be,* a symmetric design. Equally, the facades of many medieval cathedrals – such as those of St. Lo (*frontispiece*) or Chartres (*fig.* 4.1) - are flanked by towers that are so unlike one another that it is impossible to believe that their builders intended them to be, but failed to make them, (let alone, that they thought that they *had* made them) replicas of each other.

In other instances, however, the evidence is not unambiguous and we must acknowledge that, in theory at least, an artifact could have been seen by its maker as symmetric even though we recognize that it is in fact asymmetric. When we encounter this possibility, it is useful to ask whether there is any evidence that the idea of symmetry was known in the culture in which that artifact was created. Was there indeed a word or phrase for the idea in the language of that culture? When these questions cannot be answered in the affirmative, we are justified in supposing that the asymmetry of a bilateral design does *not* represent an unsuccessful attempt to create a symmetric shape.

For the rest, we can safely set aside epistemological doubts and take as our standard for deciding whether something is symmetric or asymmetric the one that guides careful observers when they look at a shape with the naked eye, or when they examine a photograph of it with the help of a divider or ruler: or, far better yet, with any of the elementary graphic software programs that are now easily obtained. I have used one of these programs to analyze the head of Leonardo da Vinci's "Vitruvian Man" drawing (*figs.* 1.3–1.5); and in this simple way have been able to demonstrate that, despite what is often supposed, its shape is asymmetric.[9] That such observations should be made meticulously deserves to be emphasized, for as I suggested in the foreword, in our culture there is undoubtedly a bias in favor of assuming that most forms – especially ones regarded as attractive or important - *are* symmetric.

I may be allowed to share another striking instance of

---

[9] Hall (2008, p.7), for example, boldly asserts the "symmetry" of Vitruvian Man, which he claims is "broken" only by the left turn of the lower body and feet. He fails to see that the entire figure is of course asymmetric.

this bias. It occurred in the decision taken in the 1970's by the University of Florence's Department of Architecture to survey only the left portion of the façade of the church of Santa Maria Novella (*fig.* 8.1), on the grounds that the purported "perfect symmetry of the façade" - *la perfetta simmetria della stessa faciata* - made it unnecessary to measure both sides. It happens however that this great and mysterious façade is markedly asymmetric, as White saw (fn. 7, *above*) and, we can be sure, was by no means the first to see. Some of its asymmetries can be detected by the naked eye, while others can be found through rudimentary analysis of photographs.[10] But of course none are likely to be discovered by an observer who is convinced *a priori* that they are not there!

Another notable instance of the same bias has recently been documented. For a long time no structure has been thought to exemplify more pleasingly the virtues of symmetric design than the Taj Mahal: and yet, as two Indian scientists have now established, the Taj's main dome is asymmetric, with opposite radii differing by as much as 5.52%.[11] The asymmetry is quite apparent in photographs once one knows enough to look for it.

There is a further point that deserves to be considered – not directly about the ideas themselves so much as about the terms we use for them. The word "asymmetry" means literally that which is "without symmetry". It carries with it the implication, not only that asymmetry lacks that which symmetry possesses (rather than, or as well as, that symmetry lacks that which asymmetry possesses!); but also that symmetry is somehow the appropriate standard for shapes, *the norm*: while asymmetry represents a defect, a falling-off from that standard.[12]

But is asymmetry really a falling-off of the standard set by symmetry? Of the two churches in Columbus, Indiana, illustrated at the beginning of this Chapter, which is the defective

---

[10] See fn.38, p.385.

[11] Ahuja and Rajani, 2016. Although the rigidly symmetric geometry of much Islamic decoration is indisputable, the role of symmetry in Islamic architecture remains unclear, as is apparent from one authority's lament about the "want of symmetry of plan" in Muslim structures (see Havell 1913, p. 19).

[12] To give just one example: Wikipedia – admittedly no *bona fida* authority on anything – defines asymmetry as "the absence or violation of symmetry". Might it not have been as – or more – appropriate to define symmetry as "the absence or violation of asymmetry"?

one? Is it the church by Saarinen *père* because it is so markedly and pervasively without *symmetry*? Or is it the church by Saarinen *fils* because it is so markedly and pervasively without *asymmetry*? Which more draws in the eye – and perhaps the soul? Which proves more elusive the longer one's gaze moves over it, and which reveals all that it has at the very first glance? Which suggests more the mysterious reality of the spirit – and which, little more than an upturned martini glass? There is, somehow, a truthfulness in the one and mere glibness in the other.

Or consider the face of Leonardo's "Vitruvian Man", mentioned on page 14. The symmetric images that consist of the mirrored left and right halves, respectitvely, of the original are flat and little more than caricatures; we would not be surprised to find them on the pages of a comic book. But when the two halves are combined asymmetrically in a single portrait – Leonardo's actual drawing - the guileless passivity of the one half and the intimidating hostility of the other interact to suggest the inner complexities and conflicts of an authentic human being.[13]

The superiority of these asymmetric forms would not be surprising, I suspect, if our perceptions were not so strongly determined by the idea that asymmetry is inherently defective. This idea, as already mentioned, is entrenched in our terminology, which implies that asymmetry is lacking that which symmetry possesses.

How unfortunate it is that we do not have a word-pair that does not convey this impression: or, perhaps better still, that suggests that (what we now call) asymmetry possesses what symmetry in today's parlance lacks: terminology that implies that symmetry is best understood somehow as a defective form of asymmetry!

---

[13] See also Close 2000, pp. 44-48 and McManus 2015

CHAPTER TWO
THE ASYMMETRIC UNIVERSE

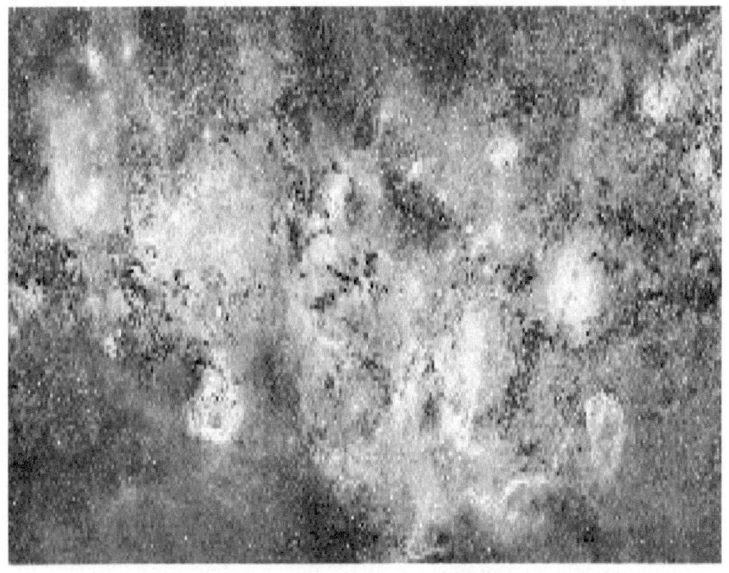

2.1 Portion of the Milky Way[1]

[1] *https://www.telegraph.co.uk/technology/2021/03/25/12-years-milky-way-mosaic-photograph/*

> *"These urbashus borders, altogether*
> *too 'ap'azard... Sloppy. That's Nature."* [1]

> *"One hundred-percent of women have*
> *breast asymmetry."* [2]

The fallacy that Nature's shapes are always symmetric originated in the middle of the 15th century.

The earliest known statement of it is by Leon Battista Alberti, (1404-1472), the Renaissance polymath. "It is of the essence of Nature (*tam ex natura est*)", he declared in *De re aedificatoria,* his treatise on architecture, "that the right should correspond in every respect to the left (*ut dextra sinistris omni parilitate correspondeant*)".[3] Alberti also declared that it was "from the lap of Nature (*ex naturae gremio*)" that the ancient Greeks deduced what he claimed was the fundamental principle that in works of art and architecture "the right must accord with the left (*dextra sinistris convenirent*) even in the most minute details (*in minutissimis*)".[4]

---

[1] Ngaio Marsh, *A Man Lay Dead* (1934; Kindle ed. at 2402).

[2] Dr Daniel Maman, board-certified plastic surgeon. *Buzzfield Health,* Jan. 6, 2017.

[3] Alberti 1966, IX, 7. For Alberti's use of *convenire* see Chap. 6, fn. 147. Alberti also claimed (*ibid,* VII, 4) that Nature delights above all (*in primis*) in circular forms and also in hexagons. Inherent in the concept of symmetry is the concept of simplicity, for a form such as a circle or hexagon whose left half contains identical information to that in the right is bound to be simpler than it would be if one of the halves contained different information. We take up the implications of this in Chapter Six, but it should be pointed out now that in the Italian Renaissance the claim that Nature's forms are symmetric was associated with the claim that they are also simple. Palladio, for instance, based his designs on what he alleged was the simplicity of things that are created by Nature – "*quella semplicita che nella cose da lei [i.e la Natura] create*" (see p. 286, fn.20, *below*).

[4] *ibid.,* VI, 3. See Gadol (1969, p. 111 *ff.*) who, in common with other Alberti scholars, failed to recognize that Alberti's belief that Nature's forms are symmetric was the foundation of his concept of *collocatio*. It is curious to note that, according to Kruft (1994, p. 418), the Russian Constructivist M. Y. Ginzburg (1892-1946), in calling for symmetric architecture, cited Alberti as his authority for claiming that Nature's forms are symmetric. Wouldn't it have been much more sensible of him if, instead of this pedantry, he had merely looked at the natural world around him and seen that everything in it is asymmetric? (In a later book, however, Ginzburg advocated asymmetric architecture, instead, on the grounds that mechanical movements are almost always asymmetric!)

This view of the shapes that Nature creates has enjoyed a long life;[5] and it continues to be held today, including by scientists. The physicist Leon Lederman, for example, a Nobel laureate, writes of "the graceful symmetry" that characterizes Nature's forms, among them "a noble tree's branches, and the veins of its leaves... a snowflake... [the] disks of the Moon and Sun".[6] Such views are not considered controversial among scientists today.[7] Indeed, I am unaware of any who repudiate them. It should be understood that the claim is not usually that Nature's forms are symmetric only in some hypothesized ideal state – that they are, as it were, "theoretically" rather than actually symmetric; or else, that they are symmetric in the metaphoric sense employed (as we saw in Chapter One) by some scientists today. *The claim, rather, is that Nature's forms in their outward, manifest, appearance – the forms, that is, that we ordinarily see when we look at them - are symmetric: that their left and right halves are mirror images of each other.*

Yet in truth Nature disdains symmetry. Draw an imaginary line, if you doubt that, between the right and left halves of any of her artifacts – down the middle of a cloud, perhaps, or the canopy of stars in the night sky; of a mountain, or a tree growing on its slopes; of a meandering river, the fish swimming in it, or the boulders resting on its bed; of a snowflake, or the cells of a honeycomb, or a spider's web; of your own face and body:[8] indeed,

---

[5] An example, from the 18th-century Scottish Enlightenment, is the statement of Lord Kames, (1696-1782), that not only the external appearance but the internal structure of organic life is characterized by "the strictest regularity ... order and symmetry". The two sides of animals, according to him, "are precisely similar ... and the two individuals of each pair are precisely accurate". Even the smallest divergence, he added, never fails to produce a "perception of deformity" (Kames 1845, pp. 161-2). These claims are repeated almost verbatim in the article on "Uniformity" in the 2nd edition (1783) of *Encyclopedia Britannica*.

[6] Lederman 2008, p. 13. See also p.43, fn. 89, *below*.

[7] Even perceptual psychologists, ironically enough, can subscribe to this fallacy – *vide* McBeath (1997): "Symmetry is a pervasive structural characteristic of 3-D objects in the [natural] world".

[8] Wyman (1867, pp. 12-13) acknowledges that in animal bodies "symmetry is rarely if ever absolute" and that "generally the symmetry is largely distorted". In other words, and more correctly, they are asymmet-

down the middle of anything animal or vegetable or mineral that
has not been tampered with by the likes of topiarists or "cosmetic"
plastic surgeons - and you will always discover that there are features on one of those halves that are not mirrored on the other half
(*figs.* 2.1 - 2.15).   And this hold true regardless of the size of the
objects we look at, for Nature's asymmetry is just as apparent in
molecules of a few Ångstroms seen through an electron microscope (*figs.* 2.6, 2.15), or in the double helix of the DNA molecule,
which is sometimes called a building block of life itself, as it is in
the Eagle nebula's colossal "Pillars of Creation" (*fig.* 2.7) and, indeed, in our own galaxy, the Milky Way (*fig.* 2.1).[9] Nature's penchant for asymmetry, moreover, could well be as old as the universe – as old as Nature – herself.[10] That, at least, is what seems to

---

ric. He claims, however, that "the limbs which in the adult are most un-
symmetrical, are quite symmetrical in the embryo" and on this basis argues that "the hypothesis that the idea of symmetry underlies their structure is rendered highly probable". That mammalian embryos are in fact
asymmetric is now generally acknowledged – see for example Shahbazi,
etc., 2019. As we will see below, Wyman was an empiricist, and resolutely
rejected the widespread belief that the cells of bees are symmetrically
shaped.   An account for the layman of why the human body is asymmetric may be found at ted.com/talks/leo_q_wan_why_are_ hu-
man_bodies_asymmetrical/transcript?language=en.   Perplexingly, the
author also repeats the fallacy that, in his words, "Symmetry is everywhere in Nature. And we usually associate it with   beauty: a perfectly
shaped leaf..." Note the blithe assumption here that a symmetric leaf (if
that were ever to be found!) is one that is beautiful and "perfectly
shaped". According to Reinberger (2016) the asymmetry of our bodies is
"primarily" determined by "the asymmetric structure of amino acids".
As a layman, I find that the most intelligible discussion of this profound
topic is by Frank Close (2000, *passim*, but particularly pp. 65–73.) Unfortunately, Close also repeats some of the conventional fallacies, such as
that the sun, the moon and the planets are perfect spheres (p. 12; but see
fn. 9, *below*); and that galaxies are symmetric (p. 19; comp. *fig.* 2.8, which
shows a galaxy that is obviously asymmetric!) Iain McGilchrist (2012)
shows how the physical asymmetry of the human brain crucially enables
the division of labor by which the brain operates. For a broader perspective on the subject of animal asymmetry, see Neville (1976).

[9] Perhaps worth noting here is that scientists using NASA's RHESSI
spacecraft have recently established that the sun has an irregularly oblate
shape. See: nasa.gov/topics/solarsystem /features/oblate_sun. html

[10] As a metaphor, asymmetry is used to identify one of the most basic
mysteries of physics: "If all the matter in existence — from miniscule atoms to gigantic galaxy clusters — strictly followed the theories of the
Standard Model of physics, then all of it would have an equal amount of
antimatter, and the two together would annihilate each other, leading to

be implied by the asymmetrically-shaped and distributed clusters of radiation left over from the Big Bang which have been mapped by the COBE, WMAP and Planck spacecraft (*fig.* 2.5). Prudence, to be sure, requires us to recognize that somewhere in the far reaches of the universe (or perhaps in the parallel universes posited by some cosmologists) there might be regions in which natural forms are always symmetric rather than asymmetric: even worlds where – who knows? – symmetrically-shaped people speak to each other in palindromes! But in the universe as we have been able to see it so far, and despite the unwillingness of many scientists to acknowledge this, we have encountered nothing, anywhere or on any scale, that challenges the notion that Nature's forms are asymmetric in appearance. Why this is so is a matter that we will explore toward the end of this Chapter.

That Nature's forms are asymmetric would have seemed self-evident to Archbishop Olaus Magnus of Uppsala (1490 - 1557), if he had been familiar - as he almost certainly was not - with the ideas of symmetry and asymmetry. Olaus is the first person we know of in the West who described the appearance of snowflakes in detail.[11] What struck him about them in particular

---

the existence of precisely nothing. But since there is something instead of nothing, scientists have long tried to understand this asymmetry between the amounts of matter and antimatter in the universe. This violation of the matter-antimatter symmetry is called the charge-parity (CP) violation..." (ibtimes.com/exploring-matter-antimatter-asymmetry-explain-why-any-thing-exists-universe-2548744). As I have already noted, it is inappropriate to call asymmetry a "violation ... of symmetry". One might well argue that what is being spoken of here is not a matter of "asymmetry" but of "discrepancy", which would have avoided using "symmetry" in a metaphoric sense that only heightens the confusion on this question.

[11] Two hundred years before Olaus was born, Albertus Magnus gave a brief description (*Meteorology* bk.I.chap.10), of what he thought snowflakes look like. Snowflakes, according to him, are made up of an unspecified number of radii reaching from the center to the circumference in a star shape ("*figura stellae*".) He seems to have believed that all snowflakes were of the same shape; and, also unlike Olaus, he did not accompany his description with drawings. It would seem that Albertus' remarks about the snowflake were unknown to Olaus.

Thanks to Joseph Needham, (1963, p.41) we now know that, for many centuries before Albertus, scholars in China had observed – and, moreover, sought to account for – the shape of snowflakes. Han Ying, in

was the great variety of their shapes. During a single day or night, he reported, one might see "fifteen to twenty distinct patterns, or sometimes more".[12] His illustration of them (*fig.* 2.10) shows 23 different forms. One looks like an arrow-head; others resemble a bell, a human hand, a crescent moon, a six-pointed star. All are asymmetric.

To Olaus the shape of snowflakes was, as he put it, "more a matter for amazement than for inquiry". Not so, a century later, to the great mathematician and astronomer, Johannes Kepler (1571- 1630). Crossing a bridge in Prague one winter's night, Kepler looked at the snow that was settling on his coat and asked himself why snowflakes always fall (*"perpetuo cadant"*) as equilateral – regular – hexagons.[13] We may suppose that this did not initially strike him as a particularly difficult question. Kepler, after all, had recently revolutionized astronomy by codifying the laws of planetary motion. In comparison to this awesome

---

the second century BCE, seems to have been the first to claim that snow-flakes are hexagonal, the reasons for that shape being something that Chinese scholars as late as the 17th century continued to ponder. Needham and Lu (1970) suggest that the "fine observations" made by some of these scholars indicate that at a certain point magnifying glasses must have become available in China. Yet it seems clear that these scholars' speculations were not based on empirical research. A 12th century text, for instance, accounts for the alleged hexagonal shape of snowflakes by the decidedly non-empirical argument that "six generated from Earth is the perfected number of Water, so snow is condensed into crystal flowers that are always six-pointed". Another text, from the late 14th or early 15th century, declares that "snow is the extreme form of Yin and so has the water-number in perfection. Thus it is that snow-flowers are always six-pointed". Indeed, the sole text - this from about 1600 – that clearly *is* based on empirical research repudiates the dogma that snowflakes are always hexagonal. As reported by Needham and Lu, this text declares that "old sayings are not always true", and claims that about one snow-flake in ten has *five* rather than six spikes. It seems that Chinese studies of the snowflake only became known in the West in the 20th century.

[12] Magnus 1996, Bk. 1, cap. 22. Needham and Lu (1970) criticize Olaus for "missing the essential unity of (snowflakes') pattern – the hexagonal symmetry present in all"; but of course, it is they who missed the truth, which is that there is no evidence (see below) that snowflakes are always hexagonal, let alone ever hexagonally symmetric. No less to the point, Needham and Lu repeatedly refer to the Chinese view of snowflakes as symmetric, even though there is no evidence that that is how the Chinese viewed them. Indeed, there appears not to have been a Chinese term or phrase that denotes symmetry, in our sense of the word, so that it is most unlikely that the concept of symmetry was known to them.

[13] Kepler 2010.

achievement, the challenge of accounting for the shape of snow-flakes may well have appeared trivial. Kepler discussed his find-ings in a book entitled *De nive sexangula* ("the hexagonal snow-flake") that was published in 1611 in Frankfurt-am-Main, two years after his work on the laws of planetary motion. The book is a very short one, of about 8,000 words. Somewhat fatefully for the history of science he began it by examining another natural form that he believed also has a regular hexagonal structure."*Quo ordine structi sint apum alveoli*" he asked: what principles guide bees as they build their honeycombs? Kepler claimed that he first approached this question empirically, with a simple look – "*ex intuitu simplici*" – at a honeycomb, but it would seem that in fact his starting point were some ideas that were probably first devel-oped by Pappus, the fourth-century mathematician of Alexan-dria. Pappus knew that, of the shapes that can be packed together without leaving any empty space between them, it is the hexagon that has the most angles and therefore the largest storage capac-ity: and so he marveled at the "geometrical foresight" that led bees to make their cells regular hexagons. (Centuries later the bee in *The Arabian Nights* would echo Pappus' assessment by boasting that Euclid himself might have learned a thing or two about ge-ometry from the bees.) We can infer from a statement by Alberti in the 15th century that the belief that bees' cells are always ("*nonnisi*") regular hexagons had continued to be accepted since Pappus' day, or perhaps even earlier.[14] Kepler, therefore, in de-scribing bee cells as regular hexagons, was perhaps repeating a commonplace.

There are two layers of cells in the honeycomb, with each facing in the opposite direction. The outward-looking faces are open so that a bee may enter and leave, while the end of the cell that is within the honeycomb is closed. Kepler focused his inves-tigation on the point where the closed end of each cell[15] meets,

---

[14] Alberti (1966, VII, 4), evidently echoing prevailing opinion, declared that bees, hornets "and insects in general" always make their cells in hex-agonal form (*nonnisi sexangulas in suis theatris cellas astruere didicere*). Three hundred years before Pappus, Ovid (*Metamorphoses*, XV, 382-3) had already described bee cells as hexagonal ("*sexangula*"). There is no evidence, however, that Ovid's view influenced Pappus, but Alberti is more likely to have been familiar with Ovid than with Pappus.

[15] He called it the *carina,* or keel.

without leaving any empty space, the ends of the cells abutting it; and this led him to determine that the shape of the cells in a honeycomb – "*haec inquam est quam effingunt in suis alvearibus*" – is that of an incomplete rhombic dodecahedron.[16] He added that this design is based on material necessity – "*rationes materialem*" – not only because (as Pappas had already noted) the hexagon is the most capacious of the space-filling solids, but also because its many obtuse angles best protect the delicate bodies of baby bees. The lattice of shared walls moreover, each reinforcing the others, imparts greater strength to the entire comb than any other form could.[17]

It is unclear how, really, Kepler's inquiry into the shape of bees' cells contributed to his understanding of the shape of snowflakes, but it did take on a life of its own. In 1712, a century after he published his book, Kepler's hypothesis about the *carina* received (or rather, *seemed* to receive) empirical validation in a paper on the structure of the honeycomb that a certain Giacomo Maraldi delivered to the Académie des Sciences in Paris. In his paper, Maraldi declared that he had measured – *measured, that is, not calculated!* - the angles of the rhombs at the base of bee cells and found that they were 70° 32', and 109° 28', respectively.[18] These are the angles of the rhombic dodecahedron, and as such they confirmed the accuracy of Kepler's mathematical calculations. Some years later a young Swiss mathematician called Samuel Koenig was asked by the great Réaumur, who was interested in the economy and efficiency of the honeycomb's design, to calculate what configuration of a hexagonal cell, terminated by three similar and equal rhombs, requires the least amount of material for its construction. Koenig's calculations produced a result which differed from the angles that Maraldi claimed to have measured by a mere 2' – that is, one-thirtieth of a degree. Koenig's comment on this finding was that the bees had long ago solved a

---

[16] Kepler 2010, p. 45.

[17] *ibid*, pp. 62 - 3. These explanations, Kepler added, are "sufficient, and thus I do not feel the need at this point to philosophize about the perfection, beauty or nobility of the rhombic figure, or to busy myself so that the essence of the little soul within the bee may be elucidated by contemplation of the shape it produced. If the shape in question were of no discernible utility, we might have had to proceed in this manner".

[18] *Mem.. Acad. des Sciences*, 1712, quoted Wyman 1866, p. 3.

problem that had been beyond the capacity of human beings before Newton and Leibnitz.[19] Upon learning of Koenig's calculations Fontenelle, the Academy's Sécretaire Perpétuel, declared that the bees, lacking intelligence as they undoubtedly did, could only have achieved this optimal result through divine guidance. Later, it was determined that the tiny discrepancy between Maraldi's alleged measurements and Koenig's calculations was caused by an error in the logarithmic tables that Koenig had used.[20] It was then said that the bees had been right all along and the mathematician wrong![21]

Maraldi's measurements, now vindicated mathematically down to the last second of a minute, would remain unchallenged for over a century. When Lord Brougham (d. 1868) – a former Lord Chancellor of England and author of *Observations, Demonstrations and Experiments upon the Structure of the Cells of Bees* – heard someone say that mathematical calculations and empirical observations "nearly agree" about the shape of the cell, he responded with some *hauteur*: "The 'nearly' is quite incorrect. There is an absolute and perfect agreement between theory and observation".[22] Among those who also accepted Maraldi's claims were the great French naturalist Buffon (d. 1788) and, a century

---

[19] Thompson 2005, p. 111.

[20] *ibid*, p. 112.

[21] A modern echo of these claims is to be found in the report on recent research findings by Klarreich (2000): "...in 1964 the Hungarian mathematician L. Fejes Tóth showed that a hexagonal cell capped off by part of a truncated octahedron would produce a tiny saving. Tóth pointed out, however, that the bees could have excellent reasons for choosing a slightly less efficient structure. Because the honeycomb walls have a definite thickness, it is not clear that Tóth's structure would indeed be an improvement. In that respect, the honeycomb is more like a wet foam than a dry foam. Recently, Weaire and Phelan undertook to construct two-layer foams with equal-sized bubbles, and they found that the dry foams did take on Tóth's pattern. But when they gradually added liquid, they wrote, something 'quite dramatic' happened: The structure suddenly switched over to the bees' configuration. It seems, then, that the bees got it right after all." It should be pointed out that these are mathematical or laboratory findings, not real-world ones; whatever their value may be, they do not obviate the need *to look at* actual honeycombs!

[22] Henry Brougham, *Natural Theology* 1856, p. 350, as quoted in Wyman 1866, pp. 4-5.

after him, Charles Darwin (d. 1882) himself. Darwin spoke for
many when he referred to "the extreme perfection of the cells of
the hive-bee", which he also declared is "absolutely perfect in
economizing labor and wax"; and he made the pronouncement
that natural selection "could not lead ... beyond this stage of perfection in architecture".[23]

But Maraldi's claims, despite their acceptance in scientific circles, are manifestly implausible. As Thompson has
pointed out: "The fact is that, were the angles and facets of the
honeycomb as sharp and smooth, and as constant and uniform,
as those of a quartz-crystal, it would still be a delicate matter to
measure the angles within a minute or two of arc, and a technique
unknown in Maraldi's time would be required to do it. The mi-
nute-hand of a clock (if it moves continuously) moves through
one degree of arc in ten seconds of time, and through an angle of
two minutes in one-third of a second; - and this last is the angle
which Maraldi is supposed to have measured!"[24] The established
view was thus based on calculations that are inherently implausible. Yet it was not until 1866 that this was at last demonstrated
empirically. In that year Jeffries Wyman, a professor of anatomy
at Harvard, published his findings about the extensive series of
measurements that he had made of bee cells.[25]

Wyman reported that, in the cells he had measured, none
had six sides of the same length and that, on the contrary, the
sides of any given cell varied in length by between 10% and 100%.
Their interior angles, moreover, far from being the 60 degrees of
a regular hexagon, were in fact "nowhere sharply defined", as we
can see from his drawing (*fig* 2.11). Wyman added that the shapes
of the rhombic faces – Kepler's "keel" – were also markedly irregular and variable; and indeed that some cells had four and not
three rhombs. Wyman summarized his findings with fine under-
statement. "The cell of the bee", he declared, "has not the strict
conformity to geometrical accuracy so often claimed for it". And
that, indeed, is what anyone who looks with reasonable care at
the cells of a honeycomb must also conclude (*fig*. 2.12).

Wyman sent a copy of his study to Darwin, and the two
men corresponded with each other about it. We should be clear
that Wyman's findings conclusively refute the claims that had
been made about the shape of the bee's cell since at least the time

---

[23] Thompson 2005, p. 114.

[24] *ibid*, p. 112.

[25] Wyman 1866.

of Pappus. They are also supported by common sense and by the kind of simple observation of honeycombs that anyone should be able to make. Wyman must therefore have expected – as indeed we might have expected – that, in the next edition of *The Origin of Species*, Darwin would replace his earlier description of the bee cell's purported "extreme perfection" with remarks that reflected Wyman's findings. But that is not what Darwin did. Instead, he retained his entire original text, with its admiration of the honeycomb's construction, and to it merely added a single sentence: I hear from Prof. Wyman, who has made numerous careful measurements, that the accuracy of the workmanship of the bee has been greatly exaggerated; so much so, that whatever the typical form of the cell may be, it is rarely, if ever realised.[26] In this one sentence Darwin tacitly acknowledged that Wyman had refuted everything that he, Darwin, had previously written on the subject: yet he deleted nothing from his earlier discussion. His analysis and Wyman's refutation of it appear in subsequent editions of *The Origin of Species* as a single, ostensibly consistent, narrative in which Darwin continued to refer to the "extreme perfection" of the bees' cells, and their "absolute perfection" in economizing materials - even as he acknowledged that this view of them is incorrect!

Despite Wyman's findings (and common sense, for that matter!), the view of the bee cell's structure that originated at least as far back as Pappus, that was endorsed by Kepler, and that was seemingly validated by Maraldi's observations, has not only outlived Darwin but remains the standard account today. D'Arcy Thompson, whose *On Growth and Form* was described by Stephen Jay Gould as "the greatest work of prose in twentieth-century science", is a particularly striking case in point. Thompson himself sums up the evidence against the standard view by noting that although "all the geometrical reasoning in the case postulates cell-walls of uniform tenuity and edges which are mathematically straight", in reality "the [cell's] base is always thicker than the side-walls; its solid angles are by no means sharp but [are] filled with curving surfaces of wax ... the Maraldi angle is seldom or

---

[26] Darwin 1998, p. 342. The alert reader will have noted that Wyman had stated flatly that the bee's cell does not conform to geometrical accuracy. Darwin's "rarely, if ever" is therefore misleading in its implication that *some* cells may perhaps be geometrically accurate.

never attained".[27] But then, as if he had not just made these remarks, Thomson went on to refer to "the beautiful regularity of
the bee's architecture", to the "regularly hexagonal" sides of their
cells, and to their "all but constant" angles. He ended his discussion by quoting at length and with approval Buffon's rhapsodic
description of the bee's cell as *"toute geometrique et toute regu-
liere"*![28]

The second edition of D'Arcy Thompson's work appeared in 1942. Ten years later Hermann Weyl published his influential *Symmetry*. Weyl's lengthy discussion of the bee cell in
this book is little more than a detailed (though unattributed) pré-
cis of Thompson's own discussion. Indeed, Weyl adopted
Thompson's ambivalence about the subject by declaring that
bees' cells are "not as regular" as had been assumed and then
nevertheless admiring the "geometrical talents" that had constructed them.[29]

More recent writers, throwing even this much caution to
the winds, have reverted to what we might call the pre-Wyman
position. Thus Frisch marveled at the "astounding precision" of
the bee's cell, adding that "human craftsmen could not do
work of this nature without the use of carpenter's squares and
sliding gauges."[30] "The combs are natural engineering marvels",
a paper of the authoritative Darwin Project declares, "using the
least possible amount of wax to provide the greatest amount of
storage space with the greatest possible structural stability."[31]
And in a book published in 2008 by Marcus du Sautoy, an Oxford
mathematician, we find the statement that "symmetry is nature's

---

[27] p. 114. Actually, there is no reliable evidence of the Maraldi angle *ever*
being found in a honeycomb.

[28] i.e., "altogether geometric and regular". Buffon, *Histoire naturelle,* 1753
vol. 4, p. 99, as cited by D'Arcy Thompson, *ibid.* p. 118-9.

[29] Weyl, 1952, pp. 90-92. In the preface Weyl refers to Thomson's *On
Growth and Form* with the remark, merely, that in it "symmetry is but a
side issue".

[30] Frisch (1975), quoted Bergman and Ishay 2007. Frisch even specified
standard measurements for the honeycomb: "The thickness of the cell
walls is 0.073 mm with a tolerance of no more than 0.002 mm." Bergman
and Ishay themselves refer to "the very precise hexagonal lateral symmetry" of hornets' cells, a claim the authors attempt to bolster with a photograph that – as one might expect – clearly shows them to be asymmetric.

[31] www.darwinproject.ac.uk/the-evolution-of-honey-comb, downloaded March 12, 2009.

way of being efficient and economical. For the bee, the lattice of hexagons allows the colony to pack the most honey into the greatest space without wasting too much wax on building its walls."[32] One must wonder: has Sautoy ever actually *looked at* a honeycomb?![33]

But back now to that fateful winter's night in Prague, and Kepler's observation that the snowflakes falling on his coat were "all" in the shape of a regular hexagon ("*omnes sexanguli*"). Even at first glance this statement seems implausible.[34] Street lighting in Prague at the beginning of the seventeenth century was quite rudimentary, and it seems unlikely that there would have been enough light on that stormy evening for Kepler to identify, let alone to determine precisely, the shapes of the snowflakes that were falling all around him: and this all the more so when we recall that at the age of four Kepler's eyesight was permanently impaired by small-pox.[35] Nevertheless, as far as Kepler was concerned, those snowflakes were indeed regular hexagons. Exclaiming *mehercule!* – "by Hercules" – he began to ponder the reason for that. "Why", he asked himself, "do snowflakes always come down ("*perpetuo cadant*") as hexagons?" He assumed that there must be a specific cause for their six-sidedness, for if their shape were merely a matter of chance, "why would they always fall with six corners, and not just as well with five, or with seven?" The answer, he felt, must lie in the "formative force" which "built its nest in the center" of the snowflake and from there distributed itself equally ("*equaliter*") in all directions (a hypothesis, one may note, that could apply just as well to flakes with fewer or more corners).

---

[32] Sautoy 2008, p. 13. Comp. Wilczek 2015, p. 39.

[33] Even Norman Mailer jumps into this trap. One of the protagonists of *The Castle in the Forest* (2007, p.121), Mailer's final novel, exults, regarding the bees' "hexagonal cells", that they are "a wonder to behold because they are constructed symmetrically"!

[34] One might say that it should perhaps be taken with a pinch of another crystal, that of sodium chloride.

[35] The only street lighting in Prague at this time came from coal or wood flames in iron braziers affixed to the walls of houses on street corners. See praha.eu/jnp/en/extra/light_in_streets/history.html, accessed Feb. 7, 2011.

Within days however Kepler began to discover snow-flakes that were not hexagonal at all but had "varying numbers of radii which spread in every direction", and were irregularly shaped. Kepler described them as lumpy and roundish, and thought that they were quite without beauty. He had believed that the hexagonal snowflakes he noticed initially were produced by the "supreme reason" that had existed "from the very beginning in the plan of the Creator". This suggested to him that the unattractive ones he now observed must have been "abandoned by the Master Builder". Further observations led Kepler to conclude that the hexagonal snowflakes that had originally caught his attention were clearly the "rarer" of the two kinds ("*duorum generum*"). More startlingly, he then decided that they were in fact not really six-sided at all but had "a seventh little radius bent down, like a kind of root, on which they settled as they fell, and on which they were propped up for a little while". One might have supposed that Kepler, upon discovering that most snow-flakes are not regular hexagons and that even the seemingly hex-agonal ones are not, in fact, hexagonal at all, would have recognized that the question he asked himself that wintry night in Prague was based on a false premise, and therefore should not be pursued. But Kepler never did reach this conclusion. He neither abandoned his original question – why do snowflakes always fall as regular hexagons? - nor even modified it to fit the fact, ascertained by himself, that most snowflakes are *not* hexagons. Instead, much to our bafflement, he stuck with his original delu-sional question to the very end, and concluded his essay with the bizarre comment, "I have not resolved this matter [i.e. the regu-lar, hexagonal shape of all snowflakes]. Much remains to be said on the subject before we know its cause".[36]

With these words we seem to enter a world of make-be-lieve, one in which *a priori* convictions trump unambiguous em-pirical evidence.[37] As we are about to see however, Kepler's dis-missal of the facts that he had himself ascertained is a practice

---

[36] This makes nonsense of the claim by Walker (1967) that "[Kepler] al-ways gave absolute priority to empirical evidence. If the theoretical pat-tern, however beautiful, did not fit the facts, it was discarded". See also fn. 38, *below*.

[37] Hauser (1965, p. 47) offers in this regard the very helpful insight that Kepler, as well as Copernicus, "cannot be regarded as typical represent-atives of the new, unbiased scientific spirit that based itself purely on ob-servation. In the development of their ideas and the description of their systems they allow themselves to be guided by all sorts of mystical, met-aphysical and aesthetic fancies; all sorts of fascinating geometrical and

that has continued to be followed by other scientists – even eminent ones – down to our own day. Scientific research into the
morphology of snowflakes has always been marred by a willing-
ness, one might almost call it a compulsion, to ignore or bend the
evidence in order to support a dogmatic, *a priori*, conviction that
snowflakes are shaped as regular hexagons.

Thus Descartes, in his sixth *Discourse on Meteorology*,
marveled at the snowflakes he observed in Amsterdam on the
night of February 5, 1634. They were, he persuaded himself, "perfectly-cut in hexagons, with their sides so straight and their six
angles so equal, that it would be impossible for human beings to
make anything so precise".[38] Like Kepler, he failed to explain
how he was able to determine their shape, particularly at night,
with such precision.[39] Descartes had never heard anyone speak
of such accurately-shaped snowflakes before ("*je n'avois jamais ouï
parler*"), which suggests that he had not read Kepler's work; and
he wondered "what could have formed and proportioned those
six teeth with such precision around each grain?" The next day

---

ornamental patterns hover alluringly before their mind's eye … and they
create the impression that their great discoveries were due to chance and
sudden inspiration." See also Wilczek (2015, p. 80): "Kepler's own work
[on the celestial spheres], had destroyed his model's conceptual foundation and its approximate agreement with observation did not survive in
more accurate work. *Yet Kepler never abandoned his ideal system* [my italics].
He prepared a later, much expanded, edition of the *Mysterium* in 1621.
There the accurate laws appear in footnotes, undermining the text like a
sober cross-examination that belies the words of a witness prone to fantasy. Symbol or model? Ambition or precision? In refusing to choose,
Kepler fell back into the Platonic temptation, to put his theoretical Ideal
above the reality it contradicted". See also *idem, pp. 50ff.* 112. For a
broader context see Evans 1984, *passim* and particularly p. 153.

[38] Descartes 1824, v. 5, pp. 226-239. Similarly implausible is Townsend
(1818), who reported of the snowflakes that he observed one afternoon
with his naked eye that the radii of the stars were all of equal length, that
they diverged in the same plane "at exact angles of 60 degrees", and that
the length of each radius was "about the 1/7th of an inch". Elsewhere in
his paper, however, Townsend reported his observation of snowflakes
"of irregular and unequal figure".

[39] According to *moonpage.com* the moon over Amsterdam that night was
about 86% full, but it was snowing and the cloud cover would presumably have screened out some or perhaps much of the moonlight.

however he noticed that the snowflakes which were falling were not as regularly formed (*"pas si régulièrement formées"*) as those of the previous night, though apparently all of them, too, were hexagonal. Descartes' investigations did not lead to any conclusion, but there is nothing to suggest that he ever altered his conviction that *some* flakes, at least, are "perfectly-cut" hexagons. He did not come up with a theory of how either the regularly, or the irregularly, shaped flakes are formed.

Three decades after Descartes, Thomas Hooke also was initially impressed by the regular hexagonal form of snowflakes which, as he reported in his *Micrographia*, were "all of equal length, shape and make, from the center, being each of them inclined ... by an angle of sixty degrees".[40] The branches of each flake, he added, "were for the most part in one flake exactly of the same make ... so that whatever Figure one of the branches were, the other five were sure to be of the same, very exactly". However, when Hooke looked at them under his microscope, he found that they were not after all "so curious and exactly figur'd as one would have imagin'd"; and the more he magnified their images, he reported, "the more irregularities appear'd in them". In his view, though, these irregularities were caused by "the thawing and breaking of the flake" during its descent, and so were to be construed as accidents and "not at all" as "the defect ... of Nature".[41] Hooke suggested ("I am very apt to think") that if it were possible to view snowflakes through a microscope when they are first formed in the clouds, and thus "before their Figures are vitiated by external accidents", they would be seen to exhibit "an abundance of ... neatness" (i.e., to be regular hexagons) no matter how greatly their images were magnified. However, he offered nothing to support this idea.Hooke's findings were echoed a few years later by Nehemiah Grew, the plant physiologist. "Many parts of snow are a regular feature", he declared in a paper to the Royal Society in London, and he suggested that those with irregular shapes are probably broken fragments.[42] The arctic

---

[40] Hooke 1665, observation 14. This was not a view shared by Hooke's contemporary, the great astronomer Cassini (d.1712). Two illustrations in Philip Ball, *Life's Matrix A Biography of Water* (1999, p. 193), show asymmetric snowflakes drawn by Cassini. Unfortunately, Ball does not cite (and I have not found) their source.

[41] Note Hooke's assumption that irregular natural shapes are defective.

[42] Grew 1673. According to Grew, when snowflakes deviate from the hexagonal, they always appear in dodecangular form (i.e., 2 x 6 =12).

explorer and scientist William Scoresby also confronted, in a fashion, the fact that not all snowflakes have regular shapes. His *Account of the Arctic Regions* (1820) contains a lengthy analysis, illustrated with engravings, of snowflakes which he had observed over the course of several voyages. These snowflakes, he declared, included "almost every shape of which the generating angles of 60 degrees and 120 degrees are susceptible"[43]. A pious man who would later be ordained in the Church of England, Scoresby ascribed the beauty and variety of these snowflakes to "the will and pleasure of the First Great Cause whose works, even the most minute and evanescent, and in regions the most remote from human observation, are altogether admirable".[44] In another passage, aspiring perhaps to a somewhat more scientific tone, he remarked of the "perfect geometrical figures" of snowflakes that " ... the constant regard to equality in the form and size of the six radii of the stellates; the geometrical accuracy of the different parts of the hexagons; the beauty and precision of the internal lines of the compound figures, with the proper arrangement of any attendant ramifications, and the general completion of the regular figure, compose one of the most interesting features of the Science of Crystallography."[45]

Elsewhere in his discussion, however, Scoresby acknowledged that "flakes of snow of the most regular and beautiful forms" only fell when the weather was exceptionally cold.[46] Indeed, even under those conditions it was not all but only "the greatest proportion" which were – not certainly, but - "probably

---

[43] Martin 1988, p. 40, writes that Scoresby "meticulously measured and drew the symmetry of hundreds of individual snowflakes ...Scoresby's drawings are the first accurate visual descriptions". Martin does not identify the reasons for her confidence in the accuracy of Scoresby's drawings or question whether snowflakes are ever symmetric. It should be pointed out, too, that Scoresby himself did not use the term "symmetry", and we cannot assume that he in fact recognized that the forms he drew were symmetric. Huxley in 1888 (cf. fn. 52, *below)* seems to have been the first to describe Scoresby's snowflakes as "symmetric".

[44] Scoresby, I: pp. 426-7. Kepler, it will be recalled, had attributed the shape of regularly-shaped snowflakes to "the plan of the Creator".

[45] *ibid*, pp. 431-2.

[46] Note Scoresby's equation of "regular" and "beautiful".

perfect geometrical figures".[47] At other times, when the temperature was within a degree or two of the freezing point, the snow which fell typically consisted of "large irregular flakes such as are common in Britain". The fraught question of why it was that "the will and pleasure of the First Great Cause" had not shaped these flakes, too, was not one which Scoresby chose to discuss with his readers.

At about the same time as Scoresby, a Japanese scholar by the name of Doi Toshitsura published his own drawings of snowflakes in a work entitled *Zoku sekka zusetsu*.[48] Close examination establishes that his snowflakes are not symmetric. Needham (1970, pl. IX) has published photographs of a Japanese sword-guard from the Tokugawa period (1603-1868) that is decorated with engravings of snowflakes, and these too are asymmetric.

A few comments are in order at this point regarding the view of snowflakes as symmetric, specifically, rather than as regular hexagons. The Japanese physicist Ukichiro Nakaya, who is recognized as the founder of the modern scientific study of snowflakes, criticized Olaus Magnus, the sixteenth-century archbishop of Uppsala whom we met on an earlier page, because he did "not indicate that snow crystals have hexagonal symmetry".[49] In view of the fact that there are no symmetric snowflakes, or at least none that have ever been reliably recorded, it is perhaps Nakaya rather than Olaus who is deserving of censure. We should note, though, that even if Olaus *had* believed that snowflakes have the shape of regular hexagons, it is unlikely that he would have seen them as symmetric. The concept of symmetry after all had been formulated for the first time (see Chapter Six) a mere century before Olaus wrote his work, and in Olaus' day was used largely by architects and others interested in aesthetics. Indeed, it is noteworthy that not just Kepler, but his intellectual descendants up to the beginning of the nineteenth century, did not describe the regular hexagonal configuration (as they thought of it) of snowflakes and other natural creations as symmetric. To be sure, the perfectly regular hexagons which both Kepler and Descartes imagined they had seen *are* indeed symmetric shapes, but what interested

---

[47] *ibid*, pp. 425, 431. My italics.

[48] Six of them appear in *fig*.14, p.104, in Needham and Lu (1970).

[49] Nakaya's 1954, p. 1. As we have seen, Needham and Lu level the same charge at Olaus. They seem to have been unfamiliar with Nakaya's work.

each man was the problem of accounting specifically for the hexagonal shape of snowflakes, and for the regularity of the hexagon: they were not interested in its symmetry and there is no evidence that they were even aware of it.[50] Hooke and Scoresby too referred to the regular hexagonal shape of snowflakes without describing it as symmetric. The distinction is not a trivial one. To describe something as symmetric nowadays is to associate its appearance with the many ramifications of the symmetry fallacies – i.e., with the fallacy that nothing can be beautiful unless it is shaped symmetrically, and that symmetry has always, and everywhere, been the accepted standard of good design: and of course, above all, with the fallacy that all of Nature's shapes are symmetric. Not so, of course, with the view of some of Nature's creations as hexagonal. Saying that a snowflake is a regular hexagon *may* possibly say something about how Nature goes about shaping the things she makes, but it establishes nothing about the nature of beauty or about how the things we make should be (let alone always have been) formed. The symmetry fallacy would have us make everything symmetric; there is no norm that would have us make everything hexagonal!

A question to be addressed, therefore, is why it took so long for people to allege that snowflakes or bee cells are symmetric and not just regularly hexagonal. (In fact, for reasons that are not clear, it is still relatively uncommon to find bee cells described as symmetric.) Why was the symmetry fallacy so slow to infiltrate these analyses, when it had been so rapidly accepted in many other areas of intellectual and cultural life? And this question is all the more perplexing when one recognizes that the techniques

---

[50] The failure to perceive the symmetry of a symmetric form does not occur only among students of snowflakes. In Chapter Three I point to the likelihood that although the ancient Greeks knew of forms that we regard as symmetric, they did not recognize them as such (and as we have seen, they lacked a word for the concept). Plato's solids, notably, are symmetric, but neither he nor Euclid, in his mathematical elucidation of them, ever referred to them as symmetric. This distinction should not be minimized, for all that its implications are elusive. Not all symmetric forms are regular polygons, of course, and the distinction between a form whose two halves mirror each other and ones whose angles are identical seems to be a quite fundamental one. Readers should also note, as we have discussed in an earlier chapter, the rather common error of misperceiving asymmetric forms as symmetric.

for evading and manipulating facts that are associated with the claim that snowflakes and bee cells are regular hexagons, are to all intents identical (as we shall see) to those associated with the false claim that the architecture and decorative arts of mankind are and have always been symmetric, too. I raise this question here but confess that it is not one into which I am able to offer any insight.

It is only quite recently that snowflakes have come to be described specifically as symmetric, rather than with terms like "perfect geometrical figures", "regular hexagons" and so on that had been used earlier. The first person known to have referred to the symmetry of at least *some* snowflakes was the great French crystallographer René Just Haüy, (1743-1822) who, in the year 1800, wrote that snowflakes quite often ("*assez souvent*", i.e., not always) have "*un charactière particulier de symétrie*". It took English scientists three-quarters of a century to follow this lead. T. H. Huxley, (1825-1895), was evidently the first to do so, referring to Scoresby's illustrations of snowflakes as being "always true to hexagonal symmetry".[51] The first description of snowflakes as symmetric in the German language seems, curiously enough, to have been by the aesthetic theorist Gottfried Semper (1803-1879), who adduced a basic principle of design from the alleged symmetry of the snowflake.[52] There are those who, retroactively as it were, attribute the perception of snowflakes as symmetric to people in past ages. The British mathematician Ian Stewart, for example, does so with his assertion that "the sixfold symmetry of snowflakes has been remarked on for thousands of years".[53]

---

[51] Huxley 1888, pp. 60-62. It is notable that in his survey of other writers, one of whom was Haüy, and in reporting his own observations, Townsend (1818) never used the terms "symmetric" or "asymmetric", even though he described shapes that are symmetric (e.g., "the radii were all of equal length, diverged in the same place at exact angles of 60 degrees").

[52] Semper 2004, p. 94; see also Rykwert 1972, p. 31.

[53] Stewart 1995, p. 3. Other untenable claims by Stewart are that the branches of a snowflake are "all identical" though its symmetry is only "almost perfect" (p. 9, a solecism - see Chapter 1 - which means that they are in fact *asymmetric*); "the human form has approximate bilateral symmetry" (p. 32; the same solecism); "butterflies are bilaterally symmetric, each wing bearing a mirror image of the pattern on the other" (p. 32); "orchids are a glorious, exuberant example of flowers that nearly all have striking bilateral symmetry" (p. 49); and "an asymmetric snowflake would be little more than an irregular speck of ice" (p. 190). Stewart documented his claim that "bilateral symmetry has been around for a long

Stewart provides no evidence for this statement, and in view of the fact that the concept of symmetry can be traced no further back than to the middle of the fifteenth century, we may confidently dismiss it as incorrect.[54] The same can also be said of Stewart's claim that Kepler "came up with a pretty good explanation of the snowflake's six-fold symmetry", for Kepler did not use the word "symmetry" in his essay on the snowflake and, as we have already seen, he frankly acknowledged that he had not, in fact, come up with *any* – let alone "a pretty good" - explanation of the shapes of snowflakes.[55] Nor is Stewart alone in making such errors. The distinguished physicist Mario Livio claimed that Kepler had wanted "to explain the symmetry of snowflakes", while the no less distinguished physicist Ken Libbrecht claimed that both Kepler and Descartes had hoped to understand "the precise six-fold symmetry" of snowflakes. Libbrecht documented this claim, with regard to Descartes, by quoting from a mistranslation which has the Frenchman wondering "what could have formed and made so exactly *symmetrical* these six teeth" – symmetrical in this sense being a word which Descartes did not use. (Frank (1974), whose translation Libbrecht used, claimed that he had translated Descartes' French "rather literally". In fact, he thrice erroneously

---

time" by reference to "the worm Spriggina which lived some 560 to 580 million years ago" (p. 32); however, a fossil of this worm - see ts4.mm.bing.net/th?id=HN.60800761823/6319991&pid=15.1&H= 133& W= 160 - shows that it was asymmetric. Stewart acknowledges (p.56), but without accounting for it, that the human brain is shaped asymmetrically.

[54] See Chapter Six, *below*: "The History of the Concept of Symmetry".

[55] Kepler also had the concept of symmetry mistakenly attributed to him by Koestler (1963, pp. 251-2), who translated the concluding portion of the introduction to *Mysterium Cosmographicum* (1521) to read, "I saw one symmetrical solid after the other fit in so precisely between the appropriate orbits…". I speculate that Koestler had relied on the German translation in *Kepler Gesammelte Werke*, eds. Caspar and v. Eyck, (which I have not seen) and not on the Latin text, and in turn translated that into English. However, the translation bears almost no resemblance to the Latin original, the relevant portion of which reads, "*cum igitur paucis post diebus res succederet, atque ego deprehenderem quam apte unum corpus, post aliud inter suos planetas sideret*": no "symmetry" there! Koestler's error was given a further lease on life when it was quoted verbatim by Ian Glenn, *Elegance in Science* (2010), p. 24.

rendered Descartes' *compassé* ("measured", or "well-shaped") as
"symmetric". Descartes certainly knew the word "*symmetrique*",
which had been current for a long time, and if that is what he had
meant that, surely, is what he would have said.)

The description of snowflakes as symmetric was popu-
larized in the German language by the novelist Thomas Mann,
who in *The Magic Mountain*, first published in 1924, referred to
the "absolutely symmetrical, icily regular" form of "all" snowflakes. Stranded on the mountain in a storm, Mann's hero Hans
Castorp found that the flakes were "too regular, as substance
adapted to life never was to this degree – the living principle
shuddered at this perfect precision, found it deathly, the very
marrow of death". And Mann added: "Hans Castorp felt he understood now the reason why the builders of antiquity purposely
and secretly introduced minute variations from absolute symmetry in their columnar structures".[56] Mann's recognition of the
contradiction between symmetry and "the living principle" is
profound; we will return to it on a later page.

We may note parenthetically that Mann's musings are
quoted by Hermann Weyl in his book *Symmetry*. Snowflakes,
Weyl declared there, "provide the best-known specimens of hexagonal symmetry" in nature. However, the twelve photographs
(one of them is reproduced as *fig.* 2.13) with which he purported
to illustrate this claim all show snowflakes that, as it is easy to see,
are unambiguously asymmetric![57] It was not until the 1930's that
it became conventional for scientists to describe snowflakes as
symmetric.[58] Nothing did more to entrench this fallacy than the
publication in 1936 of *Snow Crystals natural and artificial* by
Ukichiro Nakaya, whom we met earlier in this chapter criticizing
Olaus Magnus for having failed to recognize that "snow crystals
have hexagonal symmetry".[59] The importance Nakaya attached

---

[56] Mann 1946, p. 480. There is little if any evidence that Classical buildings
were symmetric (see Chapter Three), but the popular belief, here echoed
by Mann, in the deliberate "breaking' of the symmetry of ancient buildings is implausible. Symmetry is a concept that was unknown to the makers of ancient as well as primitive artifacts (see also Chapters Four and
Five *passim*).

[57] Weyl 1955, pp. 63; 64-5 and fig. 38.

[58] Seligman 1980, p. 36.

[59] (English translation, 1954.) Nakaya's achievement in creating the
world's first artificial snowflakes is commemorated by a granite marker
in the shape of a perfectly hexagonal snowflake on the campus of Hokkaido University – see the illustration in Hargittai, 1986, p. 78.

to the notion that snowflakes are symmetric is evident from the very first page of his book, where he claimed (inaccurately, as the reader of these pages will now know!) that "the extraordinary symmetric nature of the six branches of a snow crystal ... has long been an object of wonder and mystery". This, and his statement that "snow crystals have hexagonal symmetry", can only be understood as applying to *all* snowflakes. It is curious, therefore, that Nakaya presented a number of photographs that he said showed snowflakes "having bilateral symmetry" or "showing a perfect symmetry of design"[60]. One might suppose these descriptions to be superfluous, for if all snowflakes "have hexagonal symmetry", why would it be necessary to single out the symmetry of some for special mention? By implication, at least, Nakaya's remarks raised the question of how many snowflakes "have hexagonal symmetry". Are all symmetric? Or only some? Or – perhaps – *none*? And indeed, it can readily be seen that his photographs of snowflakes with allegedly "perfect symmetry" *all* show snowflakes that are – asymmetric!

Nakaya's attempts to deal with this problem were hesitant and ambivalent, and bring to mind Kepler's reluctance to acknowledge that his own observations contradicted his belief that snowflakes are regular hexagons. Thus, after referring to snowflakes "having bilateral symmetry", Nakaya commented that "it is surprising that crystals of such a unique form were observed again and again", a remark that implies that symmetric snowflakes are distinctive.[61] In another passage he acknowledged that asymmetric snowflakes "are not less often observed" than symmetric ones. Then, seemingly under the momentum of this startling disclosure, he went on to declare rather strangely that, "considering the frequency of [their] occurrence, these asymmetrical crystals may be more common than the symmetrical ones". Next, and as if retreating from this acknowledgment, he added: "Of course, the question is a matter of degree ... the criterion of regular crystals and asymmetrical ones is rather arbitrary". Questionable in itself, this is not a point he had thought to make earlier when he declared that snowflakes are symmetric.

---

[60] Nakaya 1954, crystals # 512-515 and pp. 407-8; p. 103, # 208.

[61] *ibid*, p. 41.

Nakaya's two-step continued with the even bolder – but still insufficient! – acknowledgement that "a perfectly symmetrical crystal is rarely observed."[62] At no point did he provide evidence that "a perfectly symmetrical" snowflake has *ever* been observed. To the very end Nakaya remained uncertain about the conclusion to which his data pointed. His findings, he declared in his summary of them, "lead us to the conclusion that snow crystals do not always show complete hexagonal symmetry". In the very next sentence however he modified this statement with an even more candid one that, "as a matter of fact, *most* [my emphasis] of the natural snowflakes of the hexagonal plane type do not exhibit … complete hexagonal symmetry", adding that they "deviate more or less" from symmetry.[63]

This statement implies a need to reconsider what the norm for snowflakes really is, but Nakaya chose instead to declare that snowflakes (by which he presumably meant *all* snowflakes) "have a tendency to grow" symmetrically.[64] But he offered no theoretical or empirical basis for this claim, which I think is inconsistent not only with the empirical data but with some of his own observations; it also of course begs the question of how far this purported "tendency" determines their ultimate shape. All in all, then, and by his own reckoning, Nakaya failed to justify his enthusiastic reference at the outset of his book to "the extraordinary symmetric nature of the six branches of a snow crystal". Did he ever think, one wonders, that he perhaps owed old Olaus an apology?!

Later works by both popular and scientific writers also express a thoroughly ambiguous attitude to the question of what the shape of snowflakes is. Edward LaChapelle, for example, in his *Field Guide to Snow Crystals,* declared rather strangely that "most snowflakes are … often asymmetric".[65] In somewhat the same vein the organic chemists István and Magdolna Hargittai refer to "the magnificent hexagonal symmetry" of snowflakes, but then retreat to the statement that the symmetry of snowflakes is "practically perfect" – a construct that careful readers will take to mean that snowflakes are asymmetric.[66] Corydon Bell

---

[62] *ibid,* p. 38; and comp. p. 19.

[63] *ibid,* p. 106; comp. p. 37.

[64] *ibid,* p. 37.

[65] LaChapelle 1960, p. 10.

[66] Hargittai 1994, p. 128.

acknowledged that asymmetric snowflakes "probably fall in greater number and more frequently than other types of snow" but then added that "By most standards they cannot be called beautiful. They give the appearance of having been born in troublesome times or on the wrong side of the storm"[67] The physical chemist Joe Rosen, who has written extensively on symmetry, somewhat more circumspectly declared that "snowflakes *generally* [my emphasis] possess a single axis of sixfold ... symmetry", and cited in support of this claim some photographs which according to Nakaya were doctored (and which moreover do *not* show symmetric forms!) as well as a statement by - Kepler himself.[68] Other scientists however affirm the snowflake's symmetry unhesitatingly. Burke, in his study of the history of crystallography, flatly declared that "the snowflake exhibits hexagonal symmetry".[69] The physicist Leon Lederman, a Nobel laureate, wrote of the "graceful symmetry" of Nature's creations, among them the snowflake.[70] Weyl, for his part, declared he was inclined to think with Plato that "the ... mathematical laws governing nature are the origins of symmetry in Nature", one of the best known expressions of which he claimed is the hexagonal symmetry of the snowflake.[71]

The most prolific scientific writer on snowflakes today is Kenneth G. Libbrecht who, although chairman of the physics department at Caltech, has written mainly popular works on the subject. Even in his scientific publications however Libbrecht continues the long tradition of ambiguity regarding the snow-

---

[67] Bell 1967, p. 207.

[68] Rosen 1975, pp. 86-88.

[69] Burke 1966, p. 35.

[70] Lederman 2008, p. 13. "Graceful symmetry", according to Lederman, is also characteristic of a flower's petals, a radiating seashell, and "a noble tree's branches and the veins of its leaves"; he also refers to "the ideal symmetrical disks of the Moon and Sun". Of course, none of these are symmetric.

[71] Weyl 1955, pp. 7 - 8. Weyl offered nothing to support his implication here that Plato recognized, let alone explained, "symmetry in Nature" – or even knew the concept of symmetry. On Plato's evident ignorance of the concept of symmetry see p. 35, fn. 51, *above*.

flake's shape. "Most snow crystals", he wrote in *Reports on Progress in Physics*, are "usually [!] without the high degree of symmetry present in well-formed specimens". The simple meaning of this statement is of course that snowflakes are "usually" asymmetric. Further on in the same paper however Libbrecht reversed course with the statement that a "combination of complexity and symmetry" is in fact to be "seen in many specimens" – a claim that he illustrated with a photograph showing a snowflake that he claimed possessed "especially precise sixfold symmetry": but which in fact shows an *asymmetric* snowflake (*fig.* 2.14).

Libbrecht's descriptions of the growth of snowflakes extend this confusion. "The six arms of an individual crystal travel together" as they fall from the clouds, he writes, and "grow in synchrony, simply because they each experience the same growth history."[72] We need not tarry here on the point that synchrony has nothing to do with the matter – things can grow at the same time, after all, without growing in the same way. The reference to "the same growth history" is of course objectionable, not only because we do not know what their growth history is but because, if they had indeed had the same growth history, their shapes would be identical: but, as we have seen, it is a matter of fact that they are *not* identical, and therefore their growth history cannot have been the same!

Rather than pursue this point however we should note Libbrecht's acknowledgment further on in this discussion that, except when conditions are "ideal", the six arms merely develop "*roughly* [my emphasis] the same pattern". In other words, they are asymmetric. Libbrecht did not describe the "ideal" conditions which allegedly produce symmetric snowflakes, or vouchsafe his readers so much as a hint as to the frequency with which these conditions occur - or indeed his evidence for suggesting that they *ever* occur. "Symmetry is inherent in snow crystals", he declared at another point, though again without identifying either the meaning of this statement or his evidence for making it. Reversing himself again, he acknowledged that the symmetry of snowflakes is "fragile and *never perfect* [my emphasis]", and that their forms are merely versions of "imperfect ... symmetry". Must we not suppose, if their symmetry is "never perfect", that snowflakes

---

[72] Libbrecht 2006, p. 12. (Libbrecht incorrectly states that Kepler dedicated his book on snowflakes to "his patron Emperor Rudolf II". In fact, the patron to whom Kepler dedicated the book was an imperial councilor by the name of von Wackenfels.)

are always asymmetric? And is not "imperfect symmetry" merely a misleading way of saying – asymmetry?

We have seen that although Kepler retreated from his initial observation that snowflakes are "always" hexagonal, it became conventional for investigators after him to repeat that claim. This belief is sometimes based on more than (inaccurate) observation generalized to account for *all* snowflakes. Snowflakes, Nakaya and Libbrecht among others have suggested, are *necessarily* regular hexagons because their structure is determined by the regular hexagonal molecular structure of water itself.[73] The problem with this suggestion is that water molecules are *not* symmetrically-shaped – see *fig* 2.15. Even if they were, of course, it could not be their shape that determined the shapes of asymmetric snowflakes.

In 1966 two of Nakaya's former colleagues at Hokkaido University, Choji Magono and Chung Woo Lee, both meteorologists, called Nakaya's work into question with a short paper which contained the tactful but hardly accurate statement that it merely "modified and supplemented" Nakaya's study. Their research, the two men were careful to say, had been undertaken with the "agreement" of Nakaya; and they added that they "would have liked to ask Dr. Nakaya's opinion" of their work but that unfortunately he died before they completed it.

Actually, the two meteorologists' paper can be seen as a vigorous critique of Nakaya's book, and as such is an unusual departure from the deferential culture which, as I understand, prevailed at the time in the Japanese academic world. Boldly, Magono and Lee declared that Nakaya's classification of snowflakes was "too simple". In particular, they criticized his analyses of "unsymmetric, modified or rimed" snow crystals and stated

---

[73] So too Furukawa 2007, pp. 70 - 71. Comp. Mason (1992) that "the hexagonal symmetry of a snow crystal is a macroscopic, outward manifestation of the internal arrangement of the atoms in ice". The – debatable – assumption here is that the shape of a building block determines the shape of the overall structure? But the point is moot, of course, since there is no solid evidence that *any* snowflakes are symmetrically shaped: or even, that snowflakes are invariably hexagonal!

bluntly: "In actual cases most of snow crystals are irregular, unsymmetric, modified or rimed". The typology they devised more than doubled the number of types identified by Nakaya.[74]

Magono and Lee's reference to "most" snowflakes implies that some snowflakes *are* symmetric.[75] However, there is no empirical evidence that snowflakes are ever shaped symmetrically. Nakaya claimed that he had observed symmetric snowflakes "again and again" but of the more than five hundred photographs of snowflakes that he published in his book, he described only five as symmetric: and all of these are in fact, without doubt, *asymmetric*. Not one of Nakaya's photographs is of a symmetric snowflake, and we must doubt whether he in fact ever saw even a single snowflake that was symmetric.

The same is true of the photographs that Libbrecht claimed were of symmetric snowflakes. Not one of them is symmetric; and their asymmetry is readily apparent to the naked eye. Equally, not a single symmetric snowflake is to be found in that remarkable collection of photographs taken by Bentley in Vermont.[76] (Nakaya thought that Bentley had doctored at least some of his photographs).

Thus it can be said, with considerable confidence, that on the basis of what the naked eye is able to perceive, without the assistance even of simple measuring instruments, *no photograph of a symmetric snowflake has ever been published, whether in scientific or popular works.* What Wyman found regarding bees' cells is also true of snowflakes. There is no evidence that they are ever symmetric.

The underlying theme we have traced here is of a determination to establish that bees' cells and snowflakes – and beyond them, Nature's forms altogether – are shaped symmetrically. In pursuit of this goal men of great scientific eminence have ignored common sense and the rules of empirical evidence, with the result that palpable absurdities that would be the undoing of a freshman's essay have been acclaimed as contributions to science.

---

[74] Magono 1966 - for example, types P2 b and c: plates 62 - 65. Magono and Lee were evidently unaware that Kepler subsequently discovered that his "hexagonal" snowflakes in fact had *seven* branches.

[75] Magono and Lee do not explain what they mean by "modified" snowflakes, but it does not seem as if they are suggesting, as e.g. Hooke had, that irregularly-shaped snowflakes were modified from their original symmetric shape.

[76] Bentley, 1961.

Some scientists however have hinted that in a certain sense Nature's creations are simultaneously symmetric *and* asymmetric. Thus Darwin came up with the category of bees' cells of "typical form", which he distinguished from the irregular shapes reported by Wyman. Nakaya, too, contrasted the "tendency" of snowflakes to be symmetric with their *actual* asymmetric form, and it is possible that Libbrecht had something like this in mind when he claimed that symmetry is "inherent" in snowflakes, even if it is seldom (we have shown in fact never) seen in them. An earlier anticipation of these remarks may perhaps be found in Kepler's notion that some snowflakes he observed conformed to "the plan of the Creator" while others had been "abandoned by the Master Builder".[77] The notions to which these men evidently allude bring to mind the speculations of F. M. Jaeger, (1877-1945), a Dutch professor of chemistry, in his *Lectures on the principle of symmetry and its application in all natural sciences.*[78] Jaeger acknowledged that he was strongly drawn to symmetric forms, and wrote of the "splendor and fascinating beauty" that he believed symmetry confers upon "a great number of living creatures" such as radiolaria, medusa, diatomae, corals, starfish and "innumerable flowers".

Nevertheless, Jaeger was led by his instincts as a scientist to ask a question that, for all its importance, many scientists have avoided. "How far", he wanted to know, "can we really speak of true 'symmetry' with respect to the geometrical properties of objects observed in nature?"

The question is very much to the point. Jaeger's answer to it however merely resurrected fallacies that had been repeated over and over again since Kepler's time. In a section entitled "Observed disagreements between crystallographic and physical symmetry", Jaeger distinguished the ideal ("crystallographic") shape of a thing from its actual ("physical") shape. He likened the difference between them to a well-governed society in which most people are law-abiding but a minority "behave

---

[77] In what may perhaps be in much the same vein, the psychologist Puffer (1905, p. 10) claimed that there is "a hidden symmetry" in some asymmetric forms. Unfortunately, she did not explain how she detected it or how we might do so too.

[78] Jaeger 1917

not as they should do".[79] In Jaeger's analogy the well-governed society is Nature, with its laws. Most forms obey Nature's laws, one of which is the requirement that they be shaped symmetrically. A minority however flout this law and adopt asymmetric shapes.

In another section Jaeger attempted to buttress this view with a different line of thought. He began by asking how it is that we are able to identify a particular object (an oak leaf, in his example) which we have not seen previously. His answer is that we have in our minds a construct of the *ideal* image of the oak leaf – the image, that is, of what the completely developed and perfect oak leaf looks like; and we instinctively use this image to recognize any actual oak leaf that appears before us.

This is a suggestive, if not entirely original, hypothesis. Having stated it, however, Jaeger went on to declare that although Nature's *actual* forms are "never" symmetric – the one half of the oak-leaf appears never to be precisely the same as the other half; the alum crystal never has twelve accurately equal angles, etc. - in their *ideal* forms (which we never see) they are *always symmetric*.[80] This startling and highly dogmatic assertion reaches the reader unencumbered by substance. We are merely *told* that the ideal form of a thing is always symmetric. That it is so Jaeger does not demonstrate; why it should be so he does not explain. We are handed an unverified and indeed unverifiable pronouncement, no more than that.[81] That this pronouncement is contradicted by some of his other utterances does not occur (it would seem) to Jaeger. For if no symmetric shapes exist in the material world then surely nothing in it (to use his figure of speech) "behaves ...as it should do", leaving us either with a very strange view of reality or a very vulnerable understanding of it. Yet we have also learned from Jaeger of the joy he experiences in the "splendor and fascinating beauty" that, in the world of *actual* perceptions, symmetry confers upon everything from radiolaria to flowers. Surely, if symmetric forms do not exist, neither can the beautiful symmetric natural forms that give him so

---

[79] *ibid*, p. 167.

[80] *ibid,* p. 6.

[81] Jaeger (*ibid,* p. 7) justifies this: "as regards living organisms, it can hardly be hoped within a measurable space of time to connect their intimate nature with the constant occurrence of their typical external form in any direct way". He claims however, but altogether unconvincingly, to have given "a rational explanation" of the two forms "in some cases" regarding crystals.

much joy. There is not much to be gained, one suspects, from pursuing Jaeger's paradoxes further.[82]

Yet there is an even more paradoxical construct which we should note here briefly. Symmetry, the British chemical scientist J. D. Bernal declared, "is attributable to all structures endowed with a structure [!], and this holds true even if [a structure] is the negation of symmetry, asymmetry, where every part only resembles itself." Which is to say (if I understand Bernal correctly), that all structures that are structured are symmetric even when they are asymmetric... Bernal then went on to say that although the symmetry of crystals "had been assumed to be perfect", recent research has shown that "in a very large number of cases the crystal is only perfect ideally, whereas really it contained geometrical imperfections or singularities, known as dislocations, which play an essential part in crystal growth.... The dislocations are not an absence of symmetry but symmetry of a different order ... a perfect crystal could exist but it could never, as such, come into existence".[83]

These are extremely subtle observations, sounding indeed more like a Zen *koan* than a scientific statement. They leave one to wonder what kind of research could establish that crystals are "only perfect" – which is to say, symmetric – *ideally* but not actually. How do we know that perfect (=*symmetric*) crystals "could exist" but yet can never "come into existence"? What in fact does that phrase even mean? And how can an asymmetric crystal be a symmetric crystal? My non-scientific mind, at least, boggles![84] "Imperfect" symmetry however is a solecism that we

---

[82] Comp. Wyman (1866, p. 18) who, otherwise so empirical, declares that in the shape of bee cells "there is a constant approach to the perfect form" though "perfection is never reached".

[83] Bernal 1955; other fallacies of Bernal (1937) include: "Symmetry is a character of nature anteceding any human construction. It is obvious in the flower, the starfish and the snowflake..."; "Something of our appreciation of symmetry must be inborn..."; and "The conscious apprehension and understanding of symmetry first [came] to our knowledge with the Greeks and their five regular solids".

[84] Another baffling construct has been put forward by the cosmologist John Barrow, who draws a distinction between "the simplicity and economy of the laws and symmetries that govern nature's fundamental forces" and, on the other hand, the asymmetric forms that we see when we look at the world around us (Barrow, n. d.) In the latter, he says, "we

can classify. It is after all simply a reluctant way of referring to asymmetry. What we can take away from Bernals' remarks therefore is that symmetric crystals do not actually exist. All crystals, in other words, are asymmetric.

We have seen how the misperception that Nature's artifacts are symmetrically shaped can be used as evidence of transcendent forces at work.[85] Among the instances of this that we have encountered was Darwin's belief that the "perfection" he insisted on seeing in the structure of the bee's cell confirmed his ideas about the purposeful evolution of the natural order of things. Similarly Weyl, who anachronistically attributed to Plato the idea that "the mathematical laws governing Nature are the origins of symmetry in Nature" – all of whose forms, Weyl went on to say, "are inherently symmetric".[86] And the distinguished physicist Leon Lederman, a Nobel prize-winner, claimed that the symmetry of Nature's shapes "in the world around us" affirms that "perfect order and harmony" underlies "everything in the universe".[87]

---

do not observe the laws of nature; rather, we see the outcome of those laws". These outcomes, he continues, "are much more complicated than the laws that govern them ... *It is possible to have a world which displayed an unlimited number of complicated asymmetrical structures, yet is governed by a few, very simple, symmetrical laws. This is one of the secrets of the universe*" [my italics]. Barrow's distinction between a law and its outcome is not entirely clear here. If we cannot "observe" the former, how can we know – or is it merely a dogmatic belief – that the shape of a particular tree, for example, is the "outcome" of that law? And what exactly *does* "outcome" mean in this context? ("Symmetrical laws" presumably refers to symmetry in the conventional sense, and not to symmetry in the sense employed by modern science – see pp. 9-10, *above*. But then one must ask, "What *is* a symmetrical law"?)

[85] It seems important to emphasize again that although Kepler, Fontenelle, Scoresby, and doubtless many others believed snowflakes and bees' cells were in the shape of equal-sided hexagons, they did not see those forms as symmetric, specifically. This distinction should be kept in mind in any study of symmetry.

[86] Weyl, op. cit., pp. 7 - 8. Weyl leaves his readers to wonder what he could have meant by "inherent" symmetry.

[87] Lederman, 2008, p. 14. Lederman went on to say that "Humans, for thousands of years, have been drawn instinctively to equate symmetry to perfection. Ancient architects incorporated symmetries into designs and constructions. Whether it was an ancient Greek temple, a geometrical tomb of a pharaoh, or a medieval cathedral..." Having written so much nonsense he returned at one point to his role as Nobel prize-winning physicist and stated flatly (*ibid*, p. 65): "There is no center about which

Nevertheless, the facts reviewed in this chapter indicate that, no matter how perfect the order and harmony of the universe may be, they do not require that Nature's artifacts have symmetric shapes. Indeed, the opposite is true, for those artifacts, as far as we can determine, are always asymmetric... [88]

But why is this? Why does Nature seem to choose only asymmetric forms? Why does she not create only symmetric forms, instead? Or both asymmetric and symmetric ones? The following observations may help in some small degree to clarify these deep questions.

We can start by noting that the transformation of a symmetric into an asymmetric form, and the transformation of an asymmetric form into a symmetric one, are fundamentally different processes. In a symmetric shape every detail on the right half must be mirrored by every detail on the left half. For an asymmetric shape to be transformed into a symmetric one, therefore, a high level of precision is called for so that all the features on one half that do not mirror those on the other are altered to conform precisely to the latters' mirror image.

By contrast, the creation of an asymmetric shape out of a symmetric one can be achieved merely by making any change at all, whether big or small, simple or complex, on the right (or left) half, and not making the mirrored version of it on the other half.

The one transformation, accordingly, requires highly specific changes to be made; the other transformation is accomplished merely by any change at all on one half of a shape that is not mirrored on the other half.

A further distinction between the two processes is that, because of the precision required, it is only in the rarest instances that the creation of a symmetric from an asymmetric form will occur unintentionally; whereas the creation of an asymmetric form may be either intentional or unintentional.

The passage of time, which is an integral part of Nature's

---

everything turns in the universe." This is to say that the universe itself is asymmetric...

[88] Alexander, Neis, Alexander (2012, pp. 204-5), in a particularly silly assertion, would have us believe that natural forms (as well as churches and the Parthenon) are symmetric, and that the reason for this is that "symmetries occur because there is no good reason for *asymmetries* to occur".

scheme, unendingly subjects all material things to the processes of change. All forms are therefore ephemeral. A sapling grows from a seed and a tree grows from the sapling; branches extend from the tree's trunk, and twigs from them; leaves and perhaps blossoms and fruit sprout and fall; eventually the tree and all that it has put forth decays and dies. At every instant of its existence the tree undergoes changes: its detritus too is subject to ceaseless change. The myriad interactions which lead to these successive changes will only in the very rarest instances give the tree or its derivatives a symmetric form. More typically – almost invariably – they will cause the transformation of an asymmetric form into another asymmetric form. And this is how it is, I dare say, with all the trees which you and I have ever seen.

But suppose, as is not *entirely* impossible, that at some stage the forces of growth or decay naturally cause a tree to ac-quire a symmetric shape. How long would it continue to retain its symmetry? After all, the passage of time during which the tree became symmetric does not cease now that the tree is symmetric. Time will continue to pass and bring with it changes that will continue to affect the shape of the tree. Merely the smallest of these changes is likely to deprive this particular tree of its bizarre symmetry: and only the most random of coincidences would en-able the tree to retain it. Miniscule as was the likelihood of the tree becoming symmetric in the first place, the likelihood that the forces of change will continue with each ephemeral transfor-mation to create another symmetric form is virtually *nil*. The overwhelming odds are that it will return instead to an asymmet-ric configuration.

And all this is true in general and not only of trees, of course. Time causes change; arguably, the two are inseparable. It is a near certainty that naturally-occurring changes will preserve the asymmetry of a form and will make a form asymmetric again if – by a fluke – it happened to have acquired a symmetric shape. This is one of the reasons why Nature favors asymmetry. It is also perhaps why we should think of asymmetry and not of symmetry as rational. In a world without time and thus without change symmetric forms would endure; symmetry, in other words, is consistent with the cessation of time. But in the world that we inhabit, a world of time and therefore of change, symmetry is in-herently vulnerable, unstable and even unpredictable (the odds against it being as great as they are). Whereas asymmetry sur-vives almost every challenge.

A consideration having to do with our perception of a shape and not with its objective appearance is perhaps worth noting at this point. In the nature of things our glimpses of symmetry are necessarily transient and ephemeral. We are after all not often stationary and as we move, our perspectives on things change, so that what appears to be symmetric from one position will – unless it is a perfect sphere – appear to be asymmetric from any other. If we stand anywhere on a line that is at an angle of 90 degrees to the precise center of the façade of the basilica of St Peter in the Vatican, for example, we will see an impeccably symmetric structure. Yet for all the care lavished on its symmetry – a roundel containing a clock on the upper right-hand corner of the façade, for instance, is matched by an identical one in the identical position on the opposite corner – the moment we move to one side or another of that line, even if only by a single step, the façade no longer presents a symmetric view to us. Our perspective continues to be asymmetric no matter how many steps we take away from the line, and it is only by returning to any point on the line that we can recover the earlier symmetric view.[89]

There are accordingly a vast number of possible asymmetric perspectives on a symmetric object but far fewer symmetric perspectives on it.[90] It is ironic perhaps that, for all the compulsive care lavished on the design and construction of symmetric buildings, their symmetry is only seldom visible to people who happen to be looking at them, something for which we can be thankful.

We see from this that movement – a natural condition for us – is the friend of asymmetry and the enemy of symmetry. It is only in a world in which we never move in relation to the symmetric projection of every object around us that this would cease

---

[89] Cf. the observation of Kames (1845, p. 444, n): "A square field appears not such to the eye when viewed from any part of it; and the centre is the only place where a circular field preserves in appearance its regular figure".

[90] Gandy (1805, pp. vii - viii) uses this as an argument for what we would now call asymmetric buildings: "Uniform buildings have but one point of view from whence their parts are corresponding; from every other point they fall into the picturesque by the change of perspective, which is an argument drawn from Nature, that the picturesque is the most beautiful…"

to be the case. Such a world is horrible to contemplate, and virtually impossible to visualize.

We saw earlier that the symmetry of material forms is, in
an objective sense, an ephemeral condition; and now we have
also established that in a specific but significant subjective sense
symmetry is a *static* phenomenon. In both respects, symmetry
contravenes fundamental aspects of Nature, our inevitable movement in time and space.

The concept of symmetry is conjoined, as we will see in
Chapters Six and Seven, to the concept of simplicity. This brings
symmetry into conflict with another fundamental aspect of Nature, too. I am referring to that most mysterious and wonderful
penchant of Nature for variety. From the room in which I am
writing I have a vast view across the upper Sonoran desert of Arizona. In the distance I see several ranges of mountains, each with
its own unique shape; and closer in, reaching into the grounds on
which my house stands, are boulder - strewn slopes on which an
awesome variety of plants – cacti, trees, shrubs – grow in profusion: soon, Spring being almost at hand, many of these will presently adorn themselves with the luridly-colored blossoms characteristic of desert flora. A very large variety of birds, from hawks
and owls to hummingbirds, live here, as do insects and animals,
many of them quite disagreeable to human beings, such as spi-
ders, scorpions, rattlesnakes, javelinas, coyotes, Gila monsters,
bobcats and mountain lions. And it is not only that each of these
species is distinct from the others but that each animal, bird, tree,
plant, boulder – everything out there - varies in size, shape and
color from all the others of its own kind. On no two of the many
saguaro, barrel, and other cacti I can see through my windows
are the furrows and innumerable needles of their shells arranged
in the same way. The configuration of twigs and branches on each
of the Palo Verde trees I can also see, now swaying vivaciously in
the wind, or the leaves and seed pods of the eucalyptus trees, or
the fronds of the palms: each is distinctive and unique.

I understand, to be sure, that this variety reflects genetic
differences and the effects of the immediate physical habitat, but
it is not fully explained by them: and although the reasons for
Nature's almost limitless variety are unclear we sense, I think,
that it is an essential part of her scheme of things.

And here too the juxtaposition of symmetry and asymmetry is an issue, for variety is well served by asymmetry and it
is at odds with symmetry. In a symmetric shape the two halves
mirror each other; but an asymmetric shape with just one of those
halves can have an unlimited variety of shapes (except one which

is the mirror image of the other) on the other half. Asymmetry, therefore increases by a factor close to infinity Nature's ability to fill the universe with varied shapes. A universe of symmetric shapes would be far less varied and would contain much less information – and therefore would be far less interesting and attractive to us – than the universe of asymmetric shapes we are fortunate enough to inhabit.[91] It would lack the "peculiarity", as Hogarth called it, "that leads the eye a wanton kind of chase, and from the pleasure that gives the mind, entitles it to the name of beautiful".[92]

Immune to the pleasure of this wanton chase are the two authors of a scholarly paper who in all solemnity intone their conclusion that we "prefer complex figures only when they are symmetric".[93] In their argument bilaterally symmetric shapes, being simpler than asymmetric ones, are more readily comprehensible to us *and are therefore preferable*! These authors, one may suppose, would rather gaze at the all-too-intelligible façade of the Seagram building in New York than at the intricate patterns of hoar-frost on one of its windows or, perhaps, of a Jackson Pollock painting hanging inside on one of its walls. But they are surely in a very small minority.

Or so one hopes.

---

[91] Lorand 2003-4.

[92] Hogarth 2007, p. 33.

[93] Eisenman and Gellens 1968. Comp. Montesquieu (in his "Essay on Taste", the section on "the pleasures of symmetry"): "*la raison que la symetrié plaît à l'âme, c'est qu'elle lui épargne de la peine, qu'elle la soulage, et qu'elle coupe pour ainsi dire l'ouvrage par la moitié*". ("The reason that symmetry pleases the mind is that it saves it trouble, that it gives it ease, that it cuts its work, so to speak, in half." A very sad and troubling approach, surely, to beholding Creation.)

CHAPTER THREE
SYMMETRY IN
THE CLASSICAL WORLD

3.1 10th-century BCE cinerary urn amphora

> *"I cannot imagine, indeed, how those ideas of symmetry*
> *as characterizing Classical architecture… which have*
> *prevailed since the 16th century, should have ever*
> *gained currency; for I do not find a trace of it either in*
> *buildings or in authors. At Pompeii there is not a*
> *single house whose plan or elevation are subject to*
> *the rules of symmetry. Cicero and Pliny speak much in*
> *their letters of the aspect, position, and arrangements*
> *of their country houses; but of symmetry not a word."*
>
> -Viollet-le-Duc[1]

Statements to the effect that the ancient Greeks and Romans had a "passion for symmetry"[2] have been bandied about for so long

---

[1] 1987, vol. 1, pp. 160 – 161; see also *ibid*, vol. 2, p. 267. In his *Dictionnaire raisonné de l'architecture française …* (1868) Viollet-le-Duc also addressed the view that symmetry, in the modern sense, was a requirement of ancient art and architecture. "In justice to the Greeks", he wrote, they never assigned to symmetry "so flat a sense" as the one in which it is used today. Indeed, in the modern sense symmetry is "an operation so banal and insignificant that the Greeks had not even the idea of defining it". He continued: "it is certain that the Greeks did not consider what we today understand by symmetry to be an essential element of art in architecture… That men strongly adhered to the symmetry introduced in the 16th and especially in the 17th century in Italy and France is a mark of their intellectual inferiority". Viollet-le-Duc's adamancy in this regard is in striking contrast to some of his own architectural work, in which he showed himself by no means averse to symmetric design. His drawing of an "ideal" Gothic cathedral in the Dictionnaire for example,shows a structure that is clearly symmetric; and his church of Saint-Denys-de-l'Estreé is almost a caricature of symmetric design. The attempt of Pevsner (1969, p. 43) to account for Viollet-le-Duc's contradictory positions does not strike one as altogether persuasive. (As we will see in Chapter 6, symmetry was first introduced and became firmly established in the 15th century.)

[2] Gardner 2005, p. 102. Compare the claim of Sturgis (1905, article on "symmetry") that the Ancients always required their buildings and other artifacts to be "perfectly symmetrical". The standard – fallacious - view of symmetry in the Classical world can be traced no further back than to the middle of the fifteenth century, when Leon Battista Alberti (1996, VI. 3) declared that the Greeks and Romans unfailingly saw to it that the left and right halves of anything they made mirrored each other in even the minutest detail. To this day restorers routinely defer to this fallacy when they replace missing parts of ancient statues (such as the right wing of the Winged Victory of Samothrace–see Landauro 2013) - and monuments

and with such assurance that one might be forgiven for believing
that there is nothing more to be said on the subject.

But a number of considerations suggest otherwise.
Perhaps foremost among them is the fact that neither Classical
Greek nor Latin had a word or phrase that denoted the mirroring
of the right and left halves of a shape: which leads one to wonder
how people could have expressed their "passion for symmetry"
if they lacked a term for it? Are we perhaps to believe that their's
was a love that dared not speak its name? Or is it more realistic
to suppose that if people had no word or phrase for symmetry,
this was probably because they not only had no "passion" for it
but were altogether ignorant of the idea? Awareness of left-right
differences, indeed, according to James Hall, "pervaded Greek
culture".[3]

To be sure, συμμετρία, *symmetria*, is a Classical Greek
word and was later incorporated into Classical Latin. Its meaning
in those tongues, however, was not at all the same as the modern
meaning of "symmetry", that in fact only dates back to the
middle of the fifteenth century.[4] As Sitte remarks, "The word is
Greek, yet it can easily be demonstrated that all of Antiquity
understood something quite different by it than we do, and *was
not acquainted with the modern theoretical concept of symmetry – that
is, of a mirror-image likeness of right to left*" [my emphasis]. Sitte
adds, "the word expresses something for which we today lack a
term. We can therefore not translate the old word *symmetria*
without paraphrasing it".[5]

Sitte is clearly correct in asserting that the modern
concept of symmetry was unknown to the Greeks and the
Romans; but the implication in his statement that we know what
"Antiquity understood" by *symmetria* is just as clearly *not* correct.
Indeed, as we can see from two of Plato's later dialogs, the

(such as certain reliefs on the Ara Pacis in Rome, for which cf. Iacopi 2008,
pp. 15, 24). These are not "restorations", but alterations!

[3] Hall 2008, p. 15. "Most of Aristotle's theories", he adds, "supported the
primacy of the right side."

[4] See Chapter Six. The term's earliest recorded use in the modern sense is
in the mid-15th century Latin *Commentarii* of Pius II (d. 1464); at the end
of the 15th century it occurs a number of times in the Tuscan vernacular
in the novel *Hypnerotomachia Poliphili.*

[5] Sitte 2006, p. 189. This work first appeared in 1889, which is to say that
his unassailable point here has been ignored by most architectural
historians for well over 125 years.

Philebus (Φιληβος), and the *Sophist* (Σοφιστησ), the word's original meaning is almost impenetrably obscure.

The main object of the *Philebus* is to identify the relative superiority of wisdom and pleasure. Jowett, in the introduction to his translation, complained of the "confusion and incompleteness" that mar the work. Nowhere are these more evident than in Plato's analyses of the relationship between symmetry and, initially, beauty and truth (καλλει καί συμμετρία καί αληθεία). These, when they are all present, make something good (αγαθου). In another passage in the *Philebus*, however, Plato declares that symmetry and measure (μέτρου) are the necessary components; but he then goes on to say that measure and symmetry are identified with beauty and with virtue (άρετή). These combinations are confusing – they seem inconsistent with each other – and they suggest that Plato was not working with a clear idea of what he meant by *symmetria*. Yet, we need not get bogged down in those other terms if we recall that numerous discussions of beauty, truth, good and virtue are found in other works of Plato – they will be familiar to all Plato scholars - and although his thoughts on them are not always entirely consistent and show an evolution over time, it is not all that difficult to gain an impression of what Plato meant by them. However, that is not the case with *symmetria*. Unlike with beauty, truth, good and virtue, references to *symmetria* both in the *Philebus* and in Plato's other works are infrequent, and we do not have so much as an inkling of what Plato really meant by them. A single reference, also in *Philebus,* allows us to detect one, probably quite minor, aspect of its meaning. *Symmetria*, Socrates declares, is to be found in the camp of Wisdom rather than in that of Pleasure, "for nothing is more immoderate than pleasure"; and from this we may infer that one – and possibly one of the lesser – attributes of *symmetria* is moderation. That is itself a term with a range of possible meanings, of course, so that it contributes only a little to our understanding of Plato's *symmetria*. But that little is all that we are able to deduce from the *Philebus* about the meaning of *symmetria*.

We are no better off when we turn to the *Sophist*. Here, an unintelligent soul is described as one that is deformed and devoid of symmetry, a condition that Plato defines as "failure in the attainment of a mark or measure" which "arises from ignorance", and which in turn is "the aberration of the soul moving towards knowledge". *Symmetria*, we may hazard a guess, here has something to do with the possession or acquisition of knowledge: but we have no inkling of what that "something" might be.

Plato, then, used the term without attaching any consistent, let alone evident, meaning or meanings to it: and clearly had in mind nothing that was even remotely akin to our principal contemporary meaning of the term. Modern translators uphold this unsatisfactory tradition. Jowett, in his translations of these two works renders *symmetria* as "symmetry"; in the Loeb editions it is rendered as "proportion". In neither are readers vouchsafed any indication as to how they are to understand *these* terms.

The entry for συμμετρία in the standard *Greek-English Lexicon* by Liddell and Scott[6] also shows just how ambiguous the term is in Classical Greek usage. The definitions it gives include: Calculate by comparison; Calculated to produce; Calculated to suit; Convenience; Due proportion; Exactly suitable; Fitting; Fixed proportion; Harmony of life; Having a common measure; In moderation; In right measure; Moderate in size; Of the same class or standard; Suitability; Suitable relation; To compute; To limit. Some of these and other definitions suggest something almost akin to desperation on the part of the lexicographers – and never more so than when they include the word "symmetry", in English, among their definitions! For their part, Lewis and Short, in their standard *Latin Dictionary*, more modestly give only two meanings for *symmetria*. One is "proportion"; the other – "symmetry" which, as in the Greek *Lexicon*, is left undefined![7]

To some extent, the confusion surrounding the word would be resolved during the Roman era, when *symmetria* came to apply to the proportions of a physical thing, and perhaps specifically to pleasing or appropriate proportions. Pliny the Elder (*Natural History*, chap.19) wrote that Polycletus paid more attention to symmetry than other sculptors, so that he was "very accurate in the proportions of his figures". Greeks in the Roman world adopted this usage, too. In the *Morals*, Plutarch (d. *circa* 120 C.E.) referred to the "just symmetry of breadth and height"; while in the *Lives* he complained that, after the columns of the Temple of Jupiter Capitolinus were recut, they did not gain as

---

[6] I use the 9th edition, 1996, eds. Stuart Jones and Mackenzie.

[7] Pliny the Elder, in the first century, had a more just appreciation of the term's problematic nature. "The Latin language has no appropriate word for the symmetry" of Greek sculptors like Lysippus, he wrote (*Natural History*, chap.19.).

much in polish as they lost in symmetry and beauty: they now are now too slender, and look thin.[8] A century later, Galen used συμμετριασ for the proportional relationship of parts of the body, such as that of the fingers to the wrist and forearm.

To the extent that "symmetry" had an accepted meaning prior to the fifteenth century, then, it would be for appropriate or pleasing proportions. As we will see in a later chapter, it continued to have this meaning down to modern times. The "fearful symmetry" of Blake's "tyger" is an instance of this usage in late eighteenth-century England. It must be stressed that there is no evidence that the word was ever used with its modern meaning before the middle of the 15th century.[9]

Some modern scholars are evidently unaware of this history. They make the mistake of assuming that when the word appears in an ancient Greek or Latin text it refers to the mirroring of the right and left halves of a shape – that it means "symmetry" in the modern sense of the word. Tavernor, for example, writes brazenly that for Vitruvius *symmetria* "meant the mirroring of form across an axis, the modern reading of 'symmetry'".[10] In fact, Vitruvius used the word only in the sense of pleasing proportion, in his distinctive framing of *that* idea. He did not use it for the mirroring of the two lateral halves of a form, a idea that all the available evidence indicates was unknown to him.[11] The mere appearance of the word *symmetria* in a Classical text, accordingly, should not be taken as evidence that our modern idea of symmetry was known to the Greeks and Romans. We have no reason to believe that it ever was.

Nor, by the same token, would it be justified to deduce,

---

[8] Plutarch 1874, v. 1, p. 495. See also Plutarch (1932, v. 1, pp. 540 - 42).

[9] A *possible* exception may, arguably, be found in Choricius' late 5th or early 6th century description of a portion of the church of St. Sergius in Gaza: see Selzer (2021, p.31, *et seq.*)

[10] Tavernor 1998, p. 43. Others mistakenly attributing the modern concept of symmetry to Vitruvius include Lowic (1983); Eco (1986, pp. 39 - 40); and, surprisingly, Hart and Hicks (in Serlio 1996 vol. I, p. 449 fn. 72, p. 83).

[11] A point that Sitte, again, (*op. cit.,* pp. 189 - 190) understood very well. Comp. also Sitte's contemporary, Thiersch (2017, p. 34), who stated flatly that Vitruvius did *not* understand symmetry "to mean ... one side being the mirror image of the other"; and Taylor, 2003, p. 25: "Vitruvian *symmetria* is not symmetry in the modern sense but a carefully proportioned relationship among elements".

from other Classical Latin terms that Renaissance writers used to refer to symmetry (in our sense), that the idea of symmetry was known to the Ancients; for although the words were the same, their meanings were not. Thus, *collocatio* was Alberti's term for symmetry. He took the term from Cicero, who did not use it for "symmetry" but, rather, for the component of rhetorical structure that deals with organization or arrangement (see *fn.* 56, p. 300, *below*). Alberti also used the verb *convenire* for the mirroring of left and right (*"dextra sinistris ... aequatissime conveniant"*[12]). In Classical Latin this verb has the sense of "coming together", even of being "consistent", but I find no instance of it having been used in ancient times in a context of "symmetry", in our sense. Alberti introduced *respondere*, too, to indicate symmetry and was followed in this by many others; "correspondence" is still used occasionally in this sense. It is a term that also occurs in Vitruvius, but it is clear that Vitruvius did not use it with the meaning it acquired in the Renaissance. As Hon and Goldstein note, Vitruvius "in no way transforms *symmetry* into a relation of correspondence... Vitruvius uses the verb *respondere* to indicate a relation which he calls *symmetry* – proportions that express agreement, a correspondence, among the parts and between the parts and the whole ... the Vitruvius *symmetry* means precisely a kind of proportion. Alberti ... uses the same verb which he found in Vitruvius, that is, *respondere*, but divorces it from *proportion* and takes it to mean *correspondence*. In this way he has forged a new technical term".[13] Vitruvian "correspondence", accordingly, does not indicate knowledge of symmetry in the modern sense. Another fallacy is to assume that, because *we* recognize a shape as symmetric, it must have been seen as such by people in Classical times, too. Symmetric shapes such as the square or the equilateral triangle were of course known to the Ancients, but there is no evidence that they recognized their symmetry. We see Plato's solids, notably, as symmetric: yet neither Plato, nor Euclid in his mathematical elucidation of them, appear to have been aware of their symmetry. This should not be surprising, because, as we have

---

[12] *De Re Aedificatoria*, IX, 5.

[13] Hon and Goldstein 2008, pp. 111–112. More commonly, *respondere* occurs in Classical Latin with a non-technical meaning of "resembling", as in Cicero, *De Haruspicum Responsis,* 49, where Cnaeus Pompeus wished to build a piazza in Carina that resembles (*responderet*) the one on the Palatine Hill.

seen, there is no evidence that the idea of symmetry was known in the ancient world. [14]

Failure to recognize these distinctions has led some modern scholars to interpret Greek notions about the cosmos in terms of the modern meaning of symmetry, even though (to repeat) there is no reason to believe that this meaning was known to the Greeks. Kahn, for example, refers to the "all-pervading symmetry which is the stamp of Anaximander's thought", even though neither the term nor the concept appears in the few brief fragments of Anaximander's work that have come down to us.[15] Anachronistic references by modern scholars to the concept of

---

[14] The curious failure, as it strikes us nowadays, to recognize the symmetry of certain forms did not occur only in the ancient world. As we saw in the previous chapter, Kepler, and following him Descartes and many others, regarded snowflakes as regular hexagons but it was only at the beginning of the nineteenth century that those – allegedly – regular hexagons were first described as symmetric. It is also curious to note that although his much - acclaimed model of the façade of Santa Maria Novella in Florence is based on the square and a progression of halved squares, which are of course symmetric shapes, all the attention of R. Wittkower and his followers (see Chapter Eight) is on the façade's alleged *proportions*, and not on its *symmetry*. It is unclear whether Wittkower recognized that his image of the façade was a symmetric one! He goes so far, indeed, to allege – of course, quite mistakenly - that the few Renaissance figures who took any note of the concept of symmetry treated it merely as "a theoretical" idea that they "rarely applied" in their designs (see p. 333, fn.149; and p.369, *below*).

[15] Kahn (1960, p. 90, echoed by Ferguson 2008, p. 17). Anaximander had asked, "why after all should the world fall?" Kahn's answer was that "if the universe is symmetrical, there is no more reason for the earth to move down than up". This was *Kahn's* answer, of course, and not the one Anaximander gave. (We might add, too, that the few lines of Anaximander that survive do not warrant Kahn's reference to *any* "all - pervading" characteristics of Anaximander's thought.) In much the same vein Kahn (*ibid*, p. 79, fn. 3) renders *Phaedo* 108e - 109a as, "because its relationship is symmetrical it will remain unswervingly at rest." Neither the term nor the concept appears in the original text, a more reliable translation of which is that of Fowler, who uses the term "equipoise" where Kahn has "symmetry". For other examples misleadingly attributing symmetry in the modern sense to ancient Greeks see Kirk, Raven and Schofield 1983, pp. 133, 136, referring to *Iliad* viii, 13, and *Theogony*, 726; Sautoy 2008, p. 60, referring to the *Symposium*; and Weyl 1952, p. 5, referring to the Pythagoreans.

symmetry cannot, of course, be accepted as evidence that the idea was known to the ancient Greeks.[16]

* * * *

The lack of philological or literary evidence of any knowledge of (let alone of a "passion" for!) symmetry in the Classical world is matched by a lack of material evidence, too. Nowhere is symmetry more obviously *not* present than in the design and decoration of Greek pottery. About one hundred thousand vases and fragments of vases from all parts and periods of the ancient Greek world have been photographed and catalogued in the *Corpus Vasorum Antiquorum* ("CVA").[17] My extensive sampling of the CVA images produced not one that I recognized as symmetric.

In what follows, a small selection of such objects, arranged chronologically and including vessels from different parts of the Greek world, will illustrate the asymmetry that is a pervasive characteristic of these artifacts. (Because CVA images are generally of low resolution, the illustrations presented here are from other sources, although the objects themselves may also appear in CVA.)

Fig. **3.1**
A 10th-century BCE cinerary urn amphora, now in the Kerameikos Museum in Athens, makes the point even more unambiguously. The two large discs, each with a cross in the center, are obviously similar, though one cannot fail to notice that the disc on the left is closer than the one on the right to the margin of the square containing it, and that the arms of the cross on the right are broader than those on the left. The outer column on the left is decorated with four triangular devices; the matching column on the right, however, has six-and-a-half diamond-shaped ones. In the center the outer bands

---

[16] Another instance: Dackyns, in his translation of Xenophon (*Oeconomicus,* Bk. IV), has Lysander marvel at the "symmetry" with which the trees in Cyrus' garden are laid out, that being a word that does not occur in Xenophon's Greek text.

[17] The *CVA* can now be accessed online at cvaonline.org/cva/default.htm

on either side differ in the number of their diagonal lines; and the bands between them are manifestly different. One band consists of six triangles; the other, of three sets of opposing diagonals that meet at a vertical line; in each of these sets the number of diagonals on one side of that line differs from the number on the other. Clearly, the painter of this vase had no intention of decorating it with a symmetric design.

Fig. 3.2

A Minoan vessel of the "old Palatial period" (2100 - 1700 BCE), now in the museum of Heraklion, has a wonderfully vivacious floral or sunburst design.[18] The outline of the body of this vessel as seen in the photograph is asymmetric. The roundels from which the petals or rays emerge are irregular in shape and are well to the right of the vessel's center. The petals or rays on the right are noticeably shorter than those on the left and do not match the shapes of the ones opposite them.

Fig. 3.3

A vase from the cemetery of ancient Thera, in the Cyclades, is dated between the 9th and 7th centuries BCE.[19] The handle on the right appears to be set higher than its opposite. The upper band contains a series of triangles, each of which is a different size; the triangle on the right is incomplete and, unlike the others, is not decorated internally. In the band below this one, two panels each differing in size, shape and interior decoration, flank stylized images of birds. The birds both face outwards but their sizes and shapes differ from each other; the roundel above each is placed to the right, i.e., asymmetrically. In the lower band, too, the designs are markedly asymmetric. The eight lozenges differ in size and as a group are not centered on the designs above them.

Fig. 3.4

Also from the geometric period is an 8th. century BCE prosthesis vase only a portion of which is shown in this

---

[18] en.wikipedia.org/wiki/File:AMI_ - _Kamaresvase_2.jpg

[19] www.eidola.eu/images/1274

image.[20] The meanders in the upper band face in the
same direction; they do not, in other words, mirror each
other across a central axis. The small rectangles in the
checkered panel above the bed vary randomly in size
and shape. The bed's leg on the left is shorter and thicker
than the one on the right. The child - size mourner on
the right is not matched by a similar figure on the left.
The adult figure closest to the bier on the right is taller
but narrower than its opposite number on the left. The
two figures below the bier on the right are seated; those
on the left are kneeling. The seated figures are flanked
by vertical arrangements of an inverted "V" - like
designs, those on the left by "M" - like ones. These "M"
- like designs also flank the taller figures standing on
both sides of the bier, but while there are 11 of these in
each of the arrangements on the right, those on the left
have 14, 15, 15 and 14, respectively. Surely, there was no
intention to arrange the elements of this vase in a
symmetric design.

Fig. 3.5
An 11th century BCE Gorgon from Rhodes. In no
instance are the elements on one lateral side of this
elaborate design mirrored on the other side. The entire
decoration is asymmetric.

Fig. 3.6
An amphora from Euboea is dated *circa* 570 - 560 BCE.
There are two lions on the vessel's neck. The middle
point between them is far to the right of the center of the
space between the handles, and the two circular patterns
are not positioned one above the other (the dots in these
patterns are arranged asymmetrically). Although both
lions are resting on their haunches, the lion on the right
has one foreleg well in front of the other, whereas the
legs of the lion on the left seem to adjoin each other. The
body of the lion on the right is considerably shorter than
that of the other lion, and the dimensions and curves of
their tails are quite different.

---

[20] Richter 1946, p. 281, pl. 400.

Fig. 3.7

An Attic neck amphora is dated *circa* 540 BCE and attributed to Exekias. The eight scrolls in this design differ irregularly from each other in size, and are set at irregularly different heights; the top-right scroll has one less revolution than its opposite number. In each of the two pairs on the upper row the scrolls face away from each other. On the right-hand side the scrolls are extended from the upper to the lower pair by tendrils that curve in toward each other, thus ensuring that the directions of the lower scrolls mirror each other as well as those directly above them. On the left - hand side, however, the scrolls are extended in such a way that the tendrils follow approximately the same direction – i.e., they do not face each other as they do on the right – so that the scroll on the bottom right faces in the same direction as the scrolls beside and above it. The cluster on the lower right is smaller but higher than the one on the lower left; it contains four leaves, while that on the left has three.

Fig. 3.8

A krater from Apulia is dated *circa* 330 to 320 BCE. The decorative band at the bottom of the vessel is not divided across a central axis, and therefore is asymmetric. The scrolls of the band above it are juxtaposed but are irregular in shape and size. The two pillars at the front of the pavilion differ from each other both in height and breadth; the two diagonal members of the pediment are of different lengths. The elaborate botanical arrays on the vessel's neck are arranged in a bilateral manner but without any attempt at making them symmetric. Note, for example, how the scroll - like tendril on the right is considerably higher up than the one on the left, and that while we see the former principally from within, we see the latter principally from without. The bust on the vase's neck is shown asymmetrically in profile and is not centered on the neck.

It is clear from merely the few examples presented here that the ancient Greeks did *not* insist on symmetric design for their pottery. This conclusion echoes that of R. M. Cook, a noted authority, who writes that "the neglect of exact symmetry" among Greek potters and their painters, "is undeniable".[21]   In Cook's view, however, the prevalence of asymmetric design in Greek pottery does not reflect an aesthetic preference but is accounted for, rather, by the fact that pots were regarded as "cheap product[s] not worth the laborious precision that was demanded in more expensive arts".[22]   Yet we also find abundant evidence of asymmetry in the "more expensive" artifacts of Greek civilization, as we can see from the following examples:

Fig. 3.9

A bronze *mitra* or abdominal shield is from sixth - century BCE Crete.[23] The bodies of the two sphinxes face each other, though their heads look out at the spectator. The bilateralism of this design is obvious. Nevertheless, the sphinx on the right has a longer body but lower wings, and its tail has wider curves than the other; the mid - point between the two heads is to the left of the center of the field.

Fig. 3.10

A bronze Argive vase of the mid-fifth century BCE in the Metropolitan Museum of Art, New York, is inscribed on top of the mouth with the words, "One of the prizes from Argive Hera".[24] The handle on the right is nearer the center of the vase and higher than the handle on the left. The neck, including the lip, and the disks are not centered on the vertical axis but are somewhat to the left of it. The right edge of the lip is rather higher than that on the left; as is the disk on the

---

[21] Cook 1972, p. 28. "*Exact* symmetry"? As was pointed out in Chapter One, a shape is either symmetric or asymmetric; and so there is no such thing as exact or inexact symmetry, for any shape that is not symmetric is asymmetric.

[22] *ibid.*, p. 28; comp. Vickers 1987. However, the asymmetry of some of the designs quoted above, in which motifs on one side are not repeated on the other, clearly shows, not an aversion to the "laborious" challenge of creating a symmetric design but indifference to or (more likely) ignorance of symmetry.

[23] *ibid*, p. 50.

[24] Metropolitan Museum of Art, New York; Purchase, Joseph Pulitzer Bequest, 1926 26.50

right. The shoulder of the vase is higher on the left than on the right.

Asymmetric forms are commonly found in Greek sculpture, too. Given the asymmetry of the human body itself [25] this should come as no surprise:

Fig. **3.11**
The highly abstracted, one could almost say minimalist, Cycladic marble statues of women from the late third millennium BCE have among their characteristics heads with asymmetric lyre - or shield-like outlines; as well as noses that are not centered on the face and heads that are not centered on the neck. The arms of full-length Cycladic statues of women are typically folded above the waist in a distinctive manner, with one forearm directly above the other so that the upper portion of the higher arm is necessarily much shorter than the upper portion of the lower arm. This disparity is sometimes partly obscured by having one shoulder higher than the other.   On the statue shown here the left breast is markedly lower than the other, and the shoulders differ in width.[26]

Fig. **3.12**
A well-preserved statue from the first half of the sixth century depicts a young man, *kouros*. His crown of locks hangs down asymmetrically over his forehead; there are eight waves of hair hanging down to his right shoulder but seven down to his left shoulder. His left eye, eyebrow and, even more so, his left ear, are higher than their counterparts on the right; although it is impossible to be sure from the photograph, it appears that his captivating smile, so characteristic of the *kouroi* statues, is crooked, with a more pronounced rise at the right than at the left corner. His left nipple is lower than its opposite number; his left shoulder is somewhat lower

---

[25] One of those oblivious to this fact is Woodford (1988, pp. 41, 4), who refers to "the natural symmetry of the human body" and alleges (of course, without providing evidence) that Greek sculptors "valued symmetry highly".

[26] Doumas 1983, p. 142, fig. 174.

and much less broad than his right shoulder; and his right wrist, forearm and biceps are considerably thicker than those on his left.[27] His left testicle is lower than the one on the right.[28] Particularly striking is the extent to which the neck and head are far removed from the central axis of the figure as drawn through the penis and navel.

Fig. **3.13** Also from the sixth century BCE is an ivory head, about two-thirds the size of life, that was discovered under the Sacred Way of Delphi in 1939, and is said to represent Apollo.[29] The ear, eyebrow and eye on the left side of his head are noticeably higher than those on the right; the ear on the left appears to be thicker than the other; the eyes are slightly crossed; and the nose is bent to the left. The pupil of the right eye is partly concealed by the upper eyelid whereas the pupil of the left eye is not.

Figs. **3.14a and b**
The hair on the top of the famous Poseidon (or, as some believe, Zeus) of Artemision is arranged asymmetrically The asymmetry of the statue's face, too, is quite apparent – the right nostril, for instance, is considerably broader than the left; the shapes of the eyes are different, and the left eye and eyebrow are lower than the right. The strands of the beard do not mirror each other and obscure the lower part of the right (but not the left) cheek.

Turning our attention now to another type of "more expensive" Greek artifact we should note that although a number of sixth- and fifth-century engraved gems have pronounced bilateral designs, none that are

---

[27] Beazley 1966, pl. 29. Despite the evidence of this (and many other) statues, Beazley (*ibid*, p. 13) declared that sculptures of the Archaic period, and notably the *kouroi*, are symmetric. See, as other clear examples of asymmetric *kouroi*, Richter, 1946, figs. 274, 391, 394, 450, 451.

[28] This being a topic on which McManus (2004; see also the literature cited there) has developed a particular if peculiar expertise.

[29] Mattusch 1988, pp. 177 - 178.

symmetric appear in Boardman's standard work on the subject.[30] These include:

Fig. **3.15** (clockwise from the top left): [31]

- The head of a cornelian Gorgon. Note the mirrored *direction* of the snake's heads but the asymmetry of their arrangement by both size and shape. The mouth is asymmetric.
- A cornelian scarab possibly from Cyprus, on which lions flank a sacred tree surmounted by a winged disk (the two wings of which are unequal in length). Note the different arrangements of their tails, and the much thicker forelegs of the beast on the right.
- A cornelian scarab showing a sphinx with two bodies: its asymmetries include the off - centered nose and mouth, and the tails, that on the right being taller and longer than the one on the left and having a different shape from it.
- Two cornelians, each with a winged Gorgon standing with outstretched arms: note the asymmetric arrangement of hair on top of the head of the creature on the left, and of the ribs of the wings of the creature on the right.

Fig. **3.16**

Perhaps nothing casts more dramatic doubt on Cook's idea that the ancient Greeks reserved symmetric designs for their "more expensive" artifacts than an exquisite seventh-century gold libation bowl at the Museum of Fine Arts in Boston.[32] The bowl is said to have been found at Olympia, and an inscription identifies it as the donation of the royal family of Heraclea. We can be certain therefore that this was not, by any standard, a

---

[30] Boardman 1968.

[31] *ibid,* figs. IV, 68; I, 22; IX, 123; III, 46 [found in Chiusi but, according to Boardman, *op. cit.,* p. 33, probably not Etruscan]; III, 48. The description of these gems' asymmetric features is not intended to be exhaustive and the alert reader will have no difficulty finding other asymmetric details in them as well.

[32] Museum of Fine Arts, Boston 21.1843.

cheap product; and we can be just as certain that it is an asymmetric one. The bowl's interior consists of nine elongated basins that radiate from a hub and give the rim of the bowl a scalloped shape. The bases of these basins are curves of the same size, a circumstance that makes the very different shapes of the basins themselves perplexing. The narrow troughs which separate the basins from each other are of unequal lengths, suggesting that the irregular configurations of the basins themselves were original. This possibility seems strengthened by the fact that the two concentric rings at the approximate center of the bowl are irregular in shape, and that the "mound" that they enclose is not centered on them. The various asymmetric features of this bowl therefore were almost certainly part of the original design and do not represent damage done in later times.

As with the Greeks, so too with the Romans, do we frequently find scholars making claims regarding an alleged "Roman taste for bilateral symmetry".[33]   And just as the empirical evidence refutes such claims regarding the Greeks, so too do we find many examples of asymmetric design in the art, architecture and decorative arts of Rome and its possessions, too. Here are four examples from Rome's provinces:

Fig. 3.17

An Ionic capital of the first century CE is from the synagogue at Gamla, Israel. The two scrolls wind in the same direction, which is to say they are asymmetric; and the scroll on the right is substantially larger than the other. The area between the scrolls is decorated with four ovals that differ in size and shape from each other.[34]

Fig. 3.18

A votive tablet from Pergamon, *circa* 200 - 250 CE, has the shape of an isosceles triangle (the lower side is shorter than the other two).[35] The figure on the lower

---

[33] Taylor 2003, p. 250.

[34] Wilson Jones 2003, p. 12, *fig.* 0.17.

[35] Staatliche Museen Berlin Antikenabteilung inv # 8612, from: Mitten and Doeringer 1967, p. 310.

right is set closer to the apex than the figure on the left; and none of the figures are centered on the large oval (which is lower than and to the left of the triangle's center). The upper figure is shown almost full length; the figure on the lower left is shown only down to just below her chiton, and the one on the lower right only down to her shins. The upper figure and the one on the right have markedly asymmetric shoulders. The figure on the left is centered on a line that bisects the bottom left - hand angle; the other two figures however are each off to one side of the bisecting line. The figure on the top has broader shoulders than either of the other two, while that on the right has a narrower waist. Finally, as we see from the dark lines flanking the left and right sides of the figure, this triangle is not symmetric but scalene.

Fig. **3.19**
The plan of a large Roman structure – its function is unknown - recently excavated in England is entirely asymmetric:[36]

Fig. **3.20**
A Graeco-Roman votive relief from Syria of two eagles, now in the Boston Museum of the Fine Arts, is to be dated no later than about 130 CE. Carved from a yellow-orange limestone, it is free standing but may have been part of a larger architectural (or perhaps specifically funerary) setting. In later times, particularly if used for heraldic or other symbolic purposes, birds might be arranged symmetrically (as, for example, the Russian or Habsburg double-headed eagles were). That is clearly not the case with the paired eagles in this relief. Fierce as any heraldic bird, these two differ from one another in size and shape (the bird on the left has a much longer neck, but smaller head) and they are not centered in the asymmetrically-curved niche in which they stand. The base of the niche is defined by two rows of what might be understood as feathers. In each row these face in the

---

[36]   news.yahoo.com/photos/mysterious-ancient-winged-structure-dis-covered-13-27331111-slideshow/ancient-rome-photo–132-7331072. html

same direction, so that their arrangement is also asymmetric.

It could be tempting to suppose that the asymmetry of artifacts like these reflects lower standards of craftsmanship that may have prevailed in the Roman provinces. In fact, asymmetric design is readily found even in the heart of the Roman empire, and in settings that are associated with wealth and power.

Fig. **3.21a and b**
Two murals now at the Archeological Museum in Naples were from Pompeii and are dated about 20 - 10 BCE (i.e., the so - called "Third Style" period).[37] Whether or not they are accurate representations of actual *villae marittimae* cannot be determined; but what is noteworthy for our purposes is that the structures depicted in the murals are clearly the homes of wealthy persons and that they are both asymmetric. Thus, there are 14 columns on the right - hand wing of the villa in the first painting but 13 on the left - hand wing. The pillars on the right are shorter, and support a much taller architrave or cornice than those on the left. The lower, apparently enclosed, portion of the right-hand flank is lower than its equivalent on the other side, and no attempt is made to conceal the disparity between them on the central portion of the building. Moreover, the structure on the right appears to end in a façade of six columns, that on the left in a seemingly wider façade supported by four broad piers.
In the second painting the columns to the right of the central portico appear to be lower, narrower and more irregularly spaced than those on the left; there are 13 columns on the right but only 12 on the left, but the colonnade on the left is somewhat longer than the one on the right. Of the two pedimented flanking structures, the one on the left has taller columns but a lower architrave than the one on the right. Its overall height appears to be somewhat greater than that of the right - hand wing.

---

[37] Ward - Perkins 1978, p. 118, #1.

Fig. **3.22**
Detail from a lintel (?) formerly attached to the Pantheon and now lying at its base. None of the details of one bird are mirror images of the other. The caduceus between them is also asymmetric.

Fig. **3.23**
The vessel in the British Museum now known as the Portland Vase was made in Rome during the first century of the Common Era. The entire outline of the vase – its bowl, neck, handles and rim – is asymmetric, and there is no hint of symmetry in the cameos that decorate it.

Fig. **3.24** The marble "Endymion" sarcophagus at the Metropolitan Museum of Art in New York dates from the later second or early third century, and was found near Rome.[38] The quality of the material and workmanship indicate that it was made for a person of wealth. The lower portion has a lion's head near each of its two sides but they do not represent an attempt at symmetry, the head on the right being somewhat smaller and perhaps further from the outside edge than the head on the left; and the manes of the two beasts are quite different from each other. The free and vivacious arrangement of figures and other elements of the design reflects indifference to symmetry. The entire ensemble on the lid is set well to the right of the midpoint of the sarcophagus. The arcade on the right side of the lid is lower than that on the left.

Asymmetric design was evidently acceptable, indeed, even in the highest circles of Rome itself:

Fig. **3.25** Nothing illustrates this as dramatically as the façade of the Roman Senate, a building whose construction was begun by Julius Caesar – hence its designation as "Curia Julia" - and was finished some years later by Augustus. The stucco that once adorned the façade, as well as a colonnade that was probably attached to it at the ground floor, are no longer present.

---

[38] Metropolitan Museum of Art 47.100.4. Roman, c. 190 - 210. McCann 1978, 39 - 41.

The doorway into the interior and the three large
windows above it remain in place, however, and as can
be seen from these are clearly not centered on the
structure but on a point well to the left of it.

Fig. **3.26**
The layout of the Markets of Trajan in Rome is markedly
asymmetric throughout; its great hall has an asymmetric
plan, and is "half a meter wider at one end than at the
other".[39]

Fig. **3.27**
Decorations on the wall of a room in the *Palatina Domus*
of the emperor Augustus.[40] Five horizontal lines extend
from one side of the wall to the other. Except for the
topmost, these lines are each connected to the one above
them with a series of vertical lines, thus creating a set of
frames. The horizontal and vertical lines are neatly
drawn. It is not possible to tell how many of these
frames are in the upper row; the middle row, which is
by far the tallest, has five frames and the lower row has
four. Each of the five frames in the middle row is of a
different width. In the lower row the two frames at the
right appear to be the same width (it is difficult to be
sure exactly where the corner between the two walls
runs), but the two frames to their left are each of a
unique size. Of the vertical lines that delineate each
frame in the upper and lower rows, several do not
accurately bisect the middle row of frames. Clearly,
then, the configuration of neatly - drawn vertical and
horizontal lines (excluding of course the areas too
damaged to allow a determination) is asymmetric.
Perhaps it bears emphasizing that this decoration is on
a wall in the palace of the great emperor Augustus, and
not in the shack of a simple herdsman in a remote
province of the Roman empire. Asymmetric design, we

[39] Taylor 2003, p. 66. .

[40] Iacopi 2008, p. 15.

may infer from this example, was evidently acceptable at the very pinnacle of Roman power.

Fig. **3.28**

The *Ara Pacis*, a ceremonial altar in honor of Augustus completed in the year 9, is a supreme expression of the power and wealth of Rome at the height of her greatness. Although the altar appears to be intact, or almost so, it suffered extensive damage over the course of the centuries and much of what we see now is in fact a modern reconstruction carried out in the 1930's under the aegis of the Fascist government. Something of the spirit in which the reconstruction was done is conveyed by the statement in an official guide that "lost parts have been replaced with casts taken from corresponding originals symmetrically placed on the opposite sides".[41] We have of course no reason to suppose that the "opposite sides" of the structure were originally part of a symmetric scheme. To the extent that portions of it have escaped modern reconstruction we can, in fact, establish that the original was *not* symmetrically designed. The two sides of the east face – that is, the structure on either side of the doorway – are not the same width as each other (the left is wider than the right) and their foliate friezes, though they appear to follow the same scheme, juxtaposed, are in fact substantially different and on different planes, and may have escaped the improving hand of Mussolini's reconstructionists. The asymmetry is particularly apparent in the two end pieces of the crowning slab of the sacrificial table shown in the plate. One winged lion faces outward from each end. The tail of the beast in the right - hand photograph is not as high or long as the opposite one, and it has a rather different shape. The clear difference in the approximately triangular empty space between the tops of the animals' wings and the cornice also calls attention to the asymmetric design.

---

[41] Rossini 2007, p. 82. The author adds, "In the few cases when the decoration was missing on both ... sides, a new relief has been remodeled [*sic*]". It is probably safe to assume that this "remodeling" too was done symmetrically.

The distance from the bottom of the tail to the corner of
the mouth of the beast on the left is significantly greater
than it is on the other (though the distance from the
bottom of the tail to the peak of the breast bone is
identical on both animals).The head of the beast on the
right is narrower but taller than that of the beast on the
left, and the arrangement of their manes is quite
different. The botanical motifs that extend from the
scrolls at the top, although they are similar in
conception, are clearly different in execution.

Figs. **3.29, 3.30**
            Another clear example of asymmetric
design in imperial Rome are some of the decorations of
the emperor Nero's *Domus Aurea* (*figs.* 3.24, 3.25). In the
first we see a wall divided into unequal – that is,
asymmetrically-arranged - compartments, in some of
which there are decorations, all asymmetrically placed
and asymmetric in form. In the second image we see an
apsidal recess set asymmetrically in the wall, and with a
vault that appears to be asymmetric.

Fig. **3.31**
            From a somewhat later period in the history of
imperial Rome, the Arch of Constantine has a large heroic
medallion set above the triumphal relief on its eastern
side, one of several sculptures removed from older
monuments by Constantine for his own arch. The
rectangular recess into which this medallion is placed is
not centered – it is further to the right than to the left. [42]
That this is not uncharacteristic of the Constantinian era

---

[42] Bober and Rubinstein 2010, pl. 182c. Desgodetz (1682, pl. III, p. 233)
"corrects" the asymmetry by showing the recess as precisely centered.
An unsigned preface to the later, London 1723, edition of Fréart's *Parallel*
(p. 8) declares preposterously that Desgodetz' drawings of ancient
Roman structures have "a precision so delicate (and even to a hair-
breadth, as they say), so scrupulously nice, as reaches not only to a single
foot, inches and lines alone, but even to the minutest part of a part of a
line"! (A line is 1-12th of an inch.) Janet Huskinson (Boardman 1993, p.
298) writes of the "almost relentless sense of symmetry" of the relief
showing Constantine distributing money. She is mistaken. The relief is
markedly *bilateral* in its organization, but it is asymmetric.

is implied by Krautheimer, *The Constantian Basilica* (1967): "The concepts of order and correspondence, and the articulating function of colonnades and entablatures count for little. In the Lateran, nave and aisle colonnades differed in number, proportion, materials, and position".

Fig. 3.32
The main portion of a ceiling fresco in the 4th-century C.E. catacombs of St. Domitilla. Ten segments converge on the central detail. Each of these segments is of a different size, and no two align with each other across the center. Dentelle-like details line many – but not all – of the borders of these segments and vary in size and number from segment to segment. In fact, there is no evidence of symmetry in any portion of the fresco. (The central detail is not a circle but an irregular ellipse.)

* * * *

The physical evidence considered here does not *prove* the negative that the Greeks and Romans never made things that were symmetrically shaped, but it convincingly establishes that they did not insist upon symmetric design. And this holds true, evidently, for both civilizations over the long span of their histories, and in the vast regions over which they held sway, and it also holds true both for their common wares and for the most precious and prestigious artifacts they made. When the physical evidence is joined to the linguistic and literary evidence considered at the outset of this Chapter, the conclusion becomes well-nigh inescapable that the ancient Greeks and Romans did not *know* the idea of symmetry, or that, if they did know it they did not attach much importance to it and seldom if ever applied it to the shape of the things that they made. The view that they felt a "passion" for symmetry and always saw to it that their artifacts were symmetrically-shaped is without doubt completely mistaken.

There is nothing ambiguous or elusive about the asymmetry of the artifacts at which we have looked here. Their asymmetry indeed is usually quite obvious to anyone who looks at them with unprejudiced eyes. Sometimes, to be sure, it can only be established by scrutinizing an artifact closely, or perhaps by measuring a photograph of it with a ruler or divider, or a simple graphics software program: but after that has been done a person should be left in no doubt that what he has examined is indeed an asymmetric design.

The artifacts at which we have been looking represent merely a tiny and more or less random sample of the many tens of thousands that have survived from Classical times. The overwhelming majority, if not indeed all, of those artifacts, when they are sufficiently intact to make such a determination possible, prove to have asymmetric shapes.

Nevertheless, for more than five hundred years scholars have insisted that the idea of symmetry was known in the Classical world, and that it was always conscientiously employed in the shaping of every kind of material object.

That such an obvious fallacy should have persisted for so long, despite a huge and largely unambiguous body of evidence to the contrary, is indeed a very curious thing.[43]

How did this error arise in the first place? And what has enabled it to survive for so long, and among so many people?

We will explore these questions in two phases. In this chapter we will identify some of the procedures that lead people to avoid acknowledging the asymmetry of an object. Later, in Chapter Six, we will examine the more fundamental problem of *why* the need arises to deny that a form is asymmetric.

What we find most commonly is people simply asserting – the scholarly *ex cathedra* pronouncement – that something is symmetric when it obviously is not. Usually (though as we will see, not always) there is no subterfuge involved. That is, the error is made in good faith by a person who has failed to recognize that an object is asymmetric. And we may speculate that this is usually not because that person suffered from poor eyesight or some cognitive impairment, or because the object's asymmetry is so muted as to be indiscernible: but because his or her perceptions have been affected by the strong bias in our culture in favor of seeing asymmetric objects as symmetric. That there is some validity to this suggestion may be deduced from the fact that the opposite error, of misperceiving symmetric shapes as asymmetric, is very rare.

The mechanism that makes the erroneous perception of

---

[43] Readers will have noted that the parallel here between the misperceptions of students of Classical art and architecture, on the one hand, and of the scientists whose ideas about Nature's forms we discussed in Chapter Two. This parallel is no coincidence. It is discussed more fully in Chapter Six.

asymmetric shapes possible is perhaps the one that Festinger identified in his celebrated theory of cognitive dissonance.[44] According to this theory, we sometimes alter our perceptions of the actual world in such a way that they do not conflict with – are not dissonant with - the ideas or values that we already hold and wish to continue to hold. In the present context this suggests that under certain circumstances we assume that if an object before us was made in ancient Greece or Rome it must be symmetric – and therefore we see it as such. "I know that 'X' is a Classical artifact; I know that all Classical artifacts are symmetric; therefore 'X' is symmetric", is the syllogism that activates a person's defenses against cognitive dissonance.

Here are some instances of what are surely good-faith misperceptions of asymmetry by scholars in which Festinger's mechanism may be in play. Let us revisit the Greek *hydria* (*fig.* 0.1)discussed in the foreword.

Ragghianti, it will be recalled, described the jar's decoration as "symmetrically and rigidly ordered".[45] Yet, not

---

[44] Festinger, 1956.

[45] Ragghianti 1979, p. 46. Similarly, Jacobsthal (1925), who, referring to a black - figured amphora by Exekias (BM210), writes – altogether incorrectly – that "the spirals roll in strictest symmetry"; or Hurwit (1977), who describes the hanging Kerkopes of a metope of Selinus Temple C as "symmetrical verticals", though neither figure is symmetric in itself or in relation to the other. See also Webster (1939), who claimed that one of the "great principles" of decoration in the early Geometric period of Greek pottery is that of symmetry. In Geometric pots, he stated, "the great structures and the minor parts within them are held together by the echoing of motives and figures symmetrically balanced". Yet the *tondo* from Knossos that Webster used to illustrate his analysis is obviously asymmetric. Johnston (Boardman 1993, p. 33) illustrates his claim that "frontality and symmetry are hallmarks of the so-called Daedalic style" with a golden Artemis plaque that seems to lack even a single, small symmetric feature. For the Byzantine period, cf. the remarkable misrepresentation by Van Nice (1965, p. 4) of Hagia Sophia as "essentially symmetrical" .

A notable example of an art-historian's avoidance of cognitive dissonance, though not bearing on the question of symmetry itself, occurs in the discussion by Paret (1997, pp. 87 - 93) of a painting, "Frederick and His Troops at Hochkirch", by Adolph Menze. This work depicts Prussian soldiers as they rush pell-mell to respond to a surprise attack by Austrian forces. The painting is thoroughly realistic, and vividly conveys the chaos of the moment. However, Max Jordan, the Prussian director of the National Gallery in Berlin, where this picture hung, was evidently unwilling to abandon his expectation that Prussian

only the shape of the palm tree and the fronds growing out of it but even the jagged protrusions (remnants of former fronds) on either side of its trunk are markedly asymmetric. The two large plants that are growing up on either side of the base of the palm tree are asymmetric in themselves and in relation to each other. The motifs on either side of the trunk are also asymmetric, in themselves and in relation to each other. Thus, immediately below the large fronds is a small pattern: that on the left has a cross - shaped figure within a quatrefoil frame; that on the right is a star - like design. Similarly, the motif that flows from the bottom of the handle on the left is unlike the one that flows from the bottom of the handle on the right. Similar discrepancies can be discovered at many other points on the vase. The juxtaposed elements we see are not incompetently-drawn versions of each other. Rather, many of them are entirely different designs – others differ in size or relative location - and this shows that there was no intention to make the decorations of this *hydria* symmetric. Indeed, although it seems likely that the vessel was thrown on a wheel, it is itself asymmetrically shaped.

The photograph's evidence that the decoration of the *hydria* is asymmetric is incontestable and immediately obvious. Therefore, if Ms. Ragghianti had *intended* (for whatever reason) to mislead her readers into believing that the *hydria* is "symmetrically and rigidly ordered" she would almost certainly not have included a photograph of it in her book. We must assume, rather, that she *expected* the vessel to be symmetric and therefore that is how she saw it and – in good faith - described it!

What seems to be another instance of this dynamic at work can be seen in Weyl's description (in his classic work *Symmetry*) of the so-called "Praying Boy" statue in Berlin. This statue, Weyl wrote, is a "symbol [of] the great significance" that bilateral symmetry has "both for life and art".[46] However, the statue itself is certainly not symmetric. The boy's left hip juts

---

soldiers always act in disciplined unison under the command of their officers, and so he described the scene as one that depicted "an infantry battalion at attention, firing by the numbers at the enemy". Almost certainly that is how Jordan actually saw the picture. He could not imagine Prussian soldiers in any other way (among other things, none of the soldiers in the painting are at attention!), and so he saw what he expected and no doubt wanted to see.

[46] Weyl, *op. cit.*, p. 6, fig. 2. In his discussion of this object Weyl again reminds readers that the notion of bilateral symmetry has "a concrete precise meaning".

sideways to support the weight that the right leg, bent at the knee, is not carrying; the two arms are not bent at the same angle; and the left shoulder seems noticeably broader than the right shoulder.

Could it be, however, that what we are seeing (and what Weyl may have had in mind) is a symmetrically-formed body whose stance makes it *appear-* asymmetric – an asymmetric statue, in other words, of a symmetric body? In all available photographs, perspectival distortions make it impossible to determine conclusively what the statue's contours are. But a careful examination of a number of images establishes that (1) the curls on the boy's head are boldly asymmetric and (2) that the nipple on the left side of the body is lower than the other. These asymmetries alone are sufficient to disallow Weyl's description of the statue as an exemplar of bilateral symmetry.[47]

Nor are such unwitting misrepresentations only verbal. In Hopkins' report on the Mithraeum excavated at Dura Europos, a *photograph* shows that the steps lead up to the right-hand side of the recess in which the (conjectured) altar was housed (*fig.* 3.34).[48] On the facing page, however, the artist's *drawing* of the plan and elevation provides a central axis on which both altar and staircase are located. As if unconsciously acknowledging this error, the artist's *sketch* of the structure has the staircase not quite centered on the altar but slightly to the *left* of center. Neither the artist, the author, nor the editors of the book appear to have noticed these discrepancies.

Yet there are also instances when the misrepresentation of an asymmetric form is done consciously, and may even be acknowledged as such. Drawings of Classical ruins made by Renaissance architects, writes Wilson Jones, "frequently 'correct' ... lapses of symmetry".[49] Serlio for example found the

---

[47] I would also suggest, though somewhat less confidently, that the contours of the boy's lips appear to be asymmetric and that his right eye is lower than the left.

[48] Hopkins 1979, pp. 204 - 5. Comp. the gentle comment of Seton Lloyd (1980, p. 134) that the French archaeologist Victor Place, in his reconstructions of Sargon's palace, "occasionally assumed in the Assyrian architect a most un-oriental passion for symmetry".

[49] Wilson Jones 2003, p. 2; Buddenseig 1971, p. 266. (Wilson-Jones' "lapse of symmetry" is *so* telling! No one would ever refer to symmetry as a "lapse of asymmetry".) The distinguished architectural historian James Ackerman (2002, p. 201) has written that the revival of antiquity by Renaissance architects "became obligatory, but only so long as the

asymmetric placement of the windows of an ancient Roman gateway in Verona "very displeasing to the eye". Since, as he put it, "I could not bear such discordance", he presented readers of his *Third Book* with a drawing in which, as he disarmingly told them, "I placed [the windows] in an ordered way".[50] Serlio was not always so candid, however. Because he thought of the Pantheon as "the most beautiful ... and best - conceived" of all the buildings in ancient Rome, and as "an architectural exemplar", he did not voice his displeasure at the asymmetries of its interior, but simply rendered them as symmetric in the drawing of it that he published in the *Third Book*.[51] In doing so he appears to have followed the example of Francesco di Giorgio (1439-1502) who also "corrected", without comment, the Pantheon's interior and depicted it as symmetric.[52] But for Antonio Sangallo (1484 - 1546) it was not enough to represent the Pantheon's interior as symmetric in his drawings; he pointed to the structure's asymmetries (which he had corrected in his drawings) by declaring in no uncertain terms that they were *una cosa perniciosissima* – "a most pernicious thing".[53]

---

ancient models did not break Renaissance rules". But this is not quite correct. What we see, rather, is that during the Renaissance the ancient models were altered – even if only on paper – to conform to, and in this way to validate, Renaissance rules. As Wilson Jones also noted, "The Renaissance conception of antiquity was a self - fulfilling myth: theory was projected onto ancient ruins, which in turn were used as evidence to justify the theory". Summerson (1963, p. 32) stated the matter more succinctly: "the [Renaissance] Italian", he wrote, "recreated Rome in his own image". An important exception to the "correction" of the asymmetry of Classical buildings during the Renaissance is to be found in the manifest asymmetry of Roman architectural elements in Mantegna's paintings in the Eremitani Chapel – see p. 334, fn. 157, *below*.

[50] Serlio 1996, Bk. III, 113v.

[51] *ibid*, Bk. III, 50 r & v.

[52] Giorgio 1967, v.II, p. 412.

[53] Buddenseig 1971, p. 26. The editors of Sangallo's drawings make this very perceptive comment on his corrections of the asymmetry of Classical structures: "Everything seems impelled by a preordained ideology. The imagination of the observer is no longer stimulated. Instead, a uniform scheme has been imposed". Sangallo's drawing that "corrected" the asymmetry of a gateway in Turin from the era of Augustus, along with an illustration showing the structure's actual design, is reproduced in Wilson Jones 2003, pl. 3. Comp. Brown, 1983.

These "corrections" were not always confined to paper. In the eighteenth century the attic drum on which the Pantheon's dome appears to rest was completely remodeled in an attempt to mute the asymmetry of the interior. This is the interior as we see it today, but for a small portion of the original scheme that was restored in the 1930's.[54] We have already noted how the "restoration" of the Ara Pacis in Rome imposed symmetric features that we have no reason to believe were part of the original structure.

The transformation – sometimes done consciously, and sometimes done unconsciously - of asymmetric to symmetric shapes, whether in written or graphic representations or by the alteration of physical structures, is only part of the arsenal of those who wish to believe that Classical art and architecture were shaped by a "passion" for symmetry. Another weapon in this arsenal is the dogmatic assertion – i.e. one made without evidence - that the asymmetry of a shape as we see it today is the result of damage done to an artifact that originally was symmetric. The asymmetry of a building, for example, may be understood, not as reflecting the preferences or skills of its builders but as an unintended consequence of acts of natural or human violence that caused the structure to lose its original symmetric form. We learn from one scholar, for example, that the asymmetry of the Pantheon's porch was brought about by "shocks from earthquakes combined with pressure from the surrounding earth deposited in the Middle Ages".[55] Similarly, the asymmetry of the

---

[54] The remodeling in the 18th century did not change the so-called "misalignment" of the dome's coffers with the details of the attic; the latter, now with heavily pedimented windows and framed panels, were aligned, rather, with the architectural details of the first level: Wilson Jones, 2003, p. 190.

[55] Licht 1966, p. 35. The difference between the largest and smallest intercolumniation on the Pantheon porch is over 6". A shock of the kind Licht refers to would almost certainly have displaced the columns equally, or progressively. Yet the intercolumniations vary irregularly, making that hypothesis questionable. That the asymmetry of the intercolumniations was original and not the result of earthquakes or other events is also suggested by the irregular arrangement relative to the center of each capital of the dentelles at the base of the pediment, that are clustered as follows: 8, 6, 7, 7, 6, 6, 7. Palladio noticed this irregularity in his drawing of the Pantheon façade but counted the dentelles incorrectly. Perrault (1993, p. 60) was baffled by the irregularities of the

curve of the north edge of the Parthenon's stylobate is said to
have been brought about by several factors, including
earthquakes and the explosion in 1687 of ammunition that the
barbaric Turks stored in the temple.[56] In much the same vein
Penrose, having determined that the east edge of the Parthenon's
stylobate was asymmetric in plan, declared this to be the result of
later developments (he did not indicate what these were) and that
the removal of a "trifling irregularity" from its present shape
would bring about the "restoration" of the step's original "exact
and symmetrical appearance".[57] Penrose describes the reasoning
that led him to determine that asymmetric and other
irregularities in the Parthenon were unintended in these words:
"We may ... always suspect some disturbing cause to exist, when
in quantities which tend to equality or some obvious proportion,
a difference sensibly greater ... is ... found".[58] In other words: if
a measurement is not what we expected it to be, we must always
suspect that it is not the measurement that was originally
intended!

Asymmetric forms are sometimes also explained as the
unintended result of incompetence or carelessness. Perrault, for
example, attributed the asymmetries of the Pantheon to a
succession of poor decisions made by its architects.[59]

---

columns on the Pantheon's porch (none have the same circumferences)
and evidently regarded this too as a shortcoming.

[56] Stevens, 1943.

[57] Penrose (1888, pp. 29 - 30).

[58] *ibid.*, p. 12,fn.

[59] Perrault (1993, p. 157) criticized the coffering of the Pantheon's vault
for their asymmetry: "The squares in the coffers of the vault recede in
steps, like hollow pyramids, and the center of the axes of the pyramids,
rather than being near the center of the vault, is located at the center of
the temple five feet above the pavement. This results in the axes not being
perpendicular to the bases of the pyramids, which would have been
necessary in order to maintain symmetry. This alteration makes the view
of the hollow pyramids from the lower center of the temple the same as
it would be if the viewer were lifted up to the center of the vault, with
this the point where all the axes of the coffers converge. However, as soon
as one moves away from the center of the pavement the effect is
destroyed, and one becomes aware of the obliqueness of these axes and
of the defective symmetry of the pyramids, which is something much
more disagreeable to the sight than if the orientation of the receding
coffers had been straight, as it ought to be relative to the vault."

Others hold the builders and not the architect responsible. Confronted by the asymmetric distribution of the columns of Temple "C" at Selinus, Italy, Dinsmoor declared that the intercolumniations "were intended to be perfectly uniform" but that "carelessness of execution" caused them to be made irregular, instead.[60] He also suggested that variations in the width of the top step of the Parthenon "may have been the result of clerical error".[61] Robertson, too, accounted for the asymmetry and other irregularities of the Parthenon as "notable faults of execution".[62] Stevens offered six reasons, four of them having to do with human fallibility, why the north edge of the Parthenon's stylobate is, as he put it, "not quite a perfect curve".[63] Lawrence, referring to the variations of as much as 12 inches in the diameters of the columns of the Temple of Apollo in Syracuse, attributed

---

[60] Dinsmoor 1950, p. 62.

[61] *ibid*, p. 80. Dinsmoor also declared (*ibid*, p. 9) that the palace of Knossos "departs widely from the principles of symmetry and axiality", evidently implying thereby that those principles existed and were normative in Crete during the 16th or 17th century before the common era, with the palace representing a deviation from them.

[62] Robertson 1954, pp. 116 - 7.

[63] Stevens 1943. In this paper Stevens both commends the "careful measurements" of the Parthenon taken by Balanos, and declares that they could only have been "close approximations". The "error" explanation of asymmetry was also used by a distinguished archeologist of middle - eastern sites, Henry Frankfort (1969, pp. 75 - 77). Frankfort wrote of Khorsabad and Sargon's palace there: "It is clear that the planners aimed at regularity and the frequent deviation from the right angle is due to imperfect methods of surveying. It is for instance characteristic of their love of symmetry that each side of the square should have two gates..." (Actually, only one side has two, or arguably three, gates). Frankfort acknowledged that the gates are placed asymmetrically but attributed this to the builders' "miscalculations". His explanation (p. 76) of those alleged miscalculations - "In a country where paper, or even papyrus, was unknown there could not be measured drawings" - is very poor. Referring (p. 48) to the plan of Gudea's citadel, Frankfort commented on "how awkwardly square elements are fitted in" but overlooks the fact that the entire plan is boldly asymmetric, and that there is no reason to suppose that those elements really *were* intended to be square. (In fact, it is doubtful that there is a single square in the plan!) The superb finish on the statues of Gudea (note the precision with which the jewels [?] in his headband are formed and arranged) suggests that in his day craftsmanship was in fact of a very high standard.

them to the builders' "incompetence".[64] Taylor speculates that
certain asymmetries in the *natatio* of the Baths of Caracala were
intended to mask the asymmetric placement of columns
necessitated by errors in the laying of the structure's
foundations.[65] Taylor also reproduces a plan of the mausoleum at
the Villa Maxentius on the Via Appia which shows the structure
in its actual asymmetry alongside a plan that shows it as
symmetric, "as it was probably envisaged" (i.e., as symmetric),
and then speculates about the cause of the discrepancy between
the two.[66] Why he finds that the symmetric plan was the one
intended, Taylor does not explain, and many readers may well
doubt. After all, the presence of an asymmetric shape is not in
itself a reason to assume that another, symmetric, shape had been
intended![67]

A particularly curious set of irregularities occurs on the
frieze of the pronaos of the second temple of Hera at Paestum,
where the space allotted to the metope is disproportionately wide
or, alternatively, the triglyph is disproportionately low. Wilson
Jones regards this as the result of a miscalculation in the height of
the frieze.[68] Although this explanation cannot be dismissed out of
hand, another anomaly in the structure gives one reason to
question it. It is conventional in Greek architecture for a row of
(usually 6) guttae to be placed below the triglyph. This row is the
same width as the triglyph above it. On the frieze of the Hera II
pronaos, however, the rows of guttae are considerably wider than
the triglyphs and project asymmetrically to the left. If this were a
mistake, it is surely one that could have been corrected quite

---

[64] Lawrence 1967, p. 169.

[65] Taylor 2003, pp. 71 – 73, citing Delaine 1997.

[66] *ibid.,* p. 69.

[67] Taylor, it is worth noting, points out that Roman builders were capable
of extraordinary precision, so that, for example, a measured section of the
aqueduct of Nimes has "a steady fall of only seven millimeters for every
hundred meters of distance, even as it winds along the valley contours".
The ability to work with such precision, although of course not always
called for, raises the possibility that structural irregularities such as are
documented on these pages could as plausibly be explained by builders'
indifference in certain contexts to precise work: or indeed even by a
desire to mitigate – cf. Goodyear's "refinements" - the potentially
oppressive monotony of endless regular perspectives or textures in a
very large structure such as the Baths of Caracalla.

[68] Wilson Jones (2003, pp. 13-14).

easily. The extended guttae at Paestum are, as far as I can tell, unique, and I have no explanation of them. But their anomalous character, including their asymmetry, is too apparent to be thought of as anything but intentional.[69]

It is not only in the field of architecture that we encounter such attempts to explain away asymmetric design as errors; it is also offered for the asymmetry of Classical artifacts and works of art. For Cook, as we have seen, the asymmetry of Greek pots attests to the slapdash production methods employed by their makers. Pottery was "a cheap product", he wrote, that did not merit "the laborious precision demanded in more expensive arts", and one must not infer from its asymmetry that the Greeks saw any "virtue in imperfection".[70] For Richter "warped lips and sagged shoulders" on pots were "accidents or mistakes".[71] Winckelmann wrote of the asymmetric arrangements of Greek sculptures, such as the lopsidedness of an otherwise beautiful head of Venus, that they were caused by carelessness or incompetence.[72]

Most ingenious of all is the explanation, in both architecture and the arts, that certain asymmetric configurations were not the result of poor design or craftsmanship but were intentionally and very carefully calculated *to make things look symmetric.*[73]

The French architect and architectural historian August Choisy, for one, argued that the asymmetric curves he detected in the steps of the east stylobate of the Parthenon were created intentionally *so as to correct an optical effect that would otherwise*

---

[69] Almost certainly intentional, too, is a curious feature – to which Wilson Jones (*ibid*, p. 203 and *fig*. 10.8) calls attention - of the capitals that surmount the *anta* columns in the Pantheon's portico. The side of the capital that faces the pronaos is seated asymmetrically over the shaft.

[70] Cook 1972, p. 28. Note Cook's tacit equation of "asymmetry" and "imperfection"!

[71] Richter, *op. cit. ad loc.*

[72] Winckelmann 1968, p. 265. "Incorrect drawing may also be observed in a head of Venus, which is a beautiful head in other respects, in the Villa Albani; the outline of it is the most beautiful that can be imagined, and the mouth is most lovely; but one eye is awry." Among other "errors" he records: a "beautiful rilievo in the Borghese villa has one arm that is too long; a laughing Leucothea in the Campidoglio has ears which should be parallel to the nose but fall below it".

[73] Schneider 1973, p. 40. Similarly, Philipp 1999.

make the steps appear to be asymmetric![74]   Winckelmann declared
that the inequality in the size of the feet in the Laocoon group and
the longer rear leg of the Apollo Belvedere were perspectival
devices calculated to compensate for "what might apparently be
lost by the legs being drawn back". Schneider claimed that the
asymmetry of the heads of Greek sculptures – which he stated
was *"viel starker als in der Natur"*, surely an unprovable statement
– was intended to convey the impression of symmetry when
viewed from below.

The world of scholarship has treated these opinions
generously, and I know of no instance in which someone has seen
fit to ask what the evidence for them is. But that of course *is* the
question that they beg.   For how, to take one example, did
Dinsmoor know that the intercolumniations of Temple C "were
intended [my emphasis] to be perfectly uniform"? How did Cook
or Richter know that the "defects" of asymmetric Greek pots were
regarded as such by the people who made and used them? And
what is the basis for Wilson Jones' statement that "Inaccuracies
[by which I think he refers to irregularities] were usually
permitted so long as they did not attract the eye's attention"![75]

---

[74] Choisy (1996, v. I, p. 419). A sober corrective to this view is Zucker
(1959, p.29), who argues that no "kind of specific system" determined the
layout of structures on ancient acropolises. On Choisy and his influence
see the important paper of Etlin (1987). Choisy also suggested that the
asymmetric placement of the (allegedly symmetric) buildings on the
Athenian Acropolis intentionally reflected Nature's own plan – *"ainsi
procède la nature"* – whereby leaves, that he claimed are symmetric (*"les
feuilles d'une plante sont symétriques"*) are on plants that are asymmetric.
In actuality though neither leaves on plants nor the buildings on the
Acropolis are symmetric! Comp. the thesis of Trachtenberg (1997, p. 74)
that the piazzi del Duomo and della Signoria in Florence were shaped
asymmetrically in a way that was intended to create the impression that
they are symmetric. We may well doubt, however, whether any visitor
to these manifestly asymmetric piazzi, other than of course Trachtenberg
himself, has ever fallen for this trick. (Trachtenberg, as is common in this
genre, does not share with his readers how he discovered the intentions
of the builders of the two spaces.).

[75] Wilson Jones 2003, p. 72. As we have seen, of course, there is no
shortage of "inaccuracies" in Classical art and architecture that "attract
the eye's attention." Wilson Jones himself (*ibid*, p. 202), quotes William L.
Macdonald's calling into question "whether the modern mind can ever
really grasp Roman architects' intentions, since what appears faulty to us
might not have appeared so to them". Wise words, indeed, that
historians of Classical art and architecture would do well to heed! I

The answer of course is that there is no evidence to support these and many similar statements. Penrose, as we have seen, insisted that an unexpected form must have originally looked like, or must have been intended to look like, what Penrose *expected* that it should have looked like! It could very well be that something like this preposterous narcissistic solipsism inspires many attempts to explain away asymmetric designs as defective or unintended versions of symmetric designs.

Nor can one reasonably conclude, from the possibility that a building's present asymmetric shape is the result of acts of natural or manmade violence, that its *original* shape must have been a symmetric one. Those episodes may well have changed an earlier shape *that also was asymmetric*. It can by no means be assumed that they caused a symmetric structure to become asymmetric.

The perspectival theory of Winckelmann and others is also inherently problematic. In particular, Schneider's account is called into question by the asymmetric arrangement of the hair on top of the head of the statue of "Poseidon" of Artemision.[76] The statue is about 6'9" tall and the top of its head would be invisible to anyone standing on the ground – all the more so if, as is probable, the statue was mounted on a plinth. Surely, the most plausible explanation of asymmetrically-shaped statues is that human bodies, along with all of Nature's shapes, are themselves asymmetrically shaped? As for Choisy's explanation of the asymmetric curves on the Parthenon: it could only apply to the view from a single point, and the asymmetry would be evident before one arrived at that point, and again as one moved away from it. Are we justified however in supposing that at least some of the asymmetric forms we see were consciously designed to be what we now recognize as asymmetric? We have already seen that builders of Roman aqueducts were capable of extraordinarily precise work; and in the appendix to this chapter we will see suggestions that Greek builders were capable of similar exactness. But of course, the ability of some workers to

---

would mention here that while some scholars marvel at the symmetry of Hagia Sophia (though it is, in fact, a pervasively asymmetric structure), Macdonald, as well as Krautheimer – in a sense, echoing Procopius' description of it as a "perplexing spectacle" – understood that its irregularities are intentional, and are what give it, in Macdonald's words, its "special quality of grace" – see Selzer 2021, pp.128-135.

[76] Schneider (1973).

achieve remarkable precision does not mean that *all* workers were capable of it, or that such precision was called for but not achieved in some work. There are however two remarkable pieces of evidence that demonstrate unequivocally *the intention* to create structures that are asymmetric. These are plans, incised on slabs of marble and limestone, respectively, for a temple in Ostia and another in Nihi, near Baalbeck in Lebanon.[77] In the plan for the Ostia temple the column spacing along the left flank and on that part of the right flank that has been preserved is asymmetric, diminishing irregularly toward the back. In the Niha plan, the stairs leading up to the adyton occupy the left and center spaces, but not that on the right, where instead a doorway is thought to have led to the undercroft. Because these asymmetric details appear on plans used in the construction of the temples, or as part of the process of designing their final plans, we must recognize that they could not be the result of poor workmanship, or of later accidents. They can only have been intentional.[78]

　　　　We will resume this discussion in Chapter Six, where we speculate about the motives that may lead otherwise dependable observers to *avoid* acknowledging that a shape is asymmetric.

Appendix 1: Is the Parthenon Symmetric?

　　　　When the 19th century architectural historian James Fergusson declared that Greek temples "are perfectly symmetrical", so that "the one side exactly corresponds with the other" he was expressing an opinion about Classical architecture that has been widely held since the middle of the fifteenth century and remains a commonplace to this day.[79]

---

[77] Wilson Jones, op. cit., pp. 55 – 56.

[78] Wilson Jones, *op. cit.,* p. 55, suggests rather cryptically that the intercolumnial diminution in the Ostia temple was "perhaps because the depth of the site was limited". This does not however address *the irregularity* of the spaces' diminution.

[79] Fergusson 1849, pp. 397 - 8. Fergusson was not consistent, however, for he also declared that symmetry is "a property which exists only in the imagination of the moderns", and that the Greeks were not guilty of such "absurdities", which he described (*ibid*, p. 399) as "an invention of the Italian architects in the worst age of an attempted revival of Classical art" - an argument that anticipates a central theme of this book. (Comp. too the baseless insistence of Fergusson – *ibid.* p. 397 - that Gothic churches

It is not at all easy to determine how valid this consensus is. Certainly, the challenge is far greater than the one we confronted in the main body of this Chapter, where we saw that it seldom requires more than a glance, merely, to decide if the decorative artifacts and works of art we looked at were symmetric. The conclusions we reached - that the Greeks and Romans did not require symmetry in the design of their works of art and their decorative artifacts, and indeed, that they did not use (and almost certainly were unaware of) the categories of symmetry and asymmetry when they made an object or looked at one made by someone else – are ones in which we can have a high degree of confidence. Not so with the architectural evidence, however, whose ambiguities prevent, and may always prevent, a conclusive determination of what if any role symmetry played in the design of buildings in ancient Greece and Rome.[80]

To start with there is the difficulty of obtaining accurate measurements. Clearly, except where a building is obviously symmetric or asymmetric, which in Classic architecture is seldom the case, we cannot know if it is one or the other unless we know its precise measurements. Unfortunately, there are almost no reliable measurements for any ancient Greek or Roman buildings. This holds true even with regard to the dimensions of that supreme achievement of Classical architecture, the Parthenon.

The Parthenon has been measured on a number of occasions during the past 350 years - it may indeed be the most frequently surveyed of all buildings – but the quality of most surveys has been demonstrably poor, and there are good reasons to doubt the reliability of even the most recent of them. The first survey was carried out in the seventeenth century by the Englishman Francis Vernon, who reported back to the Royal Society in London that his measurements of the Parthenon were "exact to ½ foot": a degree of precision – or lack of it – that makes

---

and cathedrals "were always designed as symmetrical buildings", a view that is put to rights here in Chapter Four.)   Like Fergusson, Viollet-le-Duc (1987 v. I, pp. 57 - 8, 88 - 9, 103, and 346; and comp. the motto to the present Chapter) insisted both on the Ancients' preference for symmetric design, and on their rejection of it.

[80] In addressing these ambiguities, readers should bear in mind the point, made at the outset of this Chapter, that there are no words or phrases in ancient Greek or Latin that denote asymmetry or symmetry: *prima facie* evidence, this, that these ideas played no role in the design of Classical buildings (or of anything else, for that matter).

his measurements useless for most purposes.[81] The measurements carried out by Vernon's later contemporaries, Wheler and Spon, were hardly more useful, being accurate, or so they claimed, to the inch in some instances and to the foot in others.[82]

By contrast, other surveyors made claims to accuracy that are not believable. Stuart and Revett, in the 18th century, declared that their measuring rod gave them results that were accurate to one - thousandth *of an inch*.[83] How far they fell short of such a standard is suggested by their failure to discover the *entasis* (or swelling) of the Parthenon's columns. A century later Penrose claimed that he had achieved "an exact delineation" of the Parthenon and that *his* measurements were accurate to one - thousandth of a foot (or about one-eightieth of an inch).[84] Nevertheless, he failed to discover, among other features, the curved elevations on the Parthenon's east front.

Troubling questions arise about the accuracy of more recent surveys, too. Dinsmoor and Balanos in the early decades of the twentieth century each declared that their measurements were accurate to within 1.0 mm., and in some instances to within 0.50 mm., yet they usually differed by more than that about the length of anything that they both measured. For example, according to Dinsmoor the two flanks of the Parthenon's stylobate are the same length, as are the axial spacings of the external columns on the east and west fronts. Balanos on the other hand found that the two stylobate flanks differ from each other, with neither matching the length reported by Dinsmoor; he also found variations as large as 8 cm., or about 3 - 1/8", in the axial spacings of the columns.[85] Again, in measuring the fourteen

---

[81] Quoted Redford 2002.

[82] Quoted Stuart 2008, v. II, pp. 1 - 3.

[83] See for example such measurements as 30', 8.834" in Stuart 2008, v. II, pl. 5. The introduction by Frank Salmon to this edition – i.e., a facsimile of the original *Antiquities of Athens* - offers a brief but useful critical survey of the methods employed by Stuart and Revett.

[84] Penrose 1888, p. *v*. An example (p. 9) of Penrose's measurements is the diameter of "3.656 feet" that he gives for the naos columns.

[85] It is curious to note the report of Friedlaender (1969, p. 92) that the Irish writer Frank Harris, who was no professional in art matters, mentioned in his *Autobiography* that he had measured the intercolumniations of the Parthenon and found that they were unequal. (I have not found this passage in two editions of Harris' memoir that I have examined.)

metopes on the Parthenon's east front, Dinsmoor and Balanos agreed on the widths of only two; their measurements of the other twelve differed from each other by between 2 mm. and as much as 26 mm. (or more than one inch).[86] These differences exceed the margins of error of one, or one-half, millimeter that each man claimed for his measurements.

In all probability Balanos' measurements of the Parthenon in the first decades of the twentieth century are the most reliable yet. They are, at any rate, the only ones in which advanced optical instruments were used and in which the work was carried out by trained engineers and surveyors.[87] The measurements were made over the course of thirty years, however, and there is no assurance that consistent procedures were employed during all phases of the work. Balanos' cursory description of the procedures he followed is all the more to be regretted, therefore.[88] But even if it is conceded that his measurements are superior to earlier ones, the question of *how reliable* they are remains to be answered. That can only be done by a new survey using the extremely precise technologies, above all, photogrammetry, that are now available. Given the present, unstable, state of the world it is impossible to say when or even if such a survey will ever be conducted.

To be sure, the discrepancies between various sets of measurements are usually very small, and up to a point one can sympathize with Robertson that "when critics argue from variations of small fractions of an inch, the thing becomes

---

[86] Balanos' (1938, pl. 6e) measurements of the metopes appear on the upper of the two rows below, and those of Dinsmoor (1950, pp. 338 - 340) on the lower. Measurements are north to south, in meters; the two pairs of identical measurements are highlighted in bold type:

1.246 1.254 1.167 1.288 1.271 **1.271** 1.330 **1.317** 1.294 1.331 1.253 1.241 1.234 1.277

1.256 1.256 1.278 1.279 1.276 **1.271** 1.334 **1.317** 1.309 1.305 1.234 1.239 1.239 1.268

Dinsmoor's measurements at Bassai were not corroborated by Cooper (1996, pp. 36, 230 fn. 3), who found that his figures for certain columnar diameters erred by as much as 2 cm., or more than ¾".

[87] Balanos 1938, p. 53: "*un niveau Zeiss de grande precision*".

[88] Balanos says of his surveys merely that they conformed to "*la methode habituelle*". By contrast see the extremely detailed description of the procedures Cooper (1996, pp. 36 - 42) used in surveying the temple of Apollo Bassitas. For the context in which Balanos worked and the major controversy in which he was embroiled, see Dimacopoulos (1985).

absurd".[89] Nevertheless, the claims that have been made regarding intentional, very precise, differences are not adequately addressed with exasperated remarks. They can only be upheld or rejected on the basis of demonstrably reliable empirical data.

There are, moreover, certain other crucial issues. One of the most intractable of these is the problem of determining – on a structure whose edges and surfaces are almost invariably worn or damaged – where to set the points from and to which measurements are to be taken. This is perhaps not a serious challenge where merely *approximate* measurements are considered to be sufficient; but as we have already seen, and will see in greater detail, the claims to be tested, and the hypotheses to be validated, presuppose that measurements are accurate to within 0.5mm., or even less.

But even if *this* issue could be resolved satisfactorily, we would still have to settle the problem of which dimensions of the Parthenon today reflect the intentions of its architects, and which reflect deviations from the original plans that may be attributable to normal wear and tear; to the workmen who built the structure; or to natural and human acts of violence (earthquakes and, if nothing else, the detonation of the ammunition that the Turks had stored there) that the Parthenon has suffered over the course of two and a half millennia. That said, however, there are certain measurements to which these reservations would *not* seem to apply. One set consists of the heights of the peripteral columns that are still standing. Balanos found that they "vary erratically and obey no rules". Because it is highly unlikely that the heights of the columns would have altered over time, we can be confident that these asymmetric variations were original. We cannot know, however, whether they reflect the architects' intentions or the workmen's indifference to (or inability to attain) a higher standard of precision.

Another asymmetric arrangement, also without doubt original, is that of the metopes on the east front. The metopes within each pair (counting outward from the center) differ in width by 13, 23, 40, 35, 26, 20 and 31 mm., respectively (*fig.* 3.37).[90]

---

[89] Robertson 1954, p. 116.

[90] See upper row of figures, fn. 84 above. For whatever reason, Balanos, (pl. 6e) gave the widths of only 5 of the 15 triglyphs and of 4 of the 14 metopes on the west front – too few to determine whether their distribution by width was symmetric or not.

No pair is symmetric, therefore, and the irregularities are not distributed consistently in the series of pairs. We can be confident that these measurements reflect the metopes' original dimensions and distribution, for the panels will not somehow have shrunk or expanded, let alone exchanged places, over time.[91]

Balanos' measurements of certain elevations on the east front of the Parthenon indicate the presence of symmetric arrangements.[92] His elevation γ is a line 30.870 meters in length that runs from the north to the south ends of the last step leading up to the stylobate. This elevation was measured at ten equidistant points with the following result (in meters; north to south; and with the middle of the series in bold type:)[93]

2.602   2.610   2.634   2.657   **2.667**   **2667**   2.655   2.636   2.610   2.602

Any two of these points that are equidistant from the center may be thought of as a pair. The measurements for the curve at elevation γ show that the lengths in each of three pairs are, to the nearest millimeter, the same as each other; and that the lengths of each of the other two pairs differ by 2 mm.

The ten points at elevation γ of the east stylobate of the Parthenon thus delineate an arc that deviates from exact symmetry by a total of 4 mm. in its length of 30.87 meters (or 1/6th

---

[91] Yeroulanou (1998) suggested that these metopes may have come from the earlier temple on the site. Perhaps so; but for our purposes this would mean that the builders of *both* temples did not require the metopes to be of identical size or (perhaps) arranged symmetrically. In another structure on the Acropolis, the Pinakotheke of the northwest wing of the Propylaea, the asymmetric placement of door and windows has long caused architectural historians heartburn. The suggestion of Plommer (1960) that they were placed where they were in order to shield paintings from direct sunlight seems common-sensical.

[92] In what follows we will very tentatively *assume* that Balanos' measurements are accurate. We do so with full recognition that the reliability of those measurements remains unproven, and may well be doubted. It follows, of course, that conclusions drawn from much of Balanos' data are tentative.

[93] Balanos, Table 1. The fact that the 31 - meter curve has been measured at 10 points is problematic. The "curve" is in fact a quasi - curve made up of ten equidistant points. We don't know why ten points were chosen – there is no reason to believe that the original builders used them in making those curves - and it is entirely possible that a larger number of points would have revealed a different, and perhaps even an asymmetric, curve instead.

of an inch in 1215 inches). This can be expressed as a ratio of 1:7300.

Five other elevations on the east front are also in this sense symmetric arcs. They are elevation δ at the stylobate edge

3.153 3.161 3.186 3.208 **3.218** **3.218** 3.209 3.188 3.161 3.153

elevation ε at the innermost point of the column base –

3.203 3.224 **3.232** **3.232** 3.223 3.202 3.182

elevation ζ at the base of the step leading up to the cella –

3.218 3.236 **3.241** **3.241** 3.236 3.219

elevation η on the secos wall –

3.526 3.547 **3.558** **3.558** 3.546 3.526

and elevation θ, also on the secos wall –

3.918 3.942 **3.952** **3.951** 3.940 3.920

The deviation from a consistent curve of these elevations is 1 part in 10,600, 12,800, 19,400, 21,200, and 4700, respectively.[94]

By virtually any standard these elevations, as measured by Balanos, are symmetric curves.[95]

Perhaps because they merely confirmed the prevailing assumption that Greek temples are symmetric, Balanos' measurements on the east front did not attract much attention. Balanos also surveyed the same elevations on the other three sides of the Parthenon and, as he reports, his measurements there "*m'a montré que ces courbes presentent une symmetrié parfait*". He allowed himself the triumphant observation that he had thereby "disproved" theories that the curves of the Parthenon and other ancient monuments in general are "random".[96]

---

[94] The *shape* of each arc is different from that of the other five.

[95] Of the elevations Balanos found to be symmetric only the east stylobate edge – evidently, Balanos' elevation δ - had been measured previously, by Penrose, who found its curve to be asymmetric.

[96] Balanos pp. 54 - 55: "*Ainsi la théorie qui formule que les courbes du Parthenon, et en général celles des monuments antiques, sont un effet du hasard … est exclue*". Balanos also claimed – though his reasoning for this is not apparent - that this finding disproved the notion that the Parthenon's curves "*sont des courbes esthétiques*".

In a certain sense his elation was justified. The question of whether curves and other "deviations from ordinary rectilineal construction"[97] on the Parthenon and other Classical structures were intentional refinements has been debated since their discovery in the early nineteenth century.

Captivating though such a possibility is, however, the evidence for it is not as compelling in every instance as it is for the more modest alternative which proposes that at least some of these "deviations" are probably the result of indifference to, or of inability to attain, more precise standards of construction.

But this uncertainty applies only to irregular shapes, for in the absence of unambiguous indications we simply do not know whether their irregularity was created intentionally or unintentionally.

The creation of the Parthenon's symmetric curves, on the other hand, that are symmetric to within one part in several thousand, *must* have been intentional, for it is impossible to believe that their precision was the result of happenstance.

If this is true of a single symmetric curve it is all the more true of several curves that, like the six curved elevations on the Parthenon's east front, are in close proximity to one another but with each having its own shape. These curves can only have been created intentionally by skilled and disciplined masons working with detailed instructions and within the most exacting margins of error.

Our assurance of this overrides otherwise reasonable objections that, because the ancient Greeks (as far as is known) had no term for bilateral symmetry, they could not have had the idea of it; or that we should not assume that they intended to create symmetric shapes unless we know why they wished to do so.

But this is not to say that Balanos was justified in claiming that all the elevations he measured were symmetric. Unfortunately, Balanos never indicated the standard he used for determining whether a shape is symmetric.

To be sure, the deviations of the curves on the east - front elevations noted above are so slight that one need have no hesitation in regarding them as symmetric.

The same elevations on the west front however, for all Balanos' claims that they are marked by *"une symmetrié parfait"*, tell a very different story (*fig.* 3.36).

---

[97] Penrose, 1888, p. v.

Counting outward from the center the discrepancies within each pair are 3, 7, 14, 21, 21; 6, 6, 13, 21, 20; 4, 7, 12, 17; 7, 1, 9; 3, 8, 10; and 3, 8, 9 mm., respectively. In no pair within any of these elevations are both segments the same length; in one pair there is a discrepancy of a mere 1 mm; but the rest of the discrepancies are larger, and in most instances, substantially larger. As the table below shows, the sum of all the discrepancies within each elevation of the west front is of an entirely different order of magnitude from the ones on the east front elevations:

Elevation	East Front (mm.)	West Front (mm.)
γ	4	66
δ	3	66
ε	2	40
ζ	1	17
η	1	21
θ	5	20

Balanos' claim that the elevations on the west front and on the two flanks (insofar as the latter could be measured) are symmetric implies a notion of symmetry that is so loose as to be virtually without meaning. In point of fact, the discrepancies Balanos found fall into two categories.

In one they are almost invariably either 2 mm. or less in a pair, while in the other they are considerably larger and also more erratic than that.

The arrangements in which discrepancies are 2 mm. or less – or even non - existent – seem qualitatively different from the rest; and it is on this basis that I call them "symmetric" and the ones where the discrepancies are greater, "asymmetric".[98]

But what about the other portions of the Parthenon? Some parts of the structure are too damaged to be measured.

---

[98] It may well be that the asymmetric forms were not made by the craftsmen who made the symmetric ones. However, we cannot be completely certain that they did not regard *their own* work as symmetric. Possessed of inferior skills, they may have believed that the forms that I designate as asymmetric were in fact symmetric, instead. With such an argument however – for all that it has validity – we descend further into the mire of *obscurum per obscurius*.

Other areas, where measurements could have been made, unfortunately were ignored by Balanos, for whatever reason.

In all, Balanos provided sufficient information to make a determination of whether a form is symmetric on 106 parts of the Parthenon. They (including the features we have already noted) are listed here under the number of the table or plate in which they appear in Balanos' book; an asterisk identifies a feature as symmetric:

Table I (*various elevations of steps, stylobate, sub - basement of secos wall, entablature*) [n=20; *=7]

East: β; γ*; δ*; ε*; ζ*; η*; θ*; ς; o.
West: β; γ; δ; ε; ζ ; η; θ; ς; o; χ^{1}*; o1.

Table II (*various elevations of steps, stylobate, sub - basement of secos wall, entablature*) [n=9; *=0]

North: β; γ; δ; ε; ζ ; η; θ; ς; o.

Table III (*various elevations of steps, stylobate, sub - basement of secos wall, entablature*) [n=9; *=0]

South: β; γ; δ; ε; ζ ; η; θ; ς; o.

Folding Plate 2 (*north colonnade*) [n=4; *=0]

Distance between capital centers
Axial intercolumniations at base
Height of abacus at center
Height of abacus at ends

Folding Plate 3 (*general plan*) [n=18; *=1]

West: two horizontal curves in plan
West: projection of peristyle capitals
West: projection of opisthodomos capitals
East: two horizontal curves in plan
South: two horizontal curves in plan
North: two horizontal curves in plan
West: intercolumniations at peristyle base
East: intercolumniations at peristyle base
South: intercolumniations at peristyle base
North: intercolumniations at peristyle base
West: opisthodomos intercolumniations
East: opisthodomos intercolumniations
Sum total of intercolumniations on flanks (=total lengths)[99]

---

[99] North flank 69.466 m. vs South flank, 69.51 m., a difference of 5 mm.

Sum total of intercolumniations on ends (=total widths)*[100]

Folding Plate 4 (*columnar dimensions*) [n=5; *=1]

Entasis of column x' - 22' (east - west)
Entasis of column x' - 22' (north - south)*[101]
Entasis of column m' - 12' (north - south[102]
depths of flutes on north and south sides
Entasis of a column of the pronaos

Folding Plate 5 (*interior façade of the opisthodomos entablature*) [n=3; *=0]

Abacus widths
Distance between abacus centers
Width of frieze panels

Folding Plate 6 (*interior of the west entablature*) [n=3; *=2]

Abacus widths *
Abacus heights*
Distance between abacus centers

Folding Plate 8 (*north colonnade*)[103] [n=2; *=0]
Intercolumnar distance at base
Distance between abacus centers

Folding Plate 10 (*east face*) [n=6; *=2]

Distance between triglyph centers
Width of [spaces for] metopes
Distance between abacus centers
Abacus widths*
Abacus heights*
Interaxial distance of columns at base

---

[100] East face 30.87 m. vs West face, 30.88 m., a difference of 10 mm.

[101] Inclination of the outer profile on both sides is equal.

[102] Balanos refers to the "*cannelures symmétriques*" on the south side of this column but this is unwarranted. The varying widths of the channels at each elevation – for instance, at the point where the column is 1.687 m. tall the channels vary between 0.032 and 0.137 m. - are random and asymmetric. The depths of the channels on the north side of the column are more consistent with each other but here too their distribution by depth is asymmetric.

[103] Despite the loss of columns on the north flank enough evidence remains on either side of the center to determine that the two measures given here could not have been symmetric.

Folding Plate 11 (*east face, from interior*) [n=4; *=1]

Distance between abacus centers
Abacus widths*
Abacus heights
Interaxial distance of columns at base

Folding Plate I (*east and west face curves in elevation and plan*[104])
[n=11; *=5]

East: symmetric plans at 4 levels*****
East: asymmetric plans at 2 levels
West: symmetric plans at 3 levels
West: symmetric plans at 2 levels

Folding Plate II (*north flank in elevation and plan*) [n=7; *=0]

North: asymmetric plans at 7 levels

Folding Plate III (*south flanks in elevation and plan*) [n=5; *=0]

South: asymmetric plans at 5 levels

It can be seen from this table that, of the 106 parts of the Parthenon for which Balanos provided sufficient data to determine whether their shapes are symmetric or asymmetric, only 19 are symmetric. Of these, almost two-thirds – 12 – are on the east front; there, however, as already noted, numerous asymmetric arrangements are also to be found. The measurements Balanos provided do not concentrate on any particular portion or portions of the Parthenon's existing structure, which perhaps allows one a modest degree of confidence that additional measurements might not significantly change the relative distribution of symmetric and asymmetric arrangements from the one given here.

The examples of the metope widths and the columnar heights suggest that at least some of these asymmetries may have been part of the original structure. We can be confident too that the symmetric portions on the east front of the Parthenon and elsewhere, if Balanos did indeed measure them accurately, are original, for symmetric shapes are not created by accident.

It would seem, then, and this is a significant finding, that the Parthenon may have been built as a structure that was partly symmetric and largely asymmetric. Why this was so – and in particular, why the symmetric arrangements were concentrated

---

[104] The elevations for Folding Plates I, II and III are provided in Table 1.

on only a relatively few sections of the Parthenon and chiefly on the east front – is a puzzlement, and likely to remain so.[105]

It cannot be over-emphasized, however, that this finding is a tentative one, and might only be confirmed, or refuted, when the Parthenon is surveyed once again, but this time on the basis of established (and identified) procedures and with the use of the highly precise surveying instruments that are now readily available: as well as with a satisfactory resolution of the problem of determining the placement of the points between which measurements are taken.

Also tentatively, I would put forward the suggestion that symmetric arrangements, should their presence on the fabric of the Parthenon ever indeed be reliably established, may reflect the application of esoteric doctrines of the Pythagoreans or some other group. For now, it is surely the case that the few symmetric curves registered by Balanos – *if that is what they are!* - do not require us to suppose that the idea of symmetry was generally known in the Classical world and scrupulously applied in the making of most, let alone *all*, things. They do not indicate that there was a "passion" for symmetry among the ancient Greeks.

The predominant asymmetry of the Parthenon does not set it apart from other ancient Greek structures. There is no evidence, indeed, that any major ancient Greek buildings are entirely, or even largely, symmetric.[106] The plans of the Minoan palace at Gournia on Crete (*fig. 3.35*), for example, or those of Lerna in the Argolid, imply an utter indifference to symmetric design.[107] So does the temple of Apollo in Syracuse, the diameters of whose columns vary, as we have already seen, by as much as 12 inches.[108]

Another example is the Argive Heraion. The graphic restoration by Pfaff [109] shows a symmetric layout, but the actual state plan of the temple, and of its foundations, establish that the

---

[105] In Selzer 2021, chapters 4-6, I show that a seemingly random mix of symmetric and asymmetric elements is found in many Byzantine structures, including in Hagia Sophia.

[106] See the examples of asymmetric Greek construction, and a thoughtful analysis of some of the technical issues involved, in Goodyear 1912, Chapter VI ("Asymmetric Dimensions in Greek Temples") and appendix, pp. 161 – 204; and Chapter VII ("Asymmetric Dimensions in Greek Temples and their Optical Effect") and appendix, pp. 205 – 214).

[107] Soles 1991; Wiencke 2000, v. 4, plan 24.

[108] See p. 85, *above*.

[109] Pfaff 2003, v. 1, fig. 84.

design was asymmetric. The perimeter foundations are asymmetric on both axes as indeed are the foundations of the cella, which are not parallel; the extension at the south end is not centered on the temple foundation. From the aerial photograph it is easy to tell that the foundation walls are not parallel – a feature that the "state plan" does not state quite fully enough. In another example, the windows of the East Building of the sanctuary of Apollo Hylates at Kourion are asymmetrically arranged. Noting this, Scranton adds rather curiously that this is "perhaps of no great significance".[110] The Temple "C" at Selinus is another example of asymmetry. According to Dinsmoor, while most of its columns have sixteen flutes, three on the east front and two on the west front have twenty, and these columns are not arranged symmetrically in the rows of columns of which they are a part.[111]

Most recently, Professor Philip Sapirstein of the University of Toronto, in Canada, has made the first photogrammetric measurements of the Heraion in Olympia. His findings are of great interest. "The stone peristyle columns are highly irregular", he writes.[112] "Some shafts are monolithic, while others are built from four to 11 drums. The diameters of the shafts vary from below 1.0m. to close to 1.3 m. Among the 21 surviving capitals there are a variety of profiles. Most can be dated to the Archaic period, but others can have been carved no earlier than the Classical period. Furthermore, there is no clustering of the various types in particular parts of the peristyle. Instead, columns of different types are juxtaposed apparently at random". Sapirstein also measured the intercolumniations of the peristyle and found that they varied irregularly from 0.97m to 1.25m. Some of these features of the Heraion have been recognized for a century, but were explained, according to some claims, by the theory that the original columns were of wood and were replaced over a considerable span of time by an assortment of available stone columns.[113] Sapirstein calls this thesis into question and suggests, instead, that all the columns were part of the original

---

[110] Scranton, 1967.

[111] Dinsmoor 1950, p. 80.

[112] Sapirstein, 2016.

[113] Of course, this is not much of an explanation, for it does not address the fact that the replacement columns, if such they were, could all have been cut to a uniform size and shape if whoever was in charge of the replacement work thought that important. Nevertheless, this flimsy hypothesis held for the better part of a century!

structure. However that debate resolves itself, it is clear that the Olympia Heraion does not support the thesis that the Greeks insisted that their buildings be constructed symmetrically.

Two well - documented buildings from the same century as the Parthenon are reported by twentieth - century surveyors to have a small number of symmetric areas, but are predominantly asymmetric.

According to Koch, horizontal curves on all four sides of the stylobate of the Hephaisteion in Athens are asymmetric, as are the cella walls and the geisons, and, on the west front, the curvature in plan of the columns and the distribution by height of the abaci; while on the other hand the arrangements of metopes, triglyphs and the intercolumniations throughout the structure are symmetric.[114] According to another investigator, Korres, the orthostates of this structure are symmetric.[115] These findings are reported without information about how they were obtained: and they too must be confirmed before any firm conclusions are drawn from them. Cooper, whose measurements are possibly the most reliable ever made of an ancient Greek structure, has described in detail the complicated layering of the stylobate of the temple of Apollo Bassitas that combines both symmetric and asymmetric arrangements; he also found a few minor parts of the temple's superstructure that are symmetric.[116] His measurements (*figs.* 3.38, 3.39) show that the spacing of the

---

[114] Koch 1955, figs. 31 p. 171, 34 p. 172, 40 p. 176. Koch gives the plan of the columns and the height of abaci only for the west front.

[115] Korres 1999, p. 93.

[116] Cooper, v. 1, pp. 164 - 183. Cooper does not however address the important issue of how one can reliably place the points between which measurements are made when their surfaces are worn or damaged. Cooper writes that the crowning course of the tympanum is "a single symmetric block", (p. 251) and that the east entrance may have been symmetric though he does not explain why he thinks so (p. 211). He reports that he determined with his naked eye (for they were too small to be measured with a micrometer) that two of the six types of *cyma reversa* moldings on the pteroma coffers are symmetric (p. 365). Cooper also described (p. 185) the corner intercolumniations as "symmetric", though the other intercolumniations are not distributed symmetrically. His drawing of the north end (pl. 20) indicates that the widths of the metopes and triglyphs follow a regular distribution, but as he only gives measurements for the right-hand side of the structure we cannot determine whether or not their distribution is symmetric.

flanking columns measured at their bases is asymmetric, and that the heights of the columns also vary asymmetrically.

Naturally, there is not much that we can confidently infer about the intentions of builders who lived twenty - five hundred years ago and who left behind no written texts that might have explained their work to us.[117] When the intentions we wish to understand concern a phenomenon–symmetry–for which they had no word or phrase, and when the evidence that we can obtain by studying a few ruined buildings is both scant and ambiguous, our problems are greatly compounded. Despite these difficulties however the facts reviewed here establish that the standard view – held for over five centuries - of Greek architecture as unfailingly symmetric is clearly unfounded. Even the most exemplary of Greek buildings, the Parthenon, would seem to have at most only a few symmetric portions; the greater part of the structure appears to be asymmetric.

It should not be supposed that all the asymmetries that have been detected in the Parthenon are the result of later acts of natural or human violence. That may be true of some of the structure's asymmetries, but I am inclined to think - bearing in mind the asymmetric arrangement, undoubtedly original, of the height of the peripteral columns or of the width of the metopes - that it is likely that the "irregularities" of many portions of the temple are original and were accepted in those days (to the extent that they were even noted) as the way things were done. There is no way for us to know, and almost certainly never will be, whether they represent deviations, caused by workmen's carelessness or incompetence, from the architects' plans.

It has been suggested that the Parthenon's asymmetries were created intentionally, as part of a subtle aesthetic scheme.[118] Penrose had this to say on the subject: "It has often been noticed that the works of Nature, although usually their tendency is to be symmetrical, are seldom absolutely so [*sic!*]; and when, in

---

[117] According to Tobin (1981) the "spatial symmetry" of Doric temples, which he took for an established fact, "was always subordinated to architectural ends". Unfortunately, he failed to disclose how he knew this, and what those "architectural ends" were.

[118] Scholars in the 19th century thought "each departure from the rules of regularity and symmetry in this [Greek] architecture was a carefully reasoned decision to achieve the most ineffable beauty and awe-inspiring sublimity" (Etlin 1994, p. 87).

architecture, exact symmetry does prevail, a dry effect is not infrequently produced". The Greeks, he went on to say, produced with "extreme care and refinement" the "charm which is sometimes the result ... of irregularity of design or even of workmanship".[119] Explanations like these of the asymmetries of the Parthenon and other Greek temples are less convincing than the notion that those asymmetries were unplanned and no doubt even unrecognized, and that they manifest the deep but not necessarily conscious sympathy for the asymmetry of Nature's forms that is a mainspring of all great design.

Obviously, asymmetric shapes cannot be created intentionally by people who have no idea of symmetry. More often than not, one suspects, asymmetric shapes are created unawares, by default, without deliberate intent.[120] Symmetric shapes by contrast are created when, for whatever reason, there is a deliberate decision that the two lateral halves of a design are to mirror each other.

If the measurements that I have cited here are accurate – and I cannot stress sufficiently that we do not know whether they are – then we would have to suppose that the idea of symmetry did indeed exist in ancient Greece and perhaps Rome, but that it was applied only on a very limited basis, and only to the construction of certain parts of some buildings. The lack of linguistic and literary evidence suggests that the symmetry in these structures would have manifested an esoteric idea known only to a small number of people.

This conclusion could be refuted – or confirmed and enlarged – if measurements of unquestionable reliability were made of the Parthenon and other ancient structures. As things now stand, however, our only certain evidence comes from the non-architectural sources we considered at the outset of this Chapter. They give us the confidence to say that the ancient Greeks and Romans did not require symmetric arrangements in

---

[119] Penrose 1888, pp. 11 - 12. Goodyear (1912, pp. 205 - 6) quotes the German historian Michaelis who, like Penrose, also attributes the "fascinating effect" of the Parthenon to the "considerable variations of width; the heights of the columns; the widths of the abaci, of the triglyphs and metopes".

[120] Michaelis, (quoted Goodyear, *ibid.*,) similarly argues for the *un*intentionality of the Parthenon's asymmetric portions which, as he writes, are "mostly so complicated, and of such various characters in the different parts of the building that have been compared, that it is difficult to consider them intentional".

their works of art and in their decorated objects, and that this was probably because they were unaware of the ideas of symmetry and asymmetry. Albeit with some hesitation, we may well apply this conclusion to the architecture, too, of the ancient Greeks and Romans.

Appendix 2: Restoring the Parthenon

The "restored" Stoa of Attilos. The Parthenon is just
visible on the Acropolis, in the upper right of the picture.

Our world is becoming full of "restored" buildings, and the problem with that can be seen all too vividly in the Stoa of Attilos, restored (so to say) in the 1950's with money from the Rockefellers. It stands, shimmering in its impudence, at one end of the ancient Athenian agora, within sight of the Acropolis. One might be forgiven for calling it a *restoratio ex nihilo,* for it is not a restoration at all, but a Speer-like fantasy, on a preposterously grandiose scale, and even more vulgar than grandiose, that someone or the other had about what once stood there. One must doubt that there is a single honest stone in the entire structure. Utterly implausible in itself, utterly divorced from the awesome place in which it is set – into which it so rudely intrudes - it is also as leaden a building as one is likely to encounter this side of the Kennedy Center in Washington DC.

The only decent restoration that might be done on it today is to demolish it entirely.

Now it is the Parthenon's turn. Something called the Acropolis Restoration Service, having already done what it thought it had to do with the Propylaea and the Erechtheion, is presently engaged in a massive program, funded by the European Union, of "restoring" the Parthenon itself.

But to which Parthenon do they propose to restore the existing structure? As we have seen, there is no consensus about the appearance of the Parthenon when it was first built. Since then it has suffered many vicissitudes – not just the wear and tear that accompanies the passage of time, but, among others, earthquakes, the detonation of an ammunition dump, Lord Elgin's depredations, and Balanos' well-meaning but near-fatal effort at restoration - all but the most recent of which have altered the Parthenon's appearance in ways that can never be determined again.

There is, in other words, no known original or even early state to which today's structure can authentically be "restored". Willfully ignoring this obvious fact, today's restorers are doomed to achieve something as implausible as the pristine, but lifeless and dreary, *and painfully wrong,* purported replica of the Parthenon in (of all places) Nashville, Tennessee.

Nashville's purported replica of the Parthenon

The Parthenon can have looked as Pericles and other leading Athenians intended it to look, if at all, only for its first few decades. All the changes that happened to it thereafter are inextricably part of its history, the evidence that this is an organic, evolving, structure: they do not testify to centuries of decline and destruction, but to centuries of its encounter with – its survival in face of - the forces of both Nature and Man. The things that irritate the restorers and which they want to eradicate – as if the Parthenon had not lived through these past 2400 years! – are not defects to be regretted and erased, not catastrophes to be remedied, but the heroic scars and lines of life, without which the living image is reduced to caricature. The Parthenon is not the pristine thing that was built in the 5th century BCE but the supremely glorious, scarred, battered, near-ruin that evolved to stand in triumph on the Acropolis today.

It is of course not only buildings that we want nowadays to restore. We live in an age of dyed hair, reshaped breasts, lifted faces, even cryogenics. There is a desperate denial of nature and of truth in these things. Our restorations, whether of stone or human bodies, mostly convince only those who "look but do not see". Montaigne would have us hold fast to what the years take away; but I think that we should welcome the years, and that we should not hesitate to show our faces to the world, lined as they may be with all those scars of our living and enduring. The ruins of our great buildings are also scars which should be valued, not effaced, for they too testify to *their* living and enduring – and to the indestructibility of some of life's noblest qualities.

CHAPTER FOUR
THE QUESTION OF MEDIEVAL
SYMMETRY

4.1 Chartres Cathedral

"The confusion and messiness of Chartres"
- John James (1982, p. 21)

"If one part always answers accurately to
another part, it is sure to be a bad building;
and the greater and more conspicuous the
irregularities, the greater chances are that
it is a good one"
- John Ruskin (1903-1912, vol. X, p. 268)

The concept of symmetry was unknown in the Middle Ages. This is sometimes questioned by people who suggest that medieval eyes would have accepted as symmetric shapes that we today, with our ability to measure and perhaps to see more accurately, would regard as asymmetric. Whatever merit this suggestion may have, it cannot account for medieval structures such as the west front of the 13th century cathedral of St Lo (*frontispiece*). The building was destroyed during the Allied landings in Normandy in 1944, but as we can see from old photographs, its façade was so pervasively asymmetric that there is probably no detail on one lateral half that is mirrored on the other. Note, for example, the different height, width and design of the two side portals – and the blatantly off-center position of the "central" doorway; note the different designs and dimensions of the two great flanking windows on the second level; and note, too, the entirely different designs and sizes of the two towers and their spires. It is impossible to believe, of course, that the men who built this asymmetric structure were so inept, or suffered from such defective eyesight, that they thought that they were duplicating the details of one side on the other, or that they had placed the main portal on the façade's central axis.

Nor is asymmetric design limited to relatively obscure provincial buildings.[1] The two towers of the great cathedral of Chartres (fig. 4.1) were built at more or less the same time, but do not at all resemble one another; no one will think that they were intended to be identical. Not just their shape but their dimensions, too, are different, for the left-hand (or northern) side of the facade, from the ground to the top of the steeple, is perhaps 15% narrower than the other side, though its steeple is taller. The steeples themselves are of course of entirely different design: Henry Adams thought that the south steeple is "the most perfect piece of architecture in the world"[2] but was restrained in his appreciation of its companion. The flanks of the façade moreover - the bases of the towers - are designed quite differently. We can be sure that these differences are the result of bold and magnificent – *and certainly conscious* - decisions by the builders. Assuredly, they did not arise because the builders attempted, but failed, to duplicate a single design: let alone because they could not see any difference between the two sides.

There are however those who infer that a design of such "confusion and messiness" as the west face of Chartres must have emerged from an uneasy and reluctant accommodation to contingencies that are unknown to us today.[3] In this view, the two towers and their spires were built at different times and reflect different tastes, and the builders of the later tower were willing to put up with "messiness" if as a result at least *one* tower and spire was consistent with their taste.[4] This is on the face of it a plausible explanation – but only

---

[1] Goodyear (1905) has assembled an interesting selection of examples of medieval architectural asymmetry.

[2] Adams 1933, p. 61.

[3] James, *op. cit.*

[4] The "striking difference" in the two towers that flank the west front of the abbey church of Jumieges, write Cotman and Turner (1822, p.3) "might justly lead to the inference, that there was also a material difference in their dates, and that they were not both of them part of the original plan; but there do not appear to be any grounds for such a supposition. On the other hand, the contrary seems to be well established; *and those who are best acquainted with the productions of Norman architects, will scarcely be surprised at anomalies of this nature*" (my emphasis). This seems a more sensible understanding than that of, for example, Lovejoy (1960b, p. 146), who argues without evidence that the asymmetry of Gothic cathedrals "was partly due to the historical

until we notice that the right-hand portal is wider than that on the left; and that the window above it is rather taller and perhaps slightly wider than that above the left-hand portal, Even more startlingly we must note that the great rose window of the façade is not centered on the same axis as the middle window or in the rectangle in which it is contained: *and indeed, that it is an eccentric circle* (fig. 4.2). The row of sixteen statues a little higher up on the façade, moreover, is not centered on the rose window. I have no explanation for these irregularities, but the point to make now is that they tell us that the asymmetries of the west front, and not merely those of the towers, must have been built into the structure virtually from the outset, and that they were not the result of later contingencies. (Simson believes – unfortunately, without providing evidence - that the construction of the two towers took place more or less concurrently, in the middle of the 12th century.[5] If this could be confirmed it would further strengthen the case for the discrepancies between the two towers not being the result of unavoidable contingencies.)

The asymmetry of the great façade of Notre Dame in Paris, also makes no attempt at concealing itself, (fig. 4.3), and it too cannot possibly have been unintentional or the awkward result of later decisions. Even the casual observer must notice that the two flanking portals of the façade are quite different from one another. The portal on the left is wider and shorter than the one on the right and it is contained in a sharply-angled recess, which the other is not. Moreover, while there are eight kings lining the arcade above the left-hand portal there are only seven in the arcade above the portal on the right. This

---

accident that few ... were completed in accordance with the original design". Weyl (1952, p. 16) seems to echo this explanation when he suggested that there are "historical reasons" why the towers of Chartres are different from one another. (Unfortunately, he did not disclose what, in his opinion, those reasons may have been.) Whether or not this view is correct may be debated, but the relevant point is surely that if people had thought it important for the two towers of Chartres, Jumieges, and many other structures, to look alike they would have built the one as the facsimile of the other. That they did not do so shows that, at the least, they had no objection to having the façade flanked by two obviously different towers. Rodin, for one, (1965, p.228) was unperturbed by the difference between Chartres' two towers, calling one of them "of silver", and the other, "of gold".

[5] Simson 1962, p. 148.

indeed is a clue, for those who have not yet detected it, that *the entire left side* of the façade, from the base to the top of the tower, is wider – by about 15% - than the façade's right-hand side: the left tower, moreover, is somewhat taller than that on the right. These are variations that must have been built into the structure from the outset. Their meaning, alas, is unknown to us today.

(For Corbusier's attempt to rationalize the design of Notre Dame's façade see *Appendix One*, to the present Chapter, pp. 129-131, *below*).

Another example of asymmetry that cannot possibly have been created unintentionally is the remarkable façade of the cathedral of Rouen (*fig.* 4.4). Its two towers, both magnificent, are altogether different. The central block of the façade is clearly much closer to the south tower (that on the right). Of the flanking portals, that on the north is taller and perhaps wider than the other which, unlike it, terminates in an aedicule. The colonnade above the south portal extends to the end of the structure; that above the north portal does not.

The standard study of the great church of the royal abbey of St Denis, by Crosby, presents a confusing and largely inaccurate account of its asymmetry, which it may be appropriate to rectify here. Crosby acknowledges that early engravings show that the north tower of St Denis (demolished in the middle of the 19th century) "was not the same as ... the south tower".[6] Nevertheless, his reconstruction of the façade shows the towers as identical, "on the assumption that [the...] master mason intended to build twin towers"! Crosby does not explain this thoroughly dubious assumption, which ignores the fact that the two towers are manifestly different in their proportions as well as in their overall design (*fig.* 4.5). He acknowledges however (pp. 176; 177-179) that "the major vertical axis of the central portal is not in exact alignment with the axis of the upper rose window, and the north side is slightly narrower than the southern". (The difference between the widths of the two sides is 30 cm., or just under 12 inches: or rather a little more than just "slightly".[7]) What evidently escaped Crosby's notice was that the window at the center of the arcade immediately above the north portal is *wider* but *shorter* than its opposite. The solid panels on either side of those windows also differ. Those on the north are incised with a grid

---

[6] Crosby 1987, p. 174.

[7] *Ibid*, pp. 176; 177-179.

of squares, while those on the south have a pattern of triangles. The arcade on the south, moreover, is much more deeply set back than the arcade on the north. On these arcades, it appears that while the arches on the south tower are pointed, only the inner arch on the north tower is; the other two are rounded. The rose window is flanked on either side by a vertical recess (fig. 4.6). That on the left is capped by a pointed arch, and that on the right by a rounded one. The left recess is substantially wider than the one on the right. The black bars that divide the recesses are spaced irregularly, are not of uniform height, and do not align with their opposite number.[8]

There are some striking asymmetric medieval churches in Germany.

Fig. 4.7

The Marienkirche cathedral in Lübeck built in the second half of the 13th century and the first half of the 14th, was largely destroyed during World War II, but – perhaps somewhat unusually – its postwar reconstruction retained the very evident asymmetry of the two tall towers (one is about 6" taller than the other). The towers are surmounted by large decorative gables, whose designs are entirely and obviously different, that on the North being filled with a grid of x-shaped figures, that on the South, with a 12-spoked wheel surmounting a simple dog-tooth design. Bars that I have superimposed on the photograph show that the North tower is narrower than the South tower; but that on the second level, at least, the windows of the latter are significantly taller and narrower. The seven quatrefoils on the second level of the South tower are noticeably larger than those on the other tower.

Fig. 4.8

Construction of the Regensburg cathedral began in the late 13th century; the façade of the West front was not completed until the beginning of the 15th century. The façade is a compendium of asymmetric design, a marvel of irregularity that will allow the eye no rest, but that nevertheless and bafflingly combines all of its disparate parts into an energetic, mysterious but

---

[8] Other medieval French churches with asymmetric facades include St Etienne, Caen, and the cathedrals of Coutances, Tours and Amiens.

never implausible whole.

In England a number of important churches lost their facades to eighteenth- and nineteenth-century renovators. Among them are Westminster Abbey, whose west front is largely the work of Nicholas Hawksmoor (d. 1736), and Norwich Cathedral, the west front of which is an extraordinarily banal and unpersuasive asymmetric structure (*fig.* 4.9) by the Gothic Revival architect Anthony Salvin (d. 1881). We do not know what the earlier facades of these churches looked like.

Fig. **4.10**

The façade of Exeter Cathedral on the other hand, probably completed in the late 14th century, is preserved in its pronounced asymmetry. The flanking doors differ in height and width; the five windows to the right of the main portal are not matched on the other side; while the chapel (?) that extends from the left wall is not matched on the right, where we find a bulky buttress, instead. On the next level, the massive window is not centered on the panel in which it is set, nor on the crenellations above and below it. The angled line of crenellations (reflecting, perhaps, the pitches of the nave roof) on the left is shorter and ends at a somewhat higher point than the matching line on the right. Above these rise two short towers, each with a single crenellation. The two on the left are both wider and taller than the one on the right.

Fig. **4.11**

The fourteenth-century façade of Gloucester Cathedral is also pervasively asymmetric. The flanking window to the right is wider and taller than its opposite, which has a doorway below it that is not centered on it. The two pairs of buttresses that flank these windows differ in height and in the placement of their decorative details. The section to the left of the main window is extended past the buttress, while that to the right is not similarly extended but is supported at the end by another buttress. The diagonal roof line at the left starts higher up than its opposite, is longer, and descends at a shallower angle.

In Spain, the 13th century façade of the cathedral of El Burgo de Osma, in Oviedo, is a boldly asymmetric structure (*fig.* 4.12). Numerous other asymmetric buildings are to be found in Spain. They include some of the most notable and important structures in that country.

Fig. 4.13
The great cathedral of Leon, also of the 13th century, has a façade that is asymmetric in most details. The designs of the two towers differ from each other, including in the placement of string courses and slit windows; the spire atop the south (right) tower is taller than the other. The porch that connects the two towers at ground level contains five arched portals. That on the extreme right is taller than the one on the extreme left. Between these two arches and the middle one are two unusual narrow arches. The one on the right is shorter and perhaps narrower than that on the left.

There are still a number of important asymmetric medieval churches in Italy.

Fig. 4.14
The remarkable unfinished façade of the Fiesole Badia is one of the most memorable in Italy. A rectangle of elaborate dark-green and white marble geometric patterns is superimposed asymmetrically (it is much further to the left of the structure) on an unfinished brick wall. The two large panels on either side of the doorway – the one on the right is both wider and taller than the other - are each framed with a band of a continuous (thus asymmetric) design that differs from one side to the next. Rectangular panels set into the three white marble compartments on either side are all of different patterns (I have not determined whether they are symmetric). In the left-hand panel the two flanking compartments are of different widths. The linear designs in the three flanking semicircles in the band above the doorway are asymmetric in themselves and also in the relation of one semicircle to the other; as are the arrangements of vertical lines above this arcade. The vertical bands of the upper row, too, are asymmetric both within and between each panel.

Figs. **4.15, 4.16**

The modernization of Santa Maria Assunta, the cathedral of Padua, was completed in the middle of the eighteenth century and no trace of its medieval interior survives. The cathedral's façade, however, which probably dates back to the twelfth century, was not rebuilt. It is markedly asymmetric. The façade's two windows are not centered on an axis drawn from the peak of the roof; neither is the main portal. The two shoulders are set at visibly different heights, and the wall of the central section that reaches up to the pitched roof is much taller on the right than on the left. The door on the left is taller and wider than that on the right, and further from the center door. The structure attached to the cathedral is its baptistery also of the twelfth century. Its porch is asymmetric and is sited asymmetrically on the main body of the building. Both drums are sited asymmetrically in relation to the structures on which they rest.

Fig. **4.17**

Otranto Cathedral. The façade of this little-known but fascinating 11th-century cathedral is asymmetric in almost every detail. On the upper roof, the left is longer and ends at a lower point than the right; and the opposite is the case with the lower pitches. The right-hand wall of the nave is longer than that on the left. The lancet window on the right is taller and wider than its opposite. The columns on the right of the doorway appear to be set further apart from each other than the pair on the opposite side, which also appears to be further from the side of the doorway.

Fig. **4.18**

The façade of San Giorgio, the cathedral of Ferrara, is pervasively asymmetric. The right-hand flank is wider and taller than that on the left. The top of the three sets of columnar arcades is higher on the right than it is on the left; while the bottom row of arcades is *lower* on the right than it is on the left. The doorway on the left is closer to the center of the structure as well as narrower than the one on the right.

Fig. **4.20**
The delightful little church of S. Giusto in Lucca has a late-12th century façade that is markedly asymmetric. The round window at the top is set off-center to the right; the little arcade of two arches two levels below it is also set off to the right; the circular windows that flank it are unequally spaced from the outer edge of the façade (that on the right being too far to the right). The main portal is also not centered on the façade: it too is too far to the right; and the flanking doorways are of unequal widths, that on the left being wider.

In Italy today however there are also a number of medieval churches whose facades are symmetric. Virtually without exception these are not their original facades.[9] From the late fifteenth to the late nineteenth centuries Italian churches were refaced in order to "correct" (as it was thought) their asymmetric design and other alleged flaws.[10] I have not

---

[9] Some medieval churches, such as San Miniato in Florence and the cathedrals of Venice, Pisa and Siena, or the Badia of Fiesole (*fig.* 4.14) have facades that are encrusted with asymmetric designs, sometimes multi-colored, in marble; some also have mosaic murals. I know of no asymmetric façade of this type that was demolished, a circumstance that suggests that these facades' decorative value may have trumped their asymmetry. Indeed, it would seem that the decorative scheme on the façade of Santa Maria Novella in Florence may actually have been *extended* during the work done on it in the fifteenth century (see Chapter Eight). The façade of the old St Peter's in the Vatican was surmounted by Giotto's mosaic *Navicella*, some surviving fragments of which are now in the Renaissance basilica. It is not a counter-example of the point made here, for the destruction of the façade was part of the demolition of the entire old basilica. The marble-and-mosaic-encrusted façade of Orvieto's cathedral is evidently symmetric, but it is a very rare and possibly unique exception that still awaits a satisfactory explanation.

[10] Nathaniel Hawthorne (1874, entry for October 2, 1858) noted that the Florentines and Romans "have obliterated, as far as they could, all the interest of their medieval structures by covering them with stucco, so that they have quite lost their character, and affect the spectator with no reverential idea of age." He did not recognize, however, that transforming asymmetric features, or concealing them, was a large part of the motivation for making these changes. That this was so may be seen from the fact that the construction of the new façade of the Florence Duomo was accompanied by the moving of windows and

discovered a single rebuilt façade that is asymmetric.[11]

Two of the greatest churches in Florence are among the many in that city whose facades were refashioned. Far from being authentic, as people often seem to assume, the symmetric confections of ineffable vulgarity that are the present facades of the church of Santa Croce and of Santa Maria del Fiore, the Florence cathedral, are nineteenth-century travesties. Fortunately, we have illustrations that show what the facades looked like in their unaltered states.

Figs. **4.21, 4.22**
The church of Santa Croce is shown in a painting (*fig.* 4.21) by Giovanni Signorini, dated 1846, that documents the asymmetry of its original appearance. The buttress-like devices on either side of the circular window are different in width, and the portal on the right is lower as well as closer to the outer edge of the façade than the portal on the left. The arch enclosing the main portal seems not to be centered on the great circular window above it. That window, most strikingly, is not centered on the wall in which it is placed but is closer to the right. *Fig.* 4.22, shows the symmetric façade, completed in 1863, that replaced it.[12]

Fig. **4.23**
The asymmetries of Santa Maria del Fiore were even more striking. They are recorded in the drawing made by Bernadino Poccetti in about 1585, shortly before the

---

tombs in the cathedral's interior to create a symmetric distribution. See also Ackerman (1991, p.91) for a discussion of the contortions to which Palladio subjected himself as he attempted to impose a symmetric exterior on the medieval arcades of Vicenza's Basilica.

[11] Among the many designs submitted for a new façade proposed for San Petronio in Bologna were two – by Baldassare Peruzzi (c. 1521) and by Giulio Romano and Cristoforo Lombardo (1546) – that are clearly asymmetric (Wittkower 1974a, pls. 98, 99). I can offer no explanation of this remarkable circumstance. The old façade has never, in fact, been replaced.

[12] The story, no doubt apocryphal, persists that the prominent Star of David on the façade reflects the Jewish antecedents of the façade's architect, Niccolo Matas of Ancona.

façade was demolished.[13] The drawing, which is almost certainly an accurate depiction of the structure, shows that the two flanking portals are of different designs and dimensions; that the tall recesses that rise from either side of the main portal are not centered on the great oval window nor on the point where the façade meets the peak of the roof. Those recesses moreover are of unequal height, the one on the right being lower than the other. Most obviously, the two oval side windows are set at markedly different heights on the façade.

In decorative details of "uncorrected" medieval churches we can find innumerable instances of asymmetric design. The asymmetry is typically unselfconscious.

Figs. **4.24 - 4.26**
One interesting example are the black-and-white bands on the great piers of the nave of S. Maria Assunta, the Siena cathedral.[14] The left and right piers nearest the entrance have 23 and 22 such bands; the next pairs 21 and 22; 22 and 22; 20 and 21; and 21 and 22, respectively. Only one facing pair of piers thus has the same number of bands. The bands on these piers also vary perceptibly in height.[15] The only one of the spectacular mosaic designs on the floor of the

---

[13] Businani 1993, p. 95. The original drawing is in the archives of the *Opera* of the cathedral.

[14] On the cathedral's façade the "misalignment" of the piers framing the central portal with those of the piers framing the rose window, although not an asymmetry, has been a source of consternation for art historians. Gillerman (1999) suggests that it is offset by "a heavy emphasis on axial symmetry" of the overall façade but, in fact, the façade, and the exterior generally, of this great cathedral is asymmetric. Gillerman adds, however, and I think quite correctly, that the "misalignment" would not have seemed anomalous to people in the Middle Ages.

[15] We may mention here the finding of Hiscock (2002, p. 107 and fig. 4.1) that in the transepts of Norwich Cathedral "the wall-piers dividing each part into three bays only face each other imprecisely... The south transept overall is shorter than the north". Hiscock used electronic means to measure the cathedral, so that his results can be considered more than usually reliable.

cathedral that I have examined is contained in a frame whose design is symmetric. As can be seen from my superimposed lines (fig. 4.25) the design in the space enclosed by that frame is asymmetric.

The façade of the Siena baptistery has a number of more or less square frames recessed into its lower walls, each containing a small carving. Fig. 4.26 shows one of these frames, in which it can be seen that although three sides of the frame each have eight dentelles, the fourth side has nine. The lion's head in the frame, moreover, is set markedly off-center, and its mane and facial features are asymmetric.[16]

Fig. 4.27

No less remarkable as evidence of an easy-going acceptance of the irregularities that we now regard as asymmetric is the famous 12th century "tree of life" mosaic in the apse of the church of San Clemente in Rome.[17] Jesus' right arm is visibly shorter than its opposite; the scrolled vines on the left are higher than those on the right; the curve of the vines enclosing the two standing figures differ in shape from each other; and the cloud-like shapes above the crucifix are all asymmetric in shape and in their relationship across the central axis.

Figs. 4.28, 4.29

The medieval panels of the main door of S. Zeno in Verona are unabashedly irregular in their relation to each other. None of the panels themselves have symmetric designs and many differ in size and the angles at which they are set; the overall effect is of a slightly crazy patchwork quilt. Leisinger suggests that some of the panels may have been a added at a later date, though their designs, too, are asymmetric.[18]

---

[16] Comp. the lion head reliefs on the reverse of the North Doors of the Florence Baptistery: Pope-Hennessy 1991, pp. 45 - 46.

[17] See Lloyd 1986 for a useful discussion of the date of the apse. The arms of medieval crucifices are frequently asymmetric. Brink (1978) documents a difference of 4 cm. in the arms of the Santa Croce Cimabue crucifix.

[18] Leisinger 1957, introductory section, "Verona". The bronze doors have now been placed inside for their protection; the wooden doors that

Figs. **4.30, 4.31**
Two lunettes above doorways leading into Santa
Maria della Pieve in Arezzo. In *fig.* 4.30, the reliefs of
the lunette and the scroll below it are asymmetric.
More strikingly, the doorway and its frame, along
with the lunette above it, are not centered on the
wider arch in which they are contained. *Fig.* 4.31
shows another doorway into the Pieve. The carvings
on the lunette are boldly asymmetric; the doorway
and the circular window above it are not centered on
each other.

Asymmetric design is common in medieval book arts,
too.

Fig. **4.32**
A Hebrew prayer book for Yom Kippur from 13th -
century Germany. The design of the titlepage is
strongly bilateral, but almost every detail on the one
side differs from that on the other side. For instance,
the mythical (?) beasts at the bottom of the page are
quite different from each other: note the differences in
the set of the two heads, or the lengths of the bodies.
The chevrons on the columns, by the same token,
resemble each other but their dimensions are
different; and the same can be said of the architectural
details that surmount the arch. In the crennelated
structure on the right, for example, we see one more
arch than we see on the equivalent structure on the
left.

Fig. **4.33**
A leaf from an early 15th - century Parisian Book of
Hours that had once belonged to Burne-Jones, the pre-
Raphaelite painter.[19] The design is asymmetric
throughout. The exquisite bed of flowers is of course
asymmetric; the panel set into it is well off-center; the

---

were substituted for them have frames that are evidently supposed to
recall the bronze panels of the original door but they have been
arranged (no surprise there!) symmetrically,

[19] bonhams.com/auctions/21845/lot/9/

three-sided frame that encloses the illustration and the
text is asymmetric: note, for example, the different
sizes (and in fact different colors, too, though that is
not apparent from the illustration used here) of the
opposing sections.

Fig. 4.34

The rear cover of the ninth-century Ashburnham
Gospels, one of the greatest treasures of the Pierpont
Morgan Library in New York City. This sumptuous
binding manifests complete indifference to symmetric
design. Even the three visible cruciform arms of the
haloes of the Jesus (?) figures facing the center panel
are asymmetric. So too are the four sections into
which the cover is divided, and all the intricate
designs within them.

Finally, we must also attest to the asymmetry of secular
medieval architecture.

Fig. 4.19

The Basilica of St Mark in Venice, completed during
the thirteenth century, is a treasury of asymmetric
design.[20] What I would point out here, because less
well-known, are the asymmetries of the great and
lovely piazza in which it is set. The plan of the piazza
(fig. 4.19) is asymmetric; the axis of the basilica itself is
at a very different angle from that of the piazza;[21] the
tall campanile rises from a seemingly arbitrary spot near
the mouth of the piazza and is not centered on or
aligned with any point around it. The piazzetta that
leads to the lagoon is also asymmetric; note the
position of the two columns toward the water. The
façade of the doge's palace – Ruskin called it "the first
building" of the world – also has more asymmetries
than are usually noted. Particularly fascinating, and I

---

[20] See the foldout plates, with elevations of the basilica, in Samonà et al
(1977); and Goodyear (1905, pp. 88 - 107, and the foldout plan) for
detailed measurements of the basilica's interior.

[21] Some paintings correct this axial "error", for example the view of P.
San Marco attributed to Giambattista Cimaroli (late 18th century),
published in the Winter, 2012 edition of Bonham's Magazine.

think particularly enlivening, is the fact that the diagonal lines and rows of rectangles on it, composed of different-colored bricks, are often not set straight and are not always continued on a straight line when intersected by a window.[22]

Fig. 4.36
Another example of secular medieval asymmetric design is the awesome Palazzo Vecchio in Florence, also known as the Palazzo della Signoria. Most obviously, the immense tower is not centered on the structure, and the fenestration is distributed unevenly on the principal façade; the portal beneath them is also situated asymmetrically.

Figs. 4,35a, b
The glorious Chateau de Chambord may be regarded as the last magnificent efflorescence of medieval French architecture, the close of a period that had produced the marvels of Chartres and Notre Dame. Construction of this palace was begun in 1519, well before Serlio brought the Italian Renaissance's new aesthetic of symmetry to France. An observer must spend many minutes to comprehend the intricate asymmetries of the façade, and especially of the playful roof line (*fig.* 4.35b).

There is abundant evidence, as we have seen by now, that medieval builders and craftsmen were in general indifferent to what we think of as symmetric design. Stendhal recognized this when he wrote that "There is no exact measurement, no symmetry, in medieval buildings ... if there are arches in a straight line, the breadths are rarely equal", *etc., etc.*[23] Yet he surely overstated the matter, for in France and elsewhere there are to be found medieval structures that *are* symmetric. Examples of them include the main facades of the cathedrals at Laon (completed c. 1225); Salisbury (completed c. 1320); or Orvieto (much of it completed by c. 1350).

---

[22] Palladio had hoped to replace the existing palace with a neo-Classical structure. Fortunately, his ambition was thwarted.

[23] Stendhal 1962, p. 255.

These structures however do *not* attest to a knowledge
of the concept of symmetry in medieval times. The use of the
terms "asymmetric" and "symmetric" in a medieval context is
anachronistic, for there is no evidence that anyone who lived in
those times knew those words, in our understanding of them, or
the ideas for which they stand. The fact that *we* can identify
medieval designs as being either the one or the other thus does
not mean that people in the Middle Ages also saw them as such;
almost certainly they did not. A case in point is the ground plan
of the Palazzo Vecchio. It is a quite irregular shape, and
definitely not symmetric. But for the Florentine chronicler
Villani (d. 1348) the problem with its design was not that it was
irregular or that it was asymmetric but that it was not square.
As Villani expressed it, it was *una granda diffalta di non farlo
quadrato* – "a big mistake not to have made it square".[24]
Designing the plan of a building "on the square" (*ad quadratum*)
or its elevation on the equilateral or isosceles triangle (*ad
triangulum*) was an important principle of medieval architecture,
albeit one that is not always evident in the actual structure of a
building.[25] As a principle of *proportion* based on regular
geometric forms it led to designs that we regard as symmetric.
But, to repeat, there is no evidence that the men who designed
and built structures *ad quadratum* or *ad triangulum* intended to
create – or even recognized that they were creating – symmetric
shapes.[26] The very detailed written record of discussions during

---

[24] *Cronica*, Lib.VIII, cap. 26. Significantly, Villani did not comment on the
many asymmetric features of the building's elevations, including the
design of its principal facade. Two centuries later Vasari (in the *Life* of
Michelozzo) would also criticize the Palazzo della Signoria for its
irregular design "built out of square" and with "unequal columns in the
courtyard", but although he too did not refer to these irregularities as
asymmetric, we know that he regarded symmetry as an essential
element of good architecture. A modern scholar, Goldthwaite, however,
did not see that the palazzo's design is asymmetric. Rather
astonishingly, he declared (1993, p. 181) instead that the structure "has
been called [he does not say by whom] one of the most important
buildings in the history of Italian architecture for the influence of its ...
symmetrical façade ... on the evolution of domestic architecture in the
Renaissance"!

[25] For an overview of the issues regarding *ad quadratum* see Wu 2002 and
Shelby 1972; and the excellent paper by Bucher, 1972.

[26] For Christian Platonists in the middle ages the square was "the
geometrical representation of the Godhead and as such the source of all
aesthetic perfection"- Simson 1962, p. 49; similarly Hiscock 2000, p. 39.

the late fourteenth century of *ad triangulum* issues in the design of the cathedral in Milan, notably, include no reference to the symmetry that would result from the application of this principle.[27]

We must therefore regard the symmetry of medieval architecture and decorative work, where it occurs, as unintentional and unacknowledged - an unwitting consequence of the geometric principles that were sometimes employed by their designers. It is not evidence that the concept of symmetry existed in medieval times.[28]

(It is difficult to accept the claim of Crisp, even on the basis of the material he himself presents, that medieval gardens "were ... symmetrical".[29] Some medieval gardens, as Crisp himself acknowledged, consisted of little more than an

---

But there is no evidence that the *symmetry* of the square was thought of as an aspect of its perfection (or even that it was detected). Similarly, outlines of one-half of the north transept portal and rose window for the Clermont Cathedral, evidently intended as templates for stone masons, have been found incised on a flat roof of the structure (Davis, 2002). The use of these templates *recto* and *verso,* as it were, would have resulted in structures that are symmetric. But it seems more plausible that this was a device to ensure the equal distribution of loads by matching both sides, rather than a method intended to create symmetric structures.

[27] Ackerman 1991, pp. 211 - 268; and Wittkower 1974a, pp. 17 - 65. In Chapter Three we saw that although the ancient Greeks knew shapes that are symmetric there is no reason for us to suppose that *they* recognized those shapes' symmetry. Plato's solids, notably, are symmetric but neither he, nor Euclid in his mathematical elucidation of them, seemed to have been aware of that.

[28] That *ad quadratum* was not associated with bilateral symmetry can also be seen from the fact that in a section about the proportions of a gateway captioned "...*La Symmetria dilla Magna Porta*" Colonna (1499, c4) declared, "*la principale regula peculiare al'architecto è quadratura*". In this passage Colonna clearly intended *simmetria* in the traditional meaning of harmonious proportions, which is attained (in his view) by obeying architecture's imperative of *ad quadratum* ("quadratura"). For him, therefore, building on the square was a way to ensure that a structure's proportions would be harmonious. He nowhere associated it with bilateral symmetry, though he used *simmetria* in this sense in the *Hypnerotomachia* as frequently as he used it in the older sense of "harmonious proportion" (see pp. 305-307, *below*).

[29] Crisp 1924, p. 18, 15, 27.

untended space enclosed by a wattle fence; others were made up of "a picturesque confusion of roses, hawthorns and honeysuckles mixed with fruit-trees and shrubs, all growing in wild profusion". Others again were like the garden whose "lawn of exceedingly fine grass, of so deep a green as to seem almost black [was] dotted all over with possibly a thousand kinds of gaily-colored flowers" in which Boccaccio sets part of *The Decameron's* third day.[30] There is no evidence that the "paths of unusual width, all as straight as arrows" that crisscrossed and surrounded this garden were arranged symmetrically. On the other hand, medieval times also knew of gardens - whether idealized or actual - like that of Sir Mirth in the 13th century *Roman de la Rose*, which were (in Chaucer's translation) "by mesurying, right even and square in compasyng". Whether these reflected the geometric principles of medieval architecture, views about the design of the Garden of Eden,[31] or the desire for efficient tending and harvesting of the plants – often herbs – that grew in them, is unclear. We cannot of course infer that they were designed to be, or were regarded as, symmetric.)

But few ideas ever spring forth into the world fully formed, and one supposes that the concept of symmetry probably had some of its roots in trends that were gathering momentum in the late Middle Ages. A hint of them, possibly, may be found in the suddenly relentless and self-conscious symmetry of buildings (*fig. 4.37*) in the paintings of Giotto (d. 1337) and Taddeo Gaddi (d. 1366).[32] Another hint, perhaps,

---

[30] Boccaccio 1972, p. 252. Some of the English translations of the Decameron have the garden in the conclusion to the Sixth Day "neatly arranged and *symmetrically* composed" (see for example Thacker 1979, p. 92). In fact Boccaccio describes the garden as "*sì ben composti e sì bene ordinati.*"- i.e., well-designed and well-arranged. Nothing about symmetry here, therefore, which is hardly surprising given the fact that the concept originated about 75 years after Boccaccio's death.

[31] Prest 1981.

[32] Villani likened Gaddi to Vitruvius. Ladis (2008) states that for Giotto "symmetry [was] an aesthetic imperative … a guiding ideal in every sense". Since the concept of symmetry had not yet been formulated, the statement is anachronistic. It is also incorrect in a more direct sense, for the Arena chapel, from which Ladis claims to have derived his insight, is not arranged symmetrically – compare, for example, the two banks of angels, or saints, on either side of the Last Judgment. Indeed, Giotto seems on occasion to have gone out of his way to create what we think

could be the warning of Lorenzo Maitani (d. 1330) when he was in in charge of the construction of Siena's cathedral, that if a proposed extension of the choir were carried out the dome would no longer be centered on the crossing of the nave and the transept, as it ought rationally – *rationaliter* – to be.[33] The impetus for these ideas came in all likelihood from the *ad quadratum* and *ad triangulum* precepts, but it may well be that those precepts also provided some movement toward the concept of symmetry, itself.

Yet (to anticipate) it was not from them that the concept of symmetry emerged. It seems rather than the principal impetus behind this new concept was the need for an ordered and immediately comprehensible visual environment. This need, I will suggest in Chapter Six, was engendered by the terrible depredations of the Black Death of 1348 - 1350, and the long succession of lesser (but still appallingly deadly) plagues that followed in its wake.

Appendix 1: Le Corbusier's Rationalization of Notre Dame

In *Towards a New Architecture*, Corbusier declared that the Gothic cathedral "is not very beautiful" because it is not "based" on the geometric figures that, in his view, are fundamental requirements of good architecture.[34] Later in the

---

of as asymmetries – see for example the intentionally crossed eyes of both Madonna and Child in his *Ognissanti Madonna*, and in his St. *Peter Enthroned in the Vatican*. On the asymmetry of the *Baroncelli Polyptich* cf. McManus (2005 pp. 163 - 5), and comp. Caglioti (1992 pp. 112 - 113), who writes that crossed eyes were "intentionally introduced" by the Byzantines, though he does not offer any suggestion as to what that intention may have been. The building depicted on the counter - façade of the upper basilica in Assisi (if that is by Giotto) is asymmetric, as is the roundel containing the Dove. Note for instance the 6 dentelles on the right edge of the pediment on the left versus 7 on the other edge. None of this is to deny, of course, what we today recognize as symmetric design in some of Giotto's work; rather, I mention this to dismiss Ladis' unwarranted claim about symmetry being an "imperative" that Giotto obeyed. Almost certainly, indeed, it was not a concept he knew.

[33] Norman 1995, vol. II, p. 142; and comp. Burns 1971, and Onians 1992.

[34] Corbusier 1986, p. 30.

same book, however, he praised Notre Dame, Paris, on the grounds that "the determinant surface of the Cathedral is based on the square and the circle" (and is thus, we must infer, symmetric).

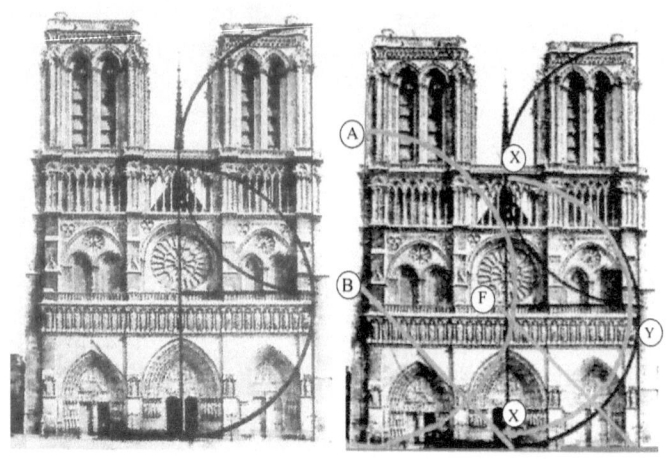

To demonstrate this point he superimposed geometric lines on a photograph of Notre Dame's west front (image above, *left*)[35]. The purported squares that Corbusier drew, one above the other, occupy the entire right half of the façade, and bisect the rose window and the central portal. A purported semi-circle sweeps from the top of the center of the façade, touches the point where the two squares meet on the extreme right of the structure, and then continues until it touches the ground at the exact center of the central portal: a seemingly nice demonstration...

...Yet one that is completely invalid, if not, indeed, actually fraudulent. In the illustration above, right, I have inscribed some additional lines on Corbusier's photograph. They show that Corbusier's lower "square" is not a square at all. "F" marks the point where the upper left-hand corner of a true square would be found, a vertical line drawn from it would bisect neither the rose window nor the central portal. Further, a circle centered, as Corbusier's purports to be centered, at the left-hand point where the two alleged squares meet, traces a very different contour ("X") from the one ("Y") depicted by

---

[35] *ibid*, p. 77.

Corbusier, and does not touch the bottom line of the façade at any point.

As indicated earlier in this chapter, the left "half" of the Notre Dame façade is considerably wider than the right (a fact of which Corbusier seems to have been unaware): and so, if a geometry of squares and circles had indeed determined the dispositions of the right side of the façade it would not have been able to determine those on the other side, too. To demonstrate this self-evident point I have indicated ("B") where a square based on the width of the *left* side of the façade would have its upper, outer corner, and where the lines of a semi-circle ("A") based on such a square would rest. None of these lines determine any feature of the left side of the façade.

Appendix 2: Ruskin on Gothic Architecture and Art

In this Chapter we have demonstrated that Gothic, or medieval, aesthetics did not include an appreciation of, let alone an insistence on, symmetric design. We have not, however, touched on the question of why this was so: or, to put the matter more usefully, what the values and attitudes were that led to the "confusion and messiness" of Gothic design, and thus precluded the clarity and orderliness of the symmetry and associated characteristics of Renaissance architecture.

There is no more profound and vivid discussion of this question than in *The Stones of Venice*. Ruskin's insights there, and the strident and sonorous tones of his prose, were informed by the passionate, and even precarious, genius of a Biblical prophet, and I can offer my readers no better service than to quote them at length here.[36]

Among the "mental tendencies" of Gothic builders, Ruskin declared - he thought of it, indeed, as its most important characteristic - is its "savagery", which he attributed in part to the harsh physical environment of northern Europe:

> There is no reproach in the word [savagery], rightly understood; on the contrary, there is a profound truth, which the instinct of mankind almost unconsciously recognizes. It is true, greatly and deeply true, that the

---

[36] Ruskin (1903 - 1912), vol. x, pp. 183 - 269, *passim*; and vol. XI, pp. 47 - 119, *passim*.

architecture of the North is rude and wild; but it is not true, that, for this reason, we are to condemn it, or despise. Far otherwise: I believe it is in this very character that it deserves our profoundest reverence ... There is, I repeat, no degradation, no reproach in this [term] but all dignity and honourableness: and we should err grievously in refusing ... to recognize as an essential character of the existing architecture of the North ... this wildness of thought, and roughness of work, this look of mountain brotherhood between the cathedral and the Alp; this magnificence of sturdy power, put forth only the more energetically because the fine finger-touch was chilled away by the frosty wind, and the eye dimmed by the moor-mist, or blinded by the hail; this out-speaking of the strong spirit of men who may not gather redundant fruitage from the earth, nor bask in dreamy benignity of sunshine, but must break the rock for bread, and cleave the forest for fire, and show, even in what they did for their delight, some of the hard habits of the arm and heart that grew on them as they swung the axe or pressed the plough."

Ruskin elaborated this thought in discussing ornament. The Greeks had demanded that their craftsmen execute all their work perfectly; the medieval Christians allowed imperfection in their work as an acknowledgment, not only of the value of every soul, but of its imperfection, too:

Neither [the Greek artisan] nor those for whom he worked could endure the appearance of imperfection in anything, and, therefore, what ornament he appointed to be done by those beneath him was composed of mere geometrical forms, – balls, ridges, and perfectly symmetrical foliage, - which could be executed with absolute precision by line and rule, and were as perfect in their way, when completed, as his own figure sculpture... The Greek gave to the lower workman no subject which he could not perfectly execute... The workman was ... a slave. But in the medieval, or especially Christian, system of ornament, this slavery is done away with altogether; Christianity having recognized, in small things as well as in great, the individual value of every soul. But it not only recognizes its value; it confesses its imperfection, in only bestowing dignity upon the acknowledgment of unworthiness. That admission of lost power and fallen nature, which the Greek ... felt to be intensely painful, and, as far as might be, altogether refused, the Christian makes daily and hourly, contemplating the fact of it without fear, as tending, in the end, to God's greater glory. Therefore, to every spirit which Christianity summons to her service, her exhortation is: Do what you can, and confess frankly what you are unable to do; neither let your effort be shortened for fear of failure, nor your confession silenced for fear of shame. And it is, perhaps, the principle admirableness of the Gothic schools of architecture, that they thus receive the results of the labour of inferior minds; and out of fragments full of imperfection, and betraying that imperfection in every touch, indulgently raise up a stately and unaccusable whole. Men were not intended to work with the accuracy of tools, to be precise and perfect in all their actions. If you will have that precision out of them, and make their fingers measure degrees like cog-wheels, and their arms strike curves like compasses, you must unhumanize them, all the energy of their spirits must be given to make cogs and compasses of themselves. All their attention and strength must go to the accomplishment of the mean act. The eye of the soul must be bent upon the finger-point, and the soul's force must fill all the invisible nerves that guide it, ten hours a day, that it may nor err from its steely precision, and so soul and sight be worn away, and the whole human being be lost at last... On the other hand, if you will make a man of the working creature, you cannot make a tool. Let him but imagine, to think, to try to do anything worth doing; and the engine-turned precision is lost at once. Out come all his roughness, all his dullness, all his incapability; shame upon shame, failure upon failure, pause after pause: but out comes the whole majesty of him, also... And now, reader, look round this English room of yours, about which you have been proud so often, because the work of it was so good and strong, and the ornaments of it so finished. Examine again all those accurate mouldings, and perfect polishings, and unerring adjustments of the seasoned wood and tempered steel. Many a time you have exulted over them, and thought how great England was, because her slightest work was done so thoroughly. Alas! If read rightly, these perfectnesses are signs of a slavery in our England a thousand times more bitter and degrading than that of the scourged African, or helot Greek ... the strength of them is given daily to be wasted into the fineness of a web, or racked into the exactness of a line. And on the other hand, go forth again to gaze upon the old cathedral front, where you have smiled so often at the fantastic ignorance of the old sculptors: examine once more those ugly goblins, and formless monsters, and stern statues, anatomiless and rigid: but do not mock at them for they are signs of the life and liberty of every workman who struck the stone; a freedom of thought, and rank in scale of being, such as no laws, no charters, no charities can secure... Rather choose rough work than smooth work, so only that the practical purpose be answered, and never imagine there is reason to be proud of anything that may be accomplished by patience and sand-paper. I shall only give one example ... Our modern glass is exquisitely clear in its substance, true in its form, accurate in its cutting. We are proud of this. We ought to be ashamed of it. The old Venice glass was muddy, inaccurate in all its forms, and clumsily cut, if at all. And the old Venetian was justly proud of it. For there is this different between the England and Venetian workman, that the former thinks only of accurately matching his patterns, and getting his curves perfectly true and his edges perfectly sharp, and becomes a mere machine for rounding curves and sharpening edges; while the old Venetian cared not a whit whether his edges were sharp or not, but he invented a new design for every glass that he made, and never moulded a handle or a lip without a new fancy for it. And therefore, though some Venetian glass is ugly and clumsy enough when made by clumsy and uninventive workmen, other Venetian glass is so lovely in its forms that no price is too great for it; and we never see the same form in it twice. Now you cannot have the finish and the varied form too. If the workman is thinking about his edges, he cannot be thinking of his design; if of his design, he cannot think of his edges. Choose whether you will pay for the lovely form or the perfect finish, and choose at the same moment whether you will make the worker a man or a grindstone.

The imperfections of Gothic workmanship, Ruskin argued, are not only a noble but an essential quality of Christian architecture. Only imperfect architecture can be truly noble, and it was the unwillingness to accept this – a new "relentless requirement of perfection" - that brought about the decline of the arts in Europe that we call the Renaissance :

... the rudeness or imperfection which at first rendered the term "Gothic" one of reproach is indeed, when rightly understood, one of the most noble characters of Christian architecture, and not only a noble but an *essential* one. It seems a fantastic paradox, but it is nevertheless a most important truth, that no architecture can be truly noble which is *not* imperfect. And this is easily demonstrable. For since the architect, whom we will suppose capable of doing all in perfection, cannot execute the whole with his own hands, he must either make slaves of his workmen in the old Greek, and present English fashion, and level his work to his slave's capabilities, which is to degrade it; or else he must take his workmen as he finds them, and let them show their weaknesses together with their strength, which will involve the Gothic imperfection, but render the whole work as noble as the intellect of the age can make it. ... Of human work none but what is bad can be perfect, in its own bad way. [Moreover], imperfection is in some sort essential to all that we know in life. It is the sign of life in a mortal body, that is to say, of a state of progress and change. Nothing that lives is, or can be, rigidly perfect; part of it is decaying, part nascent ... And in all things that live there are certain irregularities and deficiencies which are not only signs of life, but sources of beauty. No human face is exactly the same in its lines on each side, no leaf perfect in its lobes, no branch in its symmetry. All admit irregularity as they imply change; and to banish imperfection is to destroy expression, to check exertion, to paralyze vitality. All things are literally better, lovelier, and more beloved for the imperfections which have been divinely appointed... Accept this then for a universal law, that neither architecture nor any other noble work of man can be good unless it be imperfect ... the first cause of the fall of the arts of Europe [i.e. the Renaissance] was a relentless requirement of perfection.

The medieval acceptance of the inferior work of a craftsman acknowledged a duty to him, but it also rewards those who behold his work by acknowledging the deep satisfaction that variety gives us:

I have already enforced the allowing independent operation to the inferior workman, simply as a duty to *him*, and as ennobling the architecture by rendering it more Christian. We have now to consider what reward we obtain for the performance of this duty, namely, the perpetual variety of every feature of the building.

Wherever the workman is utterly enslaved, the parts of the building must of course be absolutely like each other; for the perfection of his execution can only be reached by exercising him in doing one thing, and giving him nothing else to do. The degree in which the workman is degraded may be thus known at a glance, by observing whether the several parts of the building are similar or not; and if, as in Greek work, all the capitals are alike, and all the mouldings unvaried, then the degradation is complete; if, as in Egyptian or Ninevite work, though the manner of executing certain figures is always the same, the order of design is less total; if, as in Gothic work, there is perpetual change both in design and execution, the workman must have been altogether set free. How much the beholder gains from the liberty of the labourer may perhaps be questioned in England, where one of the strongest instincts in nearly every mind is that Love of Order which makes us desire that our house windows should pair like our carriage horses, and allows us to yield our faith unhesitatingly to architectural theories which fix a form for everything, and forbid variation from it. I would not impeach love of order ... it helps us in our commerce and in all purely practical matters... Only do not let us suppose that love of order is love of art. It is true that order, in its highest sense, is one of the necessities of art, just as time is a necessity of music; but love of order has no more to do without right enjoyment of architecture or painting, than love of punctuality with the appreciation of an opera... Our architects gravely inform us that, as there are four rules of arithmetic, there are five orders of architecture; we, in our simplicity, think that this sounds consistent, and believe them. They inform us also that there is one proper form for Corinthian capitals, another for Doric, and another for Ionic.... Understanding, therefore, that one form of the said capitals is proper, and no other, and having a conscientious horror of all impropriety, we allow the architect to provide us with the said capitals, of the proper form, in such and such a quantity, and in all other points to take care that the legal forms are observed; which having done, we rest in forced confidence that we are well housed.

The aesthetic of the Middle Ages, Ruskin declared, delighted above all in change, variety and inconsistency:

But our higher instincts are not deceived. We take no pleasure in the building provided for us, resembling that which we take in a new book or a new picture. We may be proud of its size, complacent in its correctness, and happy in its convenience. We may take the same pleasure in its symmetry and workmanship as in a well-ordered room, or a skillful piece of manufacture. And this we suppose to be all the pleasure that architecture was ever intended to give us. The idea of reading a building as we would read Milton or Dante, and getting the same kind of delight out of the stones as out of the stanzas, never enters our mind for a moment. And for good reason: - There is indeed rhythm in the verses, quite as strict as the symmetries or rhythm of the architecture, and a thousand times more beautiful, but there is something else than rhythm. The verses were neither made to order, nor to match, as the capitals were; and we have therefore a kind of pleasure in them other than a sense of propriety. But it requires a strong effort of common sense to shake ourselves quit of all that we have been taught in the last two centuries, and wake to the perception of a truth just as simple and certain as it is new: that great art, whether expressing itself in words, colours, or stones, does *not* say the same thing over and over again; that the merit of architectural, as of every other art, consists in its saying new and different things; that to repeat itself is no more a characteristic of genius in marble than it is of genius in print; and that we may, without offending any laws of good taste, require of an architect, as we do of a novelist, that he should be not only correct, but entertaining. Let us then understand at once that change or variety is as much a necessity to the human heart and brain in building as in books; that there is no merit, though there is some occasional use, in monotony; and that we must no more expect to derive either pleasure or profit from an architecture whose ornaments are of one pattern, and whose pillars are of one proportion, than we should out of a universe in which the clouds are all of one shape, and the trees all of one size. .

The Gothic spirit...not only dared, but delighted in, the infringement of every servile principle; and invented a series of forms of which the merit was, not merely that they were new, but that they were *capable of perpetual novelty*. The pointed arch was not merely a bold variation from the round, but it admitted of millions of variations in itself (so too the grouped shaft, so too tracery in the windows). ... the Gothic schools exhibited [the] love of variety in culminating energy; and their influence, wherever it extended itself, may be sooner and farther traced by this character than by any other; the tendency to the adoption of Gothic types being always first shown by greater irregularity, and richer variation in the forms of architecture it is about to supersede, long before the appearance of the pointed arch or any other recognizable *outward* sign of the Gothic mind. It is one of the chief virtues of the Gothic builders, that they never suffered ideas of outside symmetries and consistencies to interfere with the real use and value of what they did. If they wanted a window, they opened one; a room, they added one; a buttress, they built one; utterly regardless of any established conventionalities of external appearance, knowing (as indeed it always happened) that such daring interruptions of the formal plan would rather give additional interest to its symmetry than injure it, so that, in the best times of Gothic, a useless window would rather have been opened in an unexpected place for the sake of the surprise, than a useful one forbidden for the sake of symmetry. Every successive architect, employed upon a great work, built the pieces he added in his own way, utterly regardless of the style adopted by his predecessors; and if two towers were raised in nominal correspondence at the sides of a cathedral front, one was nearly sure to be different from the other, and in each the style at the top to be different from the style at the bottom. These marked variations were, however, ... part of the great system of perpetual change which ran through every member of Gothic design, and rendered it as endless a field for the beholder's inquiry as for the builder's imagination: change, which in the best schools is subtle and delicate ... in the more barbaric schools is somewhat fantastic and redundant; but, in all, a necessary and constant condition of the life of the school... The vital principle is not the love of *Knowledge,* but the love of *Change.* It is that strange *disquietude* of the Gothic spirit that is its greatness: that restlessness of the dreaming mind that wanders hither and thither among the niches, and flickers feverishly around the pinnacles, and frets and fades in labyrinthine knots and shadows along wall and roof, and yet is not satisfied and shall not be satisfied. The Greek could stay in his triglyph furrow, and be at peace; but the work of the Gothic heart is fretwork still, and it can neither rest in, nor from, its labour, but must pass on sleeplessly, until its love of change shall be pacified for ever in the change that must come alike on them that wake and them that sleep.

Perhaps the most fundamental quality of Gothic architecture is its visual complexity reflecting "the fullness and wealth of the material universe":

No architecture is so haughty as that which is simple: which refuses to address the eye, except in a few clear and forceful lines; which implies, in offering so little to our regards, that all it has offered is perfect; and disdains, either by the complexity or the attractiveness of its features, to embarrass our investigation, or betray us into delight. That humility, which is the very life of the Gothic school, is shown not only in the imperfection, but in the accumulation, of ornament. The inferior rank of the workman is often shown as much in the richness, as the roughness, of his work; and if the co-operation of every hand, and the sympathy of every heart, are to be received, we must be content to allow the redundance which disguises the failure of the feeble, and wins the regard of the inattentive. There are, however, far nobler interests mingling, in the Gothic heart, with the rude love of decorative accumulation: a magnificent enthusiasm, which feels as if it never could do enough to reach the fulness of its ideal; an unselfishness of sacrifice, which would rather cast fruitless labour before the altar than stand idle in the market; and, finally, a profound sympathy with the fulness and wealth of the material universe. The sculptor who sought for his models among the forest leaves, could not but quickly and deeply feel that complexity need not involve the loss of grace, nor richness that of repose; and every hour which he spent in the study of the minute and various work of Nature, made him feel more forcibly the barrenness of what was best in that of man: nor is it to be wondered at, that, seeing her perfect and exquisite creations poured forth in a profusion which conception could not grasp nor calculation sum, he should think that it ill became him to be niggardly in his own rude craftsmanship; and where he saw throughout the universe a faultless beauty lavished on measureless spaces of broidered field and blooming mountain, to grudge his poor and imperfect labour to the few stones that he had raised one upon another, for habitation or memorial. The years of his life passed away before his task was accomplished; but generation succeeded generation with unwearied enthusiasm, and the cathedral front was at last lost in the tapestry of its traceries, like a rock among the thickets and herbage of spring.

That Ruskin's criteria for Gothic architecture remain applicable is clear from his advice to a person appraising a building:

> Observe if it be irregular, its different parts fitting themselves to different purposes, no one caring what

becomes of them, so that they do their work. If one part
always answers accurately to another part,[37] it is sure to
be a bad building; and the greater and more conspicuous
the irregularities, the greater the chances are that it is a
good one... Observe if all the traceries, capitals, and other
ornamentals are of perpetually varied design. If not, the
work is assuredly bad.

Ruskin's understanding of the character of Gothic
architecture is in many respects the mirror image of his view of
Renaissance art and architecture. He abhorred their precision,
their simplicity, their inability to grasp the essential nature of
things:

The first notable characteristic of the Renaissance ... is its
introduction of accurate knowledge into all its work, so far
as it possesses such knowledge; and its evident conviction
that such science is necessary to the excellence of the work,
and is the first thing to be expressed therein. So that all the
forms introduced, even in its minor ornament, are studied
with the utmost care ... Perspective, linear and aerial, perfect
drawing and accurate light and shade in painting, and true
anatomy in all representations of the human form, drawn or
sculptured, are the first requirements in all the work of this
school.... Considering all this in the most charitable light, as
pursued from a real love of truth, and not from vanity, it
would, of course, have been all excellent and admirable, had
it been regarded as the aid of art, and not as its essence. But
the grand mistake of the Renaissance schools lay in
supposing that science and art were the same things, and
that to advance in the one was necessarily to perfect the
other. Whereas they are, in reality, things not only different,
but so opposed that to advance the one is, in ninety-nine
cases out of the hundred, to retrograde the other. ... Science
studies the relations of things to each other: but art studies
only their relations to man: and it requires of everything
which is submitted to it imperatively this, and only this, -
what that thing is to the human eyes and human heart, what
it has to say to men, and what it can become to them: a field
of question just as much vaster than that of science, as the
soul is larger than the material creation... The whole
function of the artist in the world is to be a seeing and
feeling creature; to be an instrument of such tenderness and
sensitiveness, that no shadow, no hue, no line, no

---

[37] i.e., if the structure is symmetric.

instantaneous and evanescent expression of the visible
things around him, nor any of the emotions which they are
capable of conveying to the spirit which has been given him,
shall either be left unrecorded, or fade from the book of
record. It is not his business either to think, to judge, to
argue, or to know. His place is neither in the closet, nor on
the bench, nor at the bar, nor in the library. They are for
other men, and other work... The work of his life is to be
two-fold only; to see, to feel.... God has made every man fit
for his work; He has given to the man whom He means for a
student, the reflective, logical, sequential faculties; and to the
man whom He means for an artist, the perceptive, sensitive,
retentive faculties. And neither of these men ... can even
comprehend the way in which it is done. The student has no
understanding of the vision, nor the painter of the process;
but chiefly, the student has no idea of the colossal grasp of
the true painter's vision and sensibility. .. All the members of
Surgeons' Hall helping each other could not at this moment
see, or represent, the natural movement of a human body in
vigorous action, as a poor dyer's son did two hundred years
ago...[38] But, surely, it is still insisted, granting this particular
faculty to the painter, he will still see more as he knows
more, and the more knowledge he obtains, therefore, the
better. No: not even so. It is indeed true that, here and there,
a piece of knowledge will enable the eye to detect a truth
which might otherwise have escaped it ... but for one visible
truth to which knowledge thus opens the eyes, it seals them
to a thousand: that is to say, if the knowledge occur to the
mind so as to occupy its powers of contemplation at the
moment when the sight-work is to be done, the mind retires
inward, fixes itself upon the known fact, and forgets the
passing visible ones; and a *moment* of such forgetfulness
loses more to the painter than a day's thought can gain...
The first thing that a thinking and knowing man sees in the
course of the day, he will not easily quit. It is not his way to
quit anything without getting to the bottom of it, if possible.
But the artist is bound to receive all things on the broad,
white, lucid field of his soul, not to grasp at one. For
instance, as the knowing and thinking man watches the
sunrise he sees something the colour of a ray, or the change
of a cloud, that is new to him; and this he follows out
forthwith into a labyrinth of optical and pneumatical laws,
perceiving no more clouds nor rays all the morning. But the

---

[38] The reference is to Tintoretto.

painter must catch all the ways, all the colours that come, and see them all truly, all in their real relations and succession; therefore, everything that occupies room in his mind he must cast aside for the time as completely as may be. The thoughtful man is gone far away to seek; but the perceiving man must sit still, and open his heart to receive. The thoughtful man is knitting and sharpening himself into a two-edged sword wherewith to pierce. The perceiving man is stretching himself into a four-cornered sheet, wherewith to catch. And all the breadth to which he can expand himself, and all the white emptiness into which he can blanch himself, will not be enough to receive what God has to give him.

The shallowness of Renaissance art is the work of men who to their detriment were not "simple and unlearned":

'Well but', still answers the reader, '...the fact is that a picture of the Renaissance period, or by a modern master, does indeed represent Nature more faithfully than one wrought in the ignorance of old times'... No, not one whit. Indeed, the outside of Nature is more truly drawn; the material commonplace, which can be systematized, catalogued, and taught to all painstaking mankind, - forms of ribs and scapulae, of eyebrows, and lips, and curls of hair. Whatever can be measured and handled, dissected and demonstrated, - in a word, whatever is of the body only, - that the schools of knowledge do resolutely and courageously possess themselves of, and portray. But whatever is immeasurable, intangible, indivisible, and of the spirit, that the schools of knowledge do as certainly lose, and blot out of their sight; that is to say, all that is worth art's possessing or recording at all; for whatever can be arrested, measured and systematized, we can contemplate as much as we will in Nature herself. But what we want Art to do for us is to stay what is fleeting, and to enlighten what is incomprehensible, to incorporate the things that have no measure, and immortalize the things that have no duration. The dimly seen, momentary glance, the flitting shadow of faint emotion, the imperfect lines of fading thought, and all that by and though such things as these is recorded on the features of man, and all that in man's person and actions, and in the great natural world is infinite and wonderful; having in it that spirit and power which man may witness but not weigh; conceive but not comprehend; love but not limit; and imagine but not define; this, the beginning and the end of the aim of all

noble art, we have, in the ancient art[39] by perception; and
we have *not,* in the newer art, by knowledge. Giotto gives it
us: Orcagna gives it us; Angelico, Memmi, Pisano, – it
matters not who, - *all simple and unlearned men,* in their
measure and manner, - give it us; and the learned men that
followed them give it us not, and we, in our supreme
learning, own ourselves at this day farther from it than
ever.

The "rigid, cold, inhuman" austerity of the Renaissance
aesthetic, Ruskin continues, served the interests of domination
and power:

In the simple and meagre lines of the Renaissance ... there
is indeed an expression of aristocracy in its worst
characters; coldness, perfectness of training, incapability of
emotion, want of sympathy with the weakness of lower
men, blank, hopeless, haughty self-sufficiency. ... All other
architectures have something in them that common men
can enjoy; some concession to the simplicities of humanity,
some daily bread for the hunger of the multitude. Quaint
fancy, rich ornament, bright colour, something that shows a
sympathy with men of ordinary minds and hearts; and this
wrought out, at least in the Gothic, with a rudeness
showing that the workman did not mind exposing his own
ignorance if he could please others. But the Renaissance is
exactly the contrary of this. It is rigid, cold, inhuman;
incapable of glowing, of stopping, of conceding for an
instant. Whatever excellence it has is refined, highly-
trained, and deeply erudite; a kind which the architect well
knows no common mind can taste. He proclaims it to us
aloud. "You cannot feel my work unless you study
Vitruvius. I will give you no gay colour, no pleasant
sculpture, nothing to make you happy; for I am a learned
man. All the pleasure you can have in anything I do is in its
proud breeding, its rigid formalism, its perfect finish, its
cold tranquility. I do not work for the vulgar, only for the
men of the academy and the court".

And the instinct of the world felt this in a moment. In the
new precision and accurate law of the classical forms, they
perceived something peculiarly adapted to the setting forth
of state in an appalling manner; princes delighted in it, and
courtiers. The Gothic was good for God's worship, but this

---

[39] i.e., Gothic art

was good for man's worship. The Gothic had fellowship
with all hearts and was universal, like nature; it could
frame a temple for the prayer of nations, or shrunk into the
poor man's winding stair. But here was an architecture that
would not shrink, that had in it no submission, no mercy.
The proud princes and lords rejoiced in it. It was full of
insult to the poor in every line. It would not be built of the
materials at the poor man's hand; it would not roof itself
with thatch or shingle and black oak beams: it would not
wall itself with rough stone or brick; it would nor pierce
itself with small windows where they were needed; it
would not niche itself, wherever there was room for it, in
the street corners. It would be of hewn stone; it would have
its windows and its doors, and its stairs and its pillars, in
lordly order and stately size; it would have its wings and
its corridors, and its halls and its gardens, as if all the earth
were its own. And the rugged cottages of the mountaineers,
and the fantastic streets of the labouring burgher, were to
be thrust out of its way, as of a lower species...   The
Renaissance spirit became base both in its abstinence and
its indulgence; curtailing the bright and playful wealth of
form and thought which filled the architecture of the earlier
ages with sources of delight for their hardy spirit, pure,
simple, and yet rich as the fretwork of flowers and moss
watered by some strong and stainless mountain stream:
and base in its indulgence; as it granted to the body what it
withdrew from the heart, and exhausted, in smoothing the
pavement for the painless feet, and softening the pillow for
the sluggish brain, the powers of art which once had hewn
rough ladders into the clouds of heaven, and set up the
stones by which they rested for houses of God... The
simpletons and sophists had their way ... and the reader
can have no conception of the inanities and puerilities of
the writers who, with the help of Vitruvius, re-established
the "five orders", determined the proportions of each, and
gave the various recipes for sublimity and beauty, which
have been thenceforward followed to this day, but which
may, I believe, in this age of perfect machinery, be followed
out still farther. If, indeed, there are only five perfect forms
of columns and architraves, and there be a fixed proportion
to each, it is certainly possible, with a little ingenuity, so to
regular a stone-cutting machine as that it shall furnish
pillars and friezes, to the size ordered, of any of the five
orders, on the most perfect Greek models, in any
quantity... but if this be not so ... then let the whole system
of the orders and their proportions be cast out and

trampled down as the most vain, barbarous, and paltry deception that was ever stamped on human prejudice; and let us understand this plain truth, common to all work of man, that, if it be good work, it is not a copy, nor anything done by rule, but a freshly and divinely imagined thing. Five orders! There is not a side chapel in any Gothic cathedral but it has fifty orders, the worst of them better than the best of the Greek ones, and all new; and a single inventive human soul could create a thousand orders in an hour.[40]

---

[40] Robert Venturi sounds a bold and exciting echo of Ruskin in his *Complexity and Contradiction in Architecture* (1977): "I like elements which are hybrid rather than 'pure', compromising rather than 'clean', distorted rather than 'straightforward, ambiguous rather than 'articulated' ... I am for messy vitality over obvious unity", etc., etc. Unfortunately, Venturi's buildings, among them the annex to the London National Gallery, the Guild House in Philadelphia, the art museum at Oberlin, the Princeton student center, the theater in Hartford, etc., etc., do not live up to his verbal posturing: these structures are dull and uninteresting, and sometimes just trivially attention-grabbing. Ruskin would have dismissed them with scorn. Another, and very lovely, statement, somewhat akin to the aesthetic Ruskin would later espouse, is Herrick's poem, "Delight in Disorder" (1648). It exults in the "sweet disorder" of a woman's clothes which, he says, "do more bewitch me than when art/is too precise in every part".

ILLUSTRATIONS

(a table of illustrations may be found on the preliminary pages)

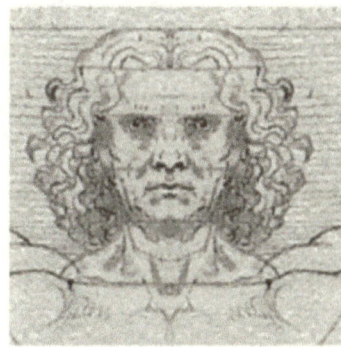

1.5 "Vitruvian Man" with right side doubled

1.3 Head of Leonardo da Vinci's "Vitruvian Man"

1.4 "Vitruvian Man" with left side doubled

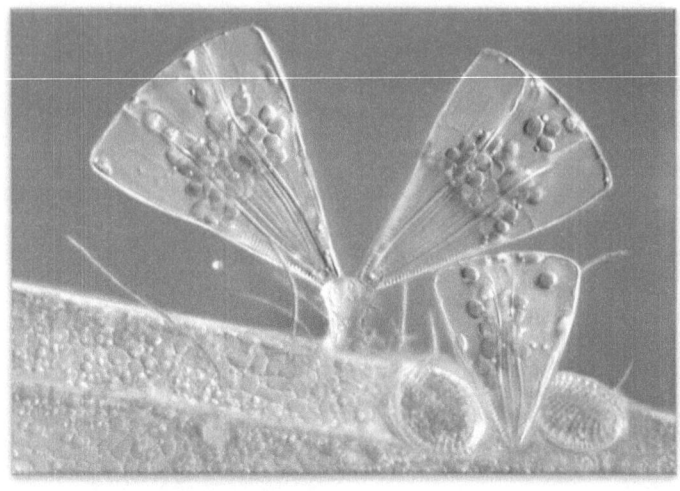

2.2 *Licmophora juegensii* on red alga An exquisite example of
the asymmetry of natural forms

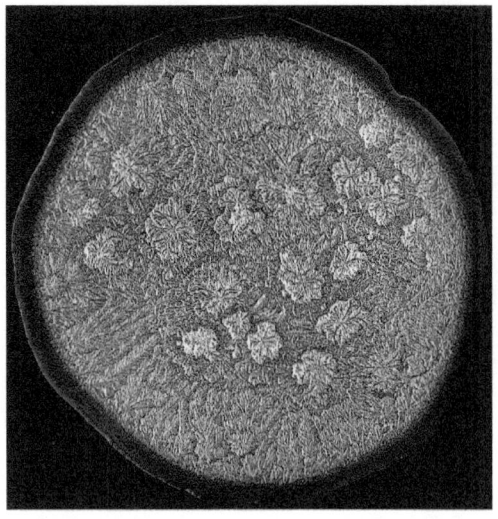

2.3 Microscopic photograph of a human tear by Maurice Mikkers

2.4 Not just the spider web but the spider itself with its uneven opposing limbs are markedly asymmetric.

2.5 Image of cosmic background radiation from the Big Bang. Note the asymmetric distribution and shapes of the forms within the (man-made) oval frame.

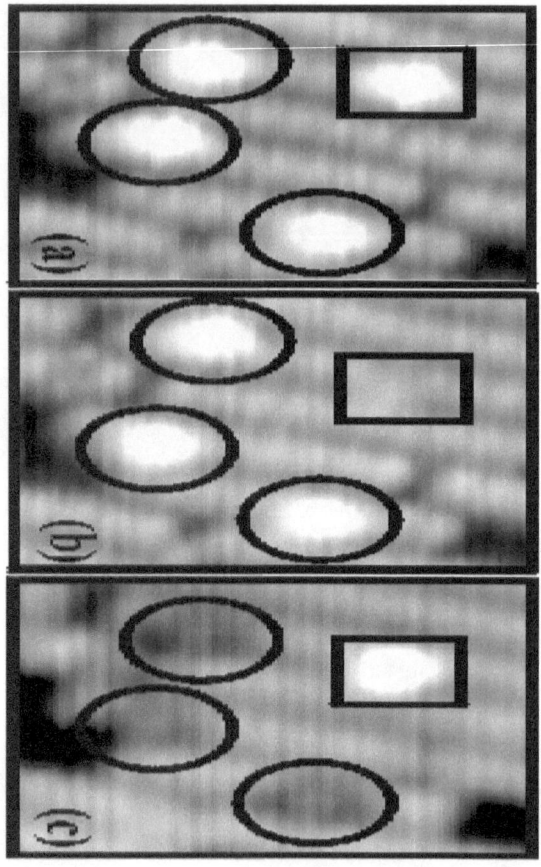

2.6 A series of 45 Å X 45Å constant current (0.1 nA) STM images of NBE molecules.[1] The molecules (within the oval and rectangular frames) are all asymmetrically shaped.

---

[1] foresight.org/Conferences/MNT7/Papers/Hersam/index.html

2.7 Eagle Nebula, "Pillars of Creation" as seen by the Hubble Telescope

2.8 M82, a "starburst" galaxy – obviously asymmetric.

2.10 Olaus Magnus' illustrations of frost (upper left) and snowflakes (right).[2] Olaus claimed that these drawings were based on observations he had made. His "snowflakes" are of course all asymmetric.

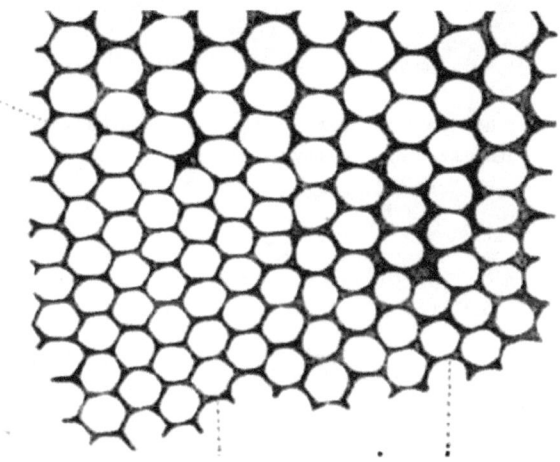

2.11 Portion of honeycomb as drawn by Wyman. He reported that none of the cells he observed had all sides of the same length, and that their interior angles were "nowhere sharply defined".

---

[2] Magnus 1996, Bk.1, cap. 22.

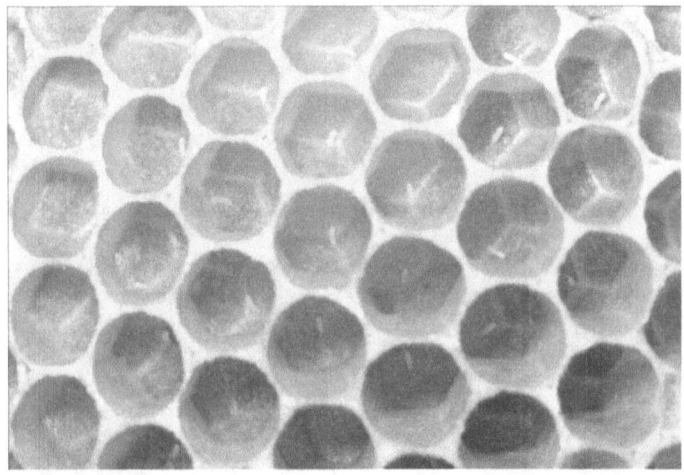

2.12 Photograph of bees' cells.[3] They are all differently and irregularly shaped; all are asymmetric.

[3]casiochulrechner.de/de/teilnehmervektoria2008/andesgymnas-iumschwabisch/gmuend/mathe _ist_alles.html

2.13   One of Weyl's purportedly symmetric snowflakes.[4] The
superimposed grid makes it evident that no two of the six major facing
branches are on the same axis, and that virtually every detail on the right
side of the crystal differs in shape and size from that on the left.

[4] Weyl 1955, *fig.* 38.

2.14 Libbrecht's "especially precise sixfold symmetry". The six sides of the notional perimeter are all of different lengths. The superimposed grid lines enable one to see how individual features on one side differ in size, shape and/or location from their equivalents.   For example, the highlighted arrows point at shapes that resemble airplanes. The one on the left has a tail-like structure behind it which the one on the right is without. The latter, moreover, is further from the central axis, and its lower "wing-tip" is higher, than the other. With the grid lines as a guide, almost every detail of the snowflake can readily be seen to be part of an asymmetric pair.

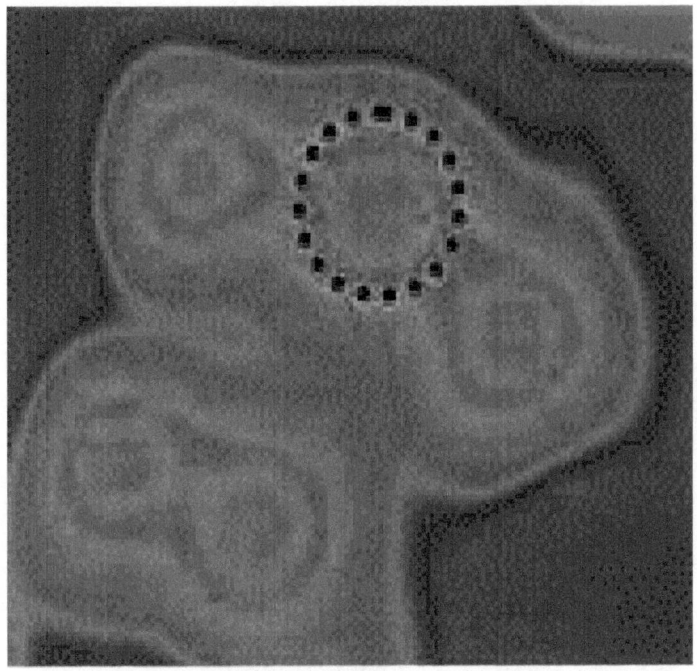

2.15 Water molecule. I do not pretend to understand this image, or why a circle of dots has been superimposed on part of it. The image does, however, demonstrate that water molecules are *not* symmetric, and that is why I am presenting it here. (Source: nature.com/nmat/journal/v9/ n5/abs/nmat 2740.html)

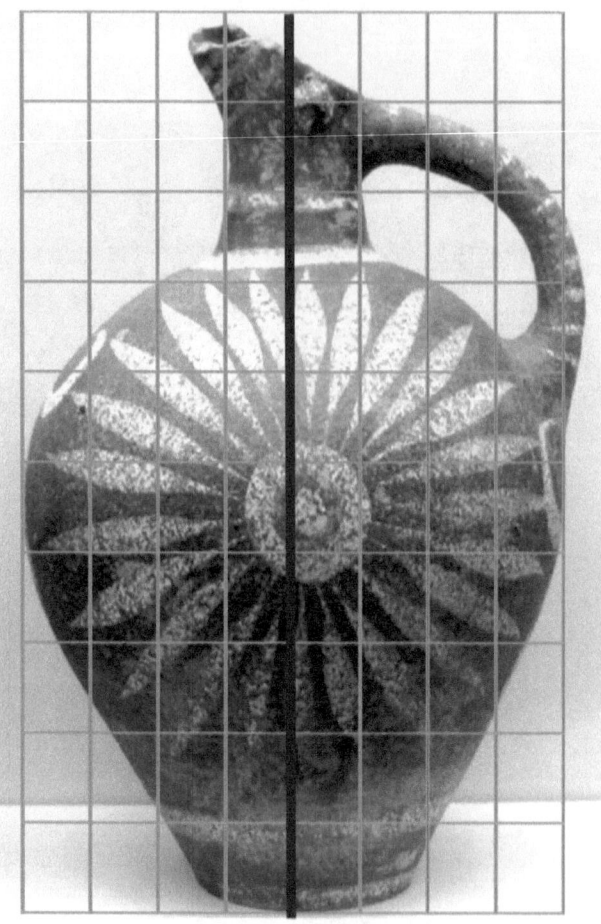

3.2 Minoan vessel, c. 2100 - 1700 BCE. The body of this object is asymmetric, though the neck is centered on the vertical axis that passes through the middle. The center of the floral or sunburst design if well to the right of that axis, and the petals or rays on the right are shorter and never quite on the same axis as those opposite them.

3.3 Vase from Thera, c.9th – 7th century BCE. None of the elements on the right half of this design are mirrored by those on the left half.

3.4 Detail of 8th - century BCE prosthesis vase. Among the many asymmetric features of this design are the differences between the figures beneath the bier. The figures on the right are flanked by columns of inverted "V"s, those on the left by "M"s. Note too the small figure next to the bier on the right, which is not matched by one on the left.

3.5 11th. cent BCE Gorgon from Rhodes

3.6 Detail of Euboean amphora, c.570 - 560 BCE. The handle on the right is considerably larger than that on the left; the lions differ in shape from each other and are not centered on the neck; the higher cluster is composed of eight; and the lower of seven dots asymmetrically arranged; and the clusters are not both on the same axis.

3.7 Attic amphora, attributed to Exekias, c.540 BCE. The most evident of the many asymmetric features of this design are the two spirals at the lower left, which not only differ in size and relative position but wind in the same direction.

3.8 Krater, Apuleia, *c.* 330 - 320 BCE.   The superimposed grid is centered on the tip of the pediment above the head of the seated figure. It is clear that the shape of the vase with its handles as well as *all* the decorations on the vase are not aligned on this axis. The artist's indifference to considerations of symmetry is particularly apparent in the various botanical embellishments on the neck. The orientation of the bust and of the seated figure also show the artist's indifference to symmetric design.

3.9 Mitra from Crete, 6th. cent. BCE. The tail and haunch of the figure on the left are higher than those on the right, but its body is considerably shorter.

3.10 Argive vase 5th cent. BCE. Almost every feature of the right (facing) side of this vessel is higher than on the left. This includes the rim, the disks on it, and the handles. The grid shows that the shape of the body of the vessel is slightly out of kilter.

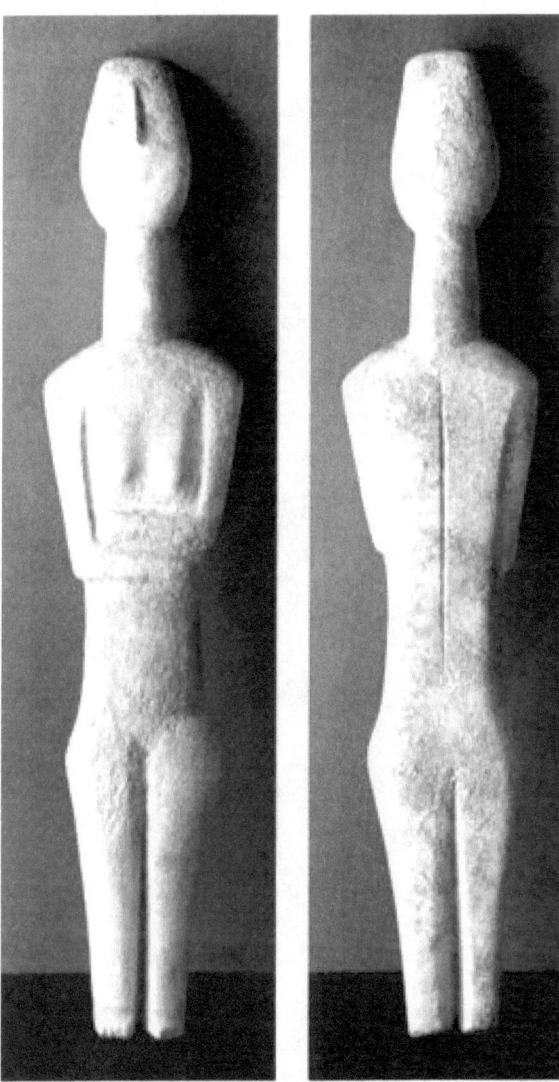

3.11 Female, Cyclades, 3rd millennium BCE. On the frontal image it is easy to see that the (facing) right shoulder is longer than the other. The left breast is lower and of course, because of the way the arms are crossed, the upper part of the left arm is much shorter than that of the right arm. It will be noticed, too, that the left forearm is considerably thicker than the right forearm. It seems too that the right leg is thicker than the left.

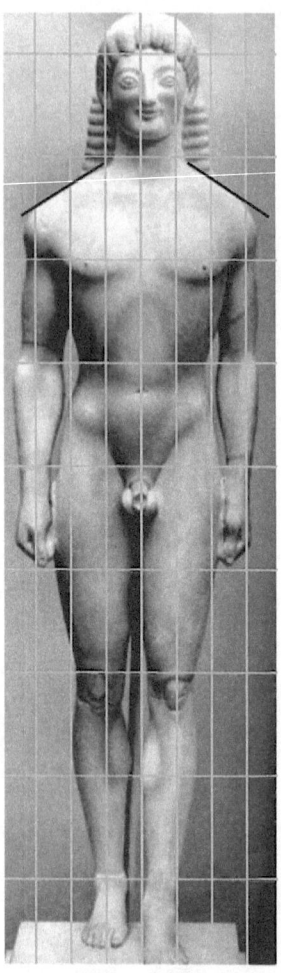

3.12 *Kouros,* 6th cent. BCE. The lines superimposed
over the shoulders show that the right shoulder is
considerably larger than the left. The grid lines also
make apparent that the right nipple is higher and
the right arm longer than those on the left. Most
notably, the central axis, which passes through the
navel and penis, shows that the neck, face and head
of the figure are not centered on its body

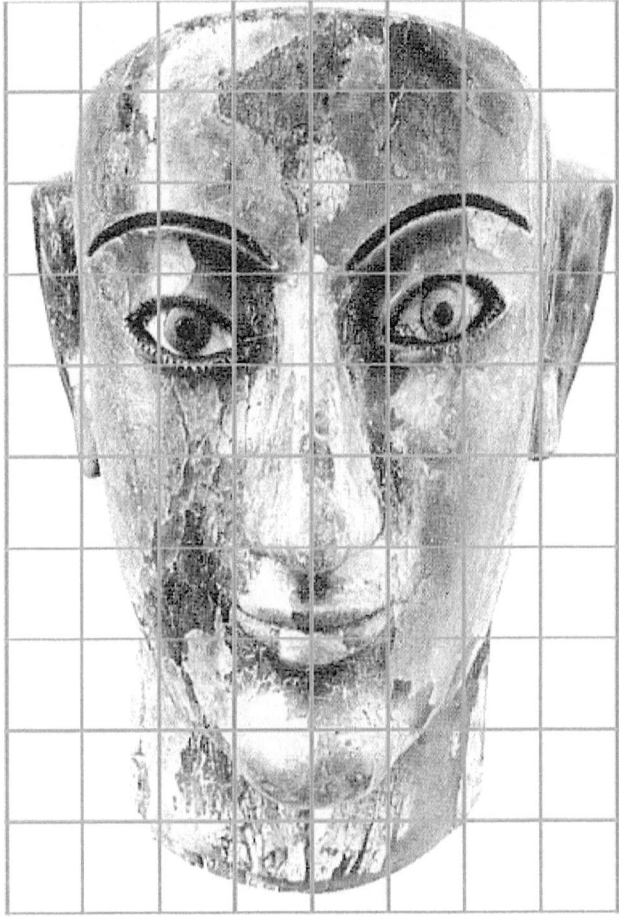

3.13 Delphi, head of Apollo (?) 6th cent. BCE. The left ear, eyebrow and eye are all higher than their opposites. The left side of the head extends further from the central axis than does the right side.

3.14a and b, Poseidon (?) of Artemision, The
superimposed bars show that the left eye socket is
broader and closer to the nose than the right eye. The
lines of the beard and of the hair on top of the head (b)
are all arranged asymmetrically.

3.15   Engraved gems, 6th - 5th century BCE. Not a symmetric detail is to be found on any of these gems.

3.16 Gold libation bowl, Olympia, 7th cent. BCE. The narrow troughs that separate the nine elongated basins of this bowl are different in length (though the curves at their base are equal in size), thus establishing that the shape of the bowl was always asymmetric.

3.17 Capital from synagogue in Gamla, Israel, 4th - 5th century C.E.
Note that the spirals both go in the same direction and that the ovals
in the center differ in location, shape and size.

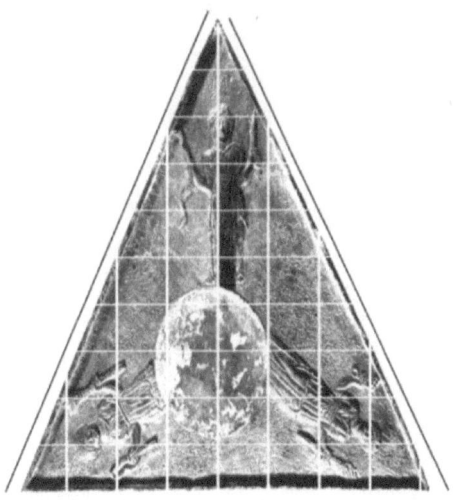

3.18 Votive tablet, Pergamon, c.200 - 250 CE The measuring lines
that flank both upper sides of the figure show that this is is a scalene
triangle. Neither the figure on the top nor the oval on which she
appears to be standing are aligned with the central axis. The figures
in the lower corners differ in size and positioning.

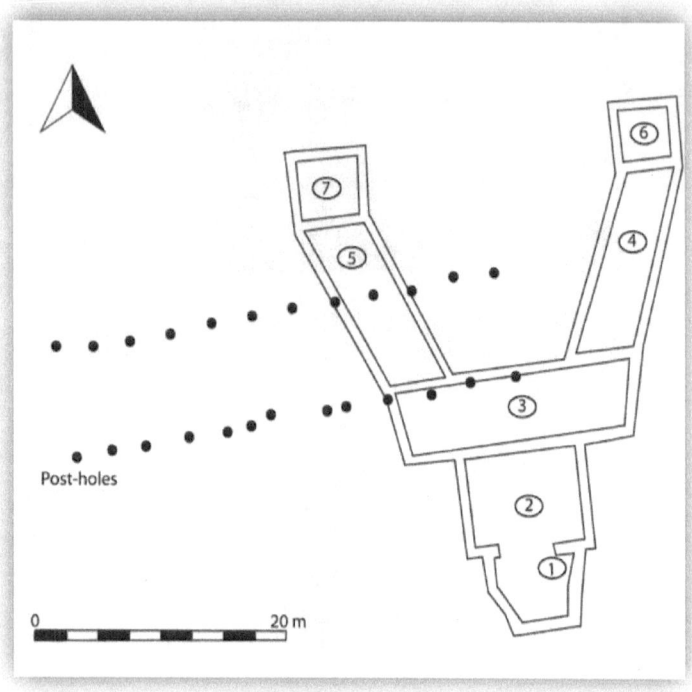

3.19 Plan of unidentified asymmetric Roman structure, England

3.20 Graeco-Roman votive relief from Syria, *circa* 130 CE. Free-standing, possibly a unit of a larger structure. Unlike the fierce heraldic birds of a much later day these eagles are obviously in an asymmetric relation to each other, and the niche in which they are placed is also asymmetric.

3.21 a and b: murals of villas, Pompeii, *c.* 20 - 10 BCE, There are fourteen columns on the right-hand wing of this villa but thirteen on the left-hand wing. The pillars on the right are shorter and support a taller architrave or cornice. The structure on the right appears to end in six, that on the left in four, columns, the former being more slender than the latter. The screen of the left side of the center section is noticeably taller than that on the right.

3.21b The thirteen columns on the right flank appear to be shorter than the twelve on the left, and are irregularly spaced. Of the two flanking structures, the one on the left appears to have taller columns but a lower architrave than that on the right.

3,22   Rome. Detail from a lintel (?) formerly attached to the Pantheon and now at its base. None of the details of one bird are mirror images of the other. The caduceus between them is asymmetric, too.

3.23 The Portland Vase. The vase, including its neck and handles, is asymmetric; so of course is the design of the cameo that decorates it.

3.24 Rome: the Curia Julia, the Roman Senate, was begun by Julius Caesar and completed by the emperor Augustus *circa* 30 BCE. The superimposed lines on this photograph establish that the windows and the doorway are not centered on the façade but are off to the right.

3.25 The Endymion Sarcophagus. Rome, 2nd - 3rd century CE. The lion's head on the right is somewhat smaller and perhaps further from the outside edge than its opposite number. The beasts' manes, too, are quite different from one another. The arcaded ensemble on the lid is set well to the right of center. The carvings show no attempt at symmetric design.

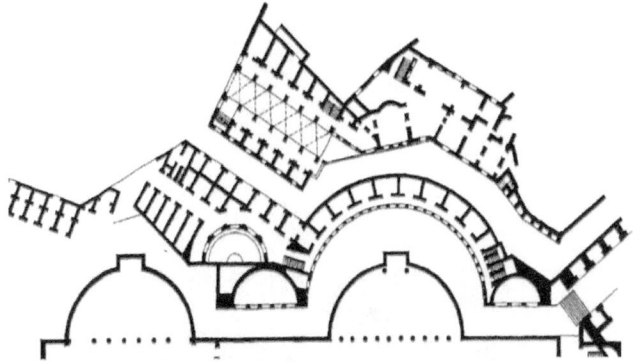

3.26 Rome: Trajan's Forum marketplace. Although many of the rooms (shops) and the smaller sem-circles and their columniations are regular in shape, every other detail   of this structure is asymmetric.

3.27   Rome, *Palatina Domus* wall decoration.. Each of the five frames in the middle is of a different width. In the row below them the two left-most frames differ from each other as well as from the other frames in that row.The vertical lines that delineate each frame in the upper and lower rows do not accurately bisect the frames in the middle row.

3.28 Rome, *Ara Pacis,* ends of sacrificial table.. The body of the beast on
the left is longer and the shape of its wing quite different from the other.
The volute on the right is larger than that on the left, and with somewhat
different leaves.

3.29 Rome: *Domus Aurea,* wall mural.The decorative motifs are neither symmetric in themselves nor centered on the frames that contain them. The frames on the right differ in widths from those on the left. The pediment above the doorway is off-center and the door frame on the right is wider than that on the left.The entire decoration, in fact, is asymmetric.

3.30 Rome, *Domus Aurea.* Note that the apse-like recess is to the right of the wall in which it is set.

3.31 Rome, Arch of Constantine (east side). The superimposed bars show that neither the circular relief nor the rectangle in which it is set are equidistant from the two sides of the structure.

3.32 Rome: Catacomb of S. Domotilla, ceiling fresco. The ten segments converging on the inner circle are all of different sizes and are not aligned on mirroring planes. The segments have unequal numbers of dentelle-like devices along some of their borders.

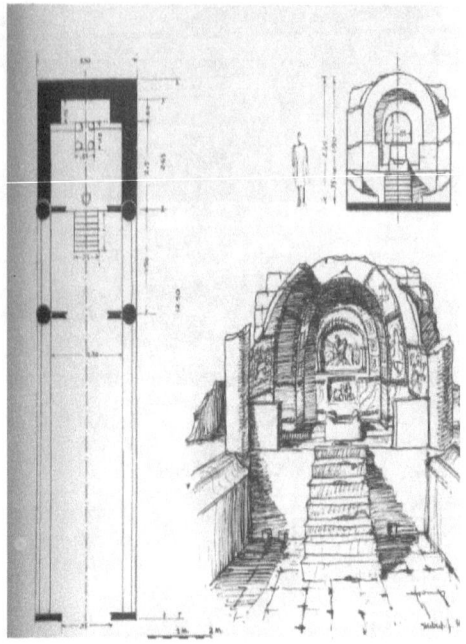

3.34 Dura Europos: *Mithraeum.* Note how the off-centered flight of steps in the photograph is centered in two of the three drawings (the sketch shows the steps to the left of center). The photograph shows six steps; the drawings, seven.

3.35 Crete: Aerial view of Minoan palace complex
Is there a symmetric detail in this entire layout?

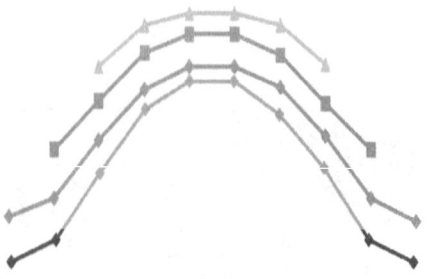

Parthenon east stylobate elevations gamma to zeta (after Balanos)

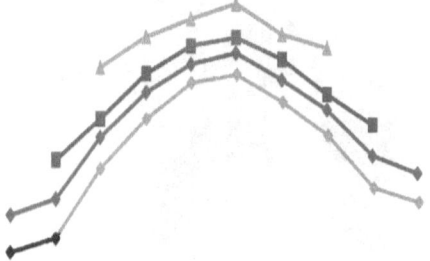

Parthenon west stylobate elevations gamma to zeta (after Balanos)

3.36 Parthenon stylobate elevations. The elevations on the east front are symmetric; those on the west front are asymmetric.

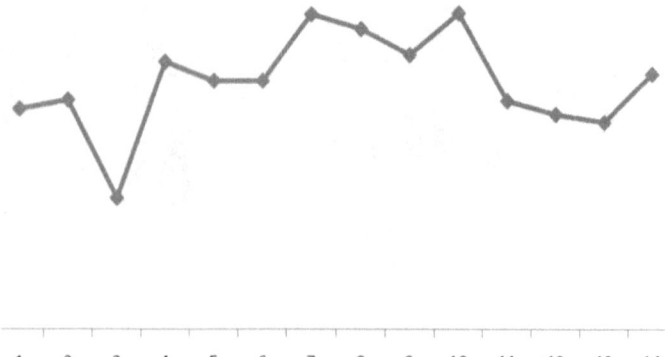

3.37 Parthenon east front metope widths (north to south) after Balanos. The arrangement by width of the metopes is asymmetric.

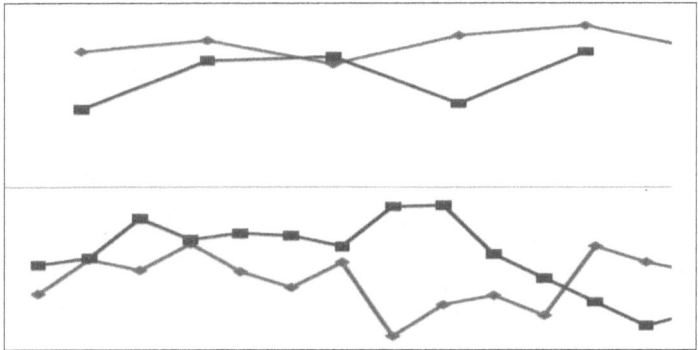

3.38 Column offsets, Temple of Apollo Bassitas: north and south flanks (upper); east and west flanks (lower).[5] The distribution of these offsets is asymmetric.

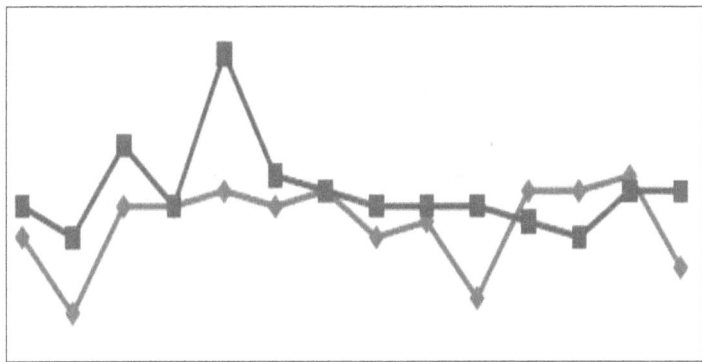

3.39 Column heights, east and west flanks, of the Temple of Apollo Bassitas.[6] The distribution by height of the columns is asymmetric.

[5] Cooper, *ibid,* v.1, 186. The variations among these offsets are not as apparent in Cooper's drawing of them in *ibid* v.4, pl. 16.

[6] Graphed from data in Cooper, *op. cit.,* v.1, 230.

4.2 Chartres rose window. The lines superimposed on this photograph show that the window is not centered in the recessed frame in which it is set (it is too far to the left, facing us). The rose window, moreover, is not centered on the window beneath it nor on the arcade of 16 figures above it. (The lower window and the arcade are however centered on the same axis.)

4.3 Paris, Notre Dame. The portal on the left is wider but shorter than that on the right, and is contained in a sharply-angled recess, which the other is not. The entire structure on the left is perhaps 15% wider than that on the right. This is apparent by a simple visual comparison of the two towers' widths; it is also apparent from the fact that the gallery above the left portal contains statues of eight kings, that on the right only seven.

4.4 Rouen Cathedral. In this visually very complex façade not one of the principal features on the right is mirrored on the left side (which is substantially wider than the right).

4.5 West front of S. Denis by Bénoit (*c.* 1844) before demolition
of north tower

4.6 St Denis west front detail with various asymmetric features

4.7 Lübeck, Marienkirche towers. The horizontal bars indicate that the South tower (right) is wider than the other, and that the windows on the second level of the South tower are narrower; the vertical bar indicates that they are also shorter. The seven quatrofoils on the second level of the South tower are substantially larger than those on the same level of the North tower.

4.8 Regensburg Cathedral, detail of West front. The North (left) and South segments of the façade differ from each other in almost every detail; the North, moreover, is 4.9% wider than the South. The lower level, with portal, of the middle segment is asymmetric (the second level however appears to be symmetric); and the pediment atop the middle segment is not aligned with the axis of the two levels beneath it.

4.9 Norwich Cathedral, West front. This is a 19th century
fabrication that reflects the Gothic Revival at its worst! The
façade's two flanks are asymmetric throughout in relation
to each other, but the asymmetry is contrived and
unpersuasive, and fails to relieve the overall impression of
dreariness. The architect has taken the glories of Gothic
design and transformed them into the dullness of a non-
Conformist chapel.

4.10   Exeter Cathedral, west front. Some asymmetries are readily apparent – the five windows to the right of the main portal are not matched on the left. The wall that extends to the left, evidently a chapel is not matched on the right, from which a buttress extends, instead. Less obviously, the large central window is not centered either in the panel in which it is set nor on the crenellatons above or below it. The upper row of crenellations pitches down at an angle on either side, but that on the left side is shorter and ends somewhat higher than its opposite. The two towers that rise above the upper crenellation also differ from one another, that on the left being wider and taller than the other.

4.11 Gloucester Cathedral, west front. Although the center portion of this structure is symmetric in itself, the buttresses on either side of it, indeed *all* of the buttresses of the façade are asymmetric in relation to those on the opposite side. On the left flank, the doorway (which is not matched by one on the other side) is not centered on the window above it, and the extension of the wall to the left of its outer buttress is also not matched on the other side. The two side windows differ in height and width from each other and the diagonal roof line on the left begins higher up, is at a shallower angle, and longer, than equivalent line on the right.

4.12 Cathedral of Oviedo (Spain), 13th – 15th centuries. Not just the major elements of the cathedral's west front but many of its decorative details, too, are asymmetric.

4.13 Leon cathedral, west front. The two towers are entirely different from each other. That on the left is wider, while the spire of that on the right reaches higher. The five arched portals in the front all differ from one another.

4.14 Fiesole, the Badia.

4.15 Padua, Santa Maria Assunta. The superimposed grid and other lines reveal the extent of the cathedral's asymmetry. The larger window and the main portal are not centered on the axis that descends from the peak of the (asymmetric) roof line. The superimposed lines show that the right portal is narrower and lower than the portal on the left; it is also closer to the central axis.

4.16 Padua, Santa Maria Assunta, baptistry. The porch is sited asymmetrically on the main body of the building and both drums are asymmetrically placed on the structures upon which they rest. The door on the right is lower and narrower than that on the left.

4.17 Otranto Cathedral. On the upper roof, the left is longer and ends at a lower point than the right; and the opposite is the case with the lower pitches. The right-hand wall of the nave is longer than that on the left. The lancet window on the right is taller and wider than its opposite. The columns on the right of the doorway appear to be set further apart from each other than the pair on the opposite side, which also appears to be further from the side of the doorway.

4.18 Ferrara Cathedral. The top and middle superimposed horizontal lines show that the right and left sections of the façade are set at different heights. The lower bars superimposed on the façade show that the right-hand section is much wider   than that on the left; because the number of arches on either side is the same, the arches on the left are rather narrower than those on the left.

4.19 Venice, Aerial view of Piazza San Marco. The piazza and the piazetta are asymmetric polygons as indeed are the basilica and the dogal palace – the two main structures on the right. The basilica is not aligned with the piazza.

4.20 The 12th-century façade of San Giusto in Lucca. Almost every feature of the façade is off-center, toward the right.

4.21 Santa Croce, Florence. Signorini's *Il Carnevale di Firenze* (1846) shows the medieval façade with its unequal protrusions on the upper level and the flanking portals of different heights and possibly widths.

4.22 The facade of Santa Croce today.

4.23 Florence, Santa Maria del Fiore. Poccetti's drawing. The tall vertical recesses are not on equidistant from the great circular window; the narrower vertical recess on the left is not matched by one on the right; the two lower circular windows are not set at the same height (and the one on the right appears to be closer to the outer edge). The portal on the left seems to be more shallow than the one on the right.

4.24   Siena cathedral, nave (first pair of piers only partly visible). Only one facing pair of piers has the same number of bands.

4.25 This mosaic panel on the floor of Siena cathedral is *not* symmetric, as can be seen from the superimposed horizontal bars (most readily, the three lower ones). The geometric design of the frame however appears to have been executed precisely.

4.26 Siena Baptistery carving. Three sides of the frame have eight, and one side has nine, dentelles. The head is set well to the left of the recess's center.

4.27 Rome, San Clemente: apse mosaic. Jesus' left arm is longer than the right. The vines on the left are higher than those on the right; the curved vines enclosing the figures on either side of the cross differ from one another; the clouds at the top are asymmetric in themselves and in relation to each other.

4.28 Verona: San Zeno, door panel. The pillars differ in height and in the angles at which they are set; the curve of the arch they support is asymmetric.

4.29 Verona, San Zeno, door. None of the panels have symmetric designs and many differ in size and the angles at which they are set.

4.30 Arezzo: Santa Maria della Pieve. The reliefs are asymmetric, as is the doorway and the elements associated with it in relation to the larger arch and columns enclosing them. The large capital on the right is lower than its opposite and the capital flanking the right side of the doorway appears to be askew.

4.31 Arezzo, Santa Maria della Pieve. The reliefs (except, possibly, the lowest band) are asymmetric as is the relationship of the entire doorway to the circular window above.

4.32 13th-century Hebrew prayer book from Germany. The chevrons on the pillars differ in height from those opposite them; the beasts at the bottom of the page differ in their dimensions. The tower at the top is not centered on the curve below it, its portal is off-center, and almost every other detail of the architectural ensemble is also asymmetric.

4.33 Leaf from early 15ᵗʰ. century, Paris, Book of Hours. The frame is not embedded in the center of the floral field which is itself highly asymmetric. The vertical sections of the frame are of different heights.

4.34 Rear cover of the Ashburnham Gospels, c. 800. The elaborate design manifests utter indifference to symmetry. Even the three cruciform arms of the halos facing the central panel are asymmetric.

4.35a Chateau de Chambord. Except for the central element, the two sides of the lower floors are asymmetric, as is the elaborate design of the roof line (see below)

4.35b Chateau de Chambord, central section of roof line.

4.36 Florence, Palazzo Vecchio. Note the asymmetric fenestration, and the un-centered portal and tower.

4.37 Giotto's symmetric architecture

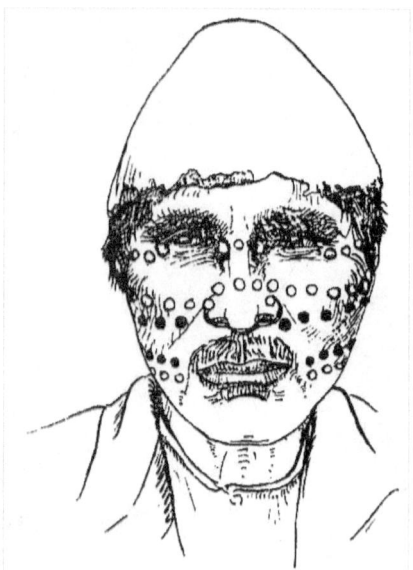

5.2 "Symmetric" Fuegan facial painting. In fact the arrangement of the circles is asymmetric. Note that there are five black circles on the left cheek, butfour on the right, etc., etc.

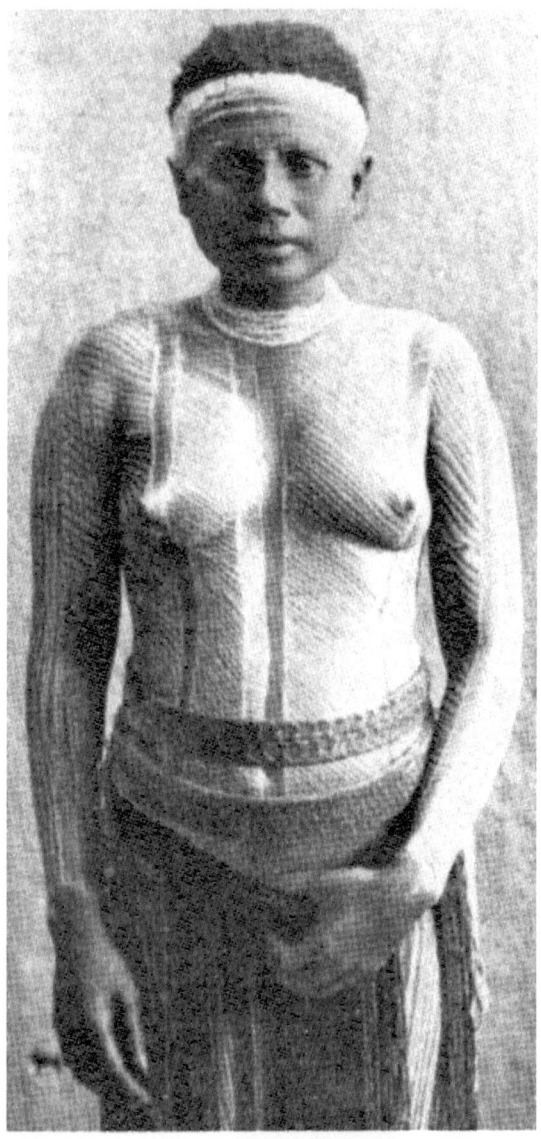

5.3 Andaman Islander with allegedly symmetric body decorations. The more or less central dark line is flanked on one but not on the   other side by a broad, light-colored band. Another dark band runs vertically down the right breast; its equivalent on the other side flanks the left armpit. The diagonal striations the arms run in the same direction.

5.4   Haida dish, described by Boas (1955, p.246) as "perfectly symmetrical".
In fact, *no* part of this design is symmetric.

5.5 Wooden mask, Urua, Congo. The two eyes are markedly different in size and shape. None of the triangles or clusters of hatch marks match their opposites. The overall design is strongly bilateral, but just as strongly asymmetric.

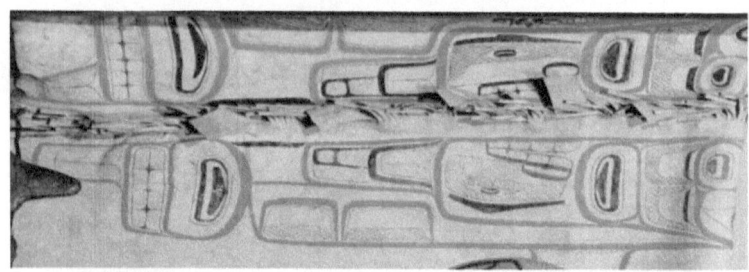

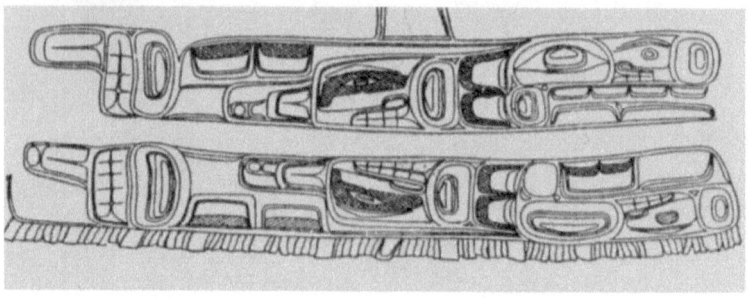

5.6 a (upper) and b (lower). The two halves of a Tlingit blanket (American Museum of Natural History E-1502). Upper image is from Boas 1955, p. 230, *fig.* 234; the lower image, an American Museum of Natural History photograph.

5,7a (upper) and 5.7b (lower). Chilkat pattern board and blanket based on it.

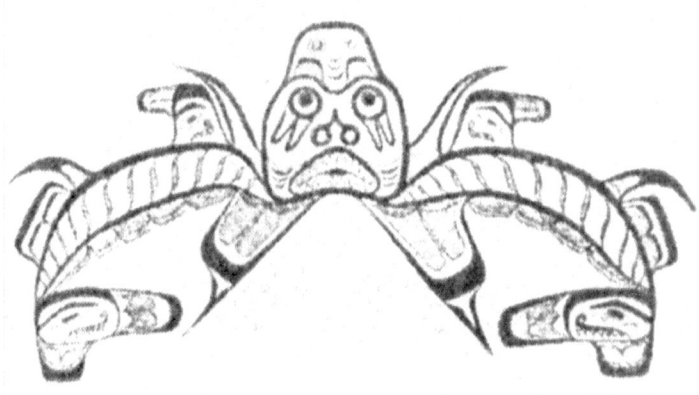

5.8 Haida Dog-Fish painting. All parts of this strongly bilateral design are asymmetric.

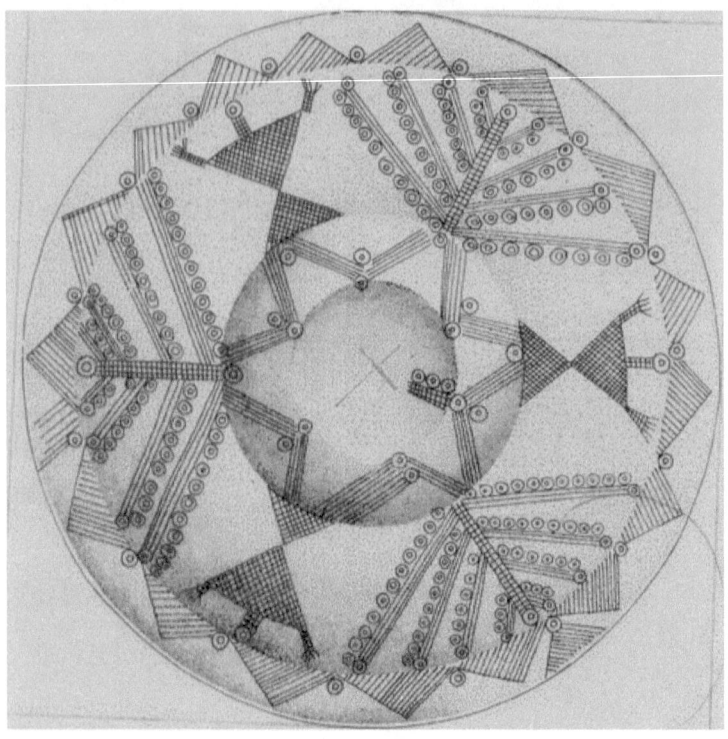

5.9 Terracotta bowl (6th or 7th cent. BCE). Not just the overall design but each of its components are asymmetric.

5.10 Venus of Vestonice. Note, among other details,   that the cleavage between the two breasts is not centered on the navel, nor on the vagina.

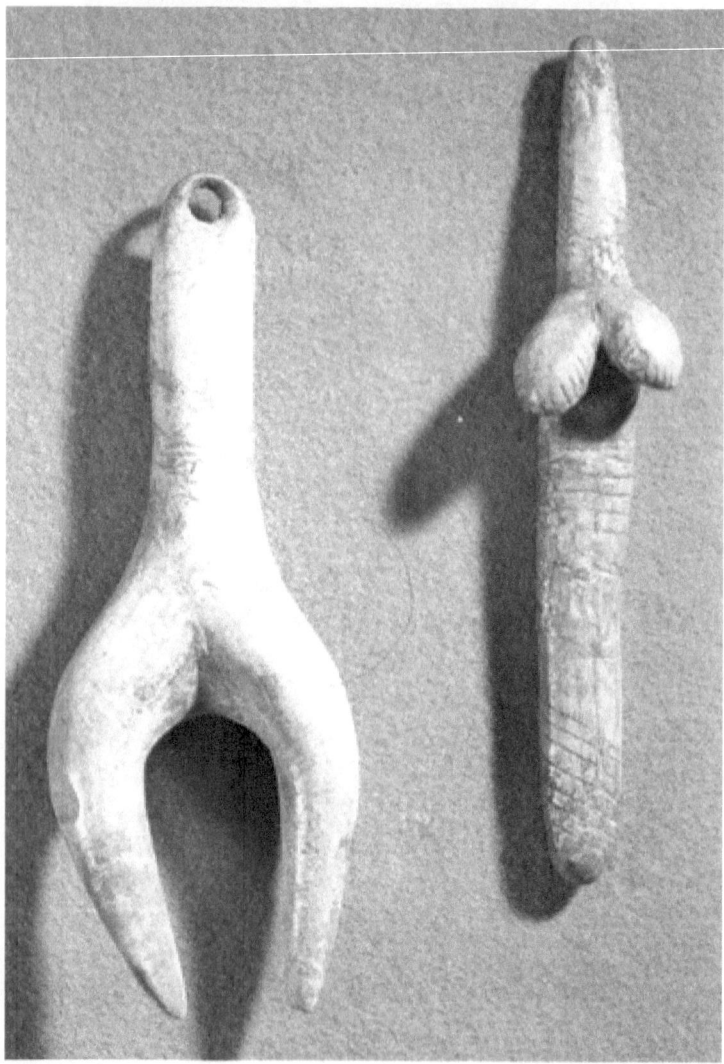

5.11 Female ivory idols from Dolni. The legs of the figure on the left differ in thickness and the torso is on a different axis from the legs. The breasts of the figure on the right are markedly different in size and shape.

5.12   Rock drawings in Skavberg. The outlines of the two bodies, as well as the shapes and sizes of their limbs, are asymmetric.

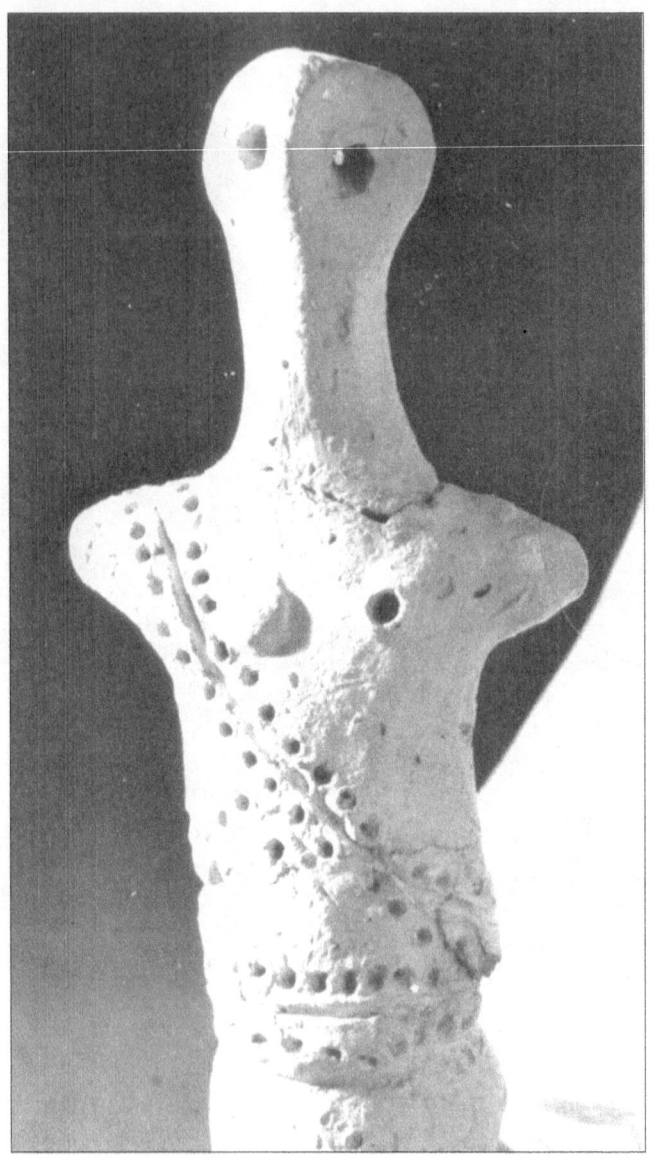

5.13   Female idol from Cucuteni-Baiceni, Among the other asymmetric details of this figure are the shape of the head and the size and location of the eyes.

5.14 Bronze belt hook from Hoelzelsau. A strongly bilateral but asymmetric design: note among others that the heads of the two higher bird-like figures differ in size and shape, and that the left-hand bar at the top is longer and appears to be thicker than its opposite.

5.15 Bronze mirror from Desborough. The sinous designs are asymmetric in themselves and do not mirror each other on either side of the mirror's central axis. Both handle and frame are asymmetric.

5.16 Tribal house in Togoland. The cylindrical towers and their rooves differ in sixe and shape, and the entrance is irregularly shaped and not centered on the cylinder in which it is placed.

5.17 Bronze king and attendants, Benin. One of the flanking figures is noticeable shorter than the other. The objects held by the king differ in size and shape. His face too is asymmetric. The nose and mouth are not centered on it. His right eye is larger and lower than the other, though its pupil is smaller. The left nostril is smaller than the other.

5.18 Dance cap, Cameroons

5.19 Congo dance mask

5.20 Royal graves, Borneo

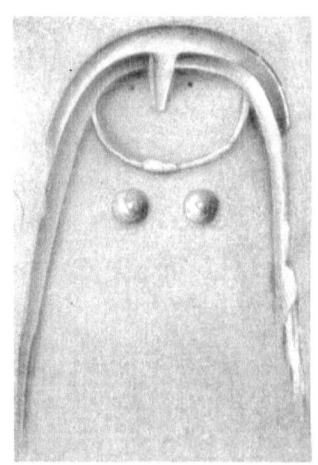 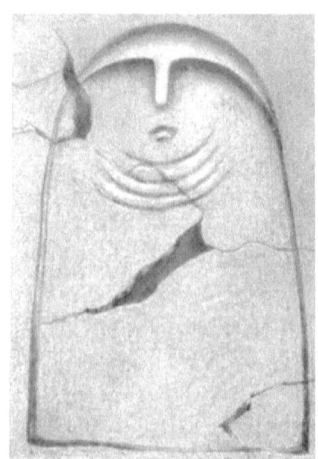

5.21 Tomb reliefs, Croizard

5.22 Silver buckle, Fornass

6.2 Late - medieval garden

6.3 Two villas by Palladio. 6.3a (*above*) Agugliaro: Villa Saraceno; and 6.3b (*below*) Lugo di Vicenza: Villa Godi

6.4 Rome: Palazzo Farnese

6.5 Windsor Castle gardens, *c.* 1725.

6.6 Villa d'Este

6.7 Villa di Castel Pulci [7]

---

[7] Zocchi 1757, pl.8

6.8   Botticelli's "dense and living forest"

6.9a "The Hunt" by Paolo Uccello (1397 - 1475)

6.9b "The Hunt" (*detail*)

6.9c Detail of 6.9b

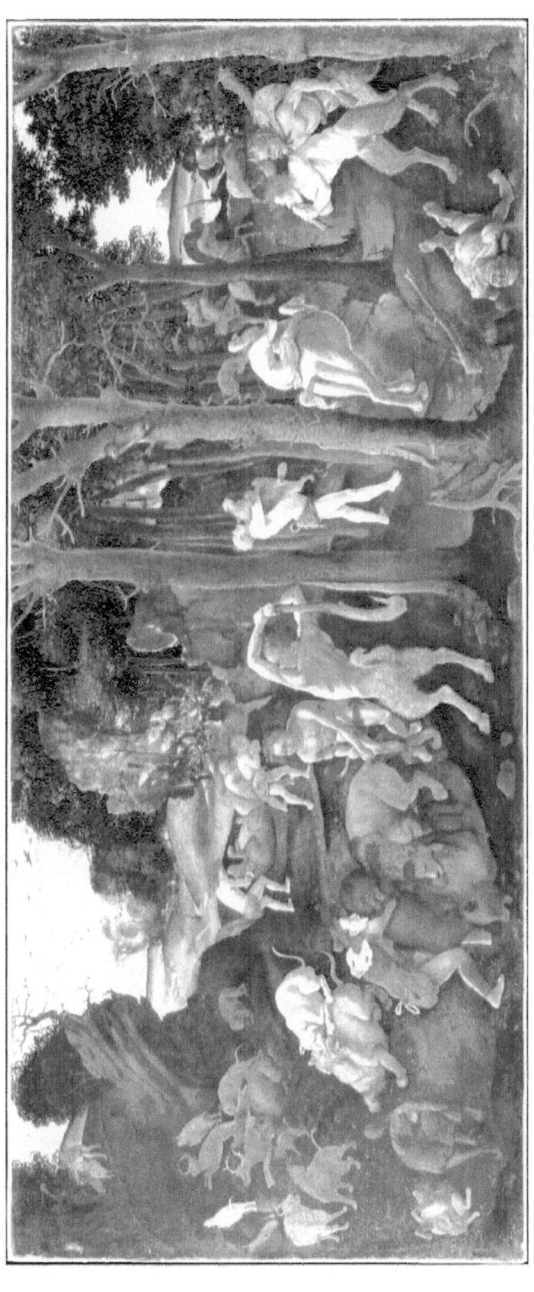

6. 10 "The Hunt" by Piero di Cosmo (1462-1522)

6.11 "Combat" by Antonio Pollaiuolo (1433 - 1498)

6.12 "Death on Horseback" by Giovanni di Paolo (1395 - 1482)

6.13 Giovanni del Biondo (active 1356 - 1399): Florence during the
1374 plague

6.14a Baccio del Bianco (1604 - 57): The Plague in Florence in 1630 (*detail*)

6.14b Pierart dou Tielt (active 1340-1360) Burying Black Death victims

6.15 Pienza, the cathedral and (*right*) the papal palace, the "*simmetrie*" of which Pius II wrote has been disturbed by the bricking - up of the inner area of the false doorway.

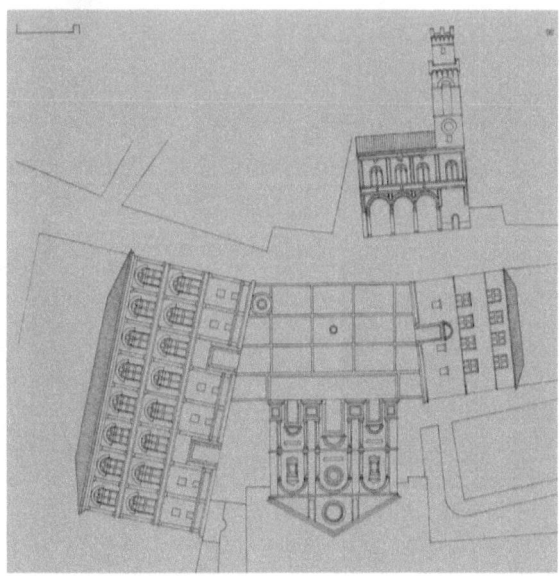

6.16 Pienza, plan showing (clockwise) town hall (top), episcopal palace, cathedral and papal palace.

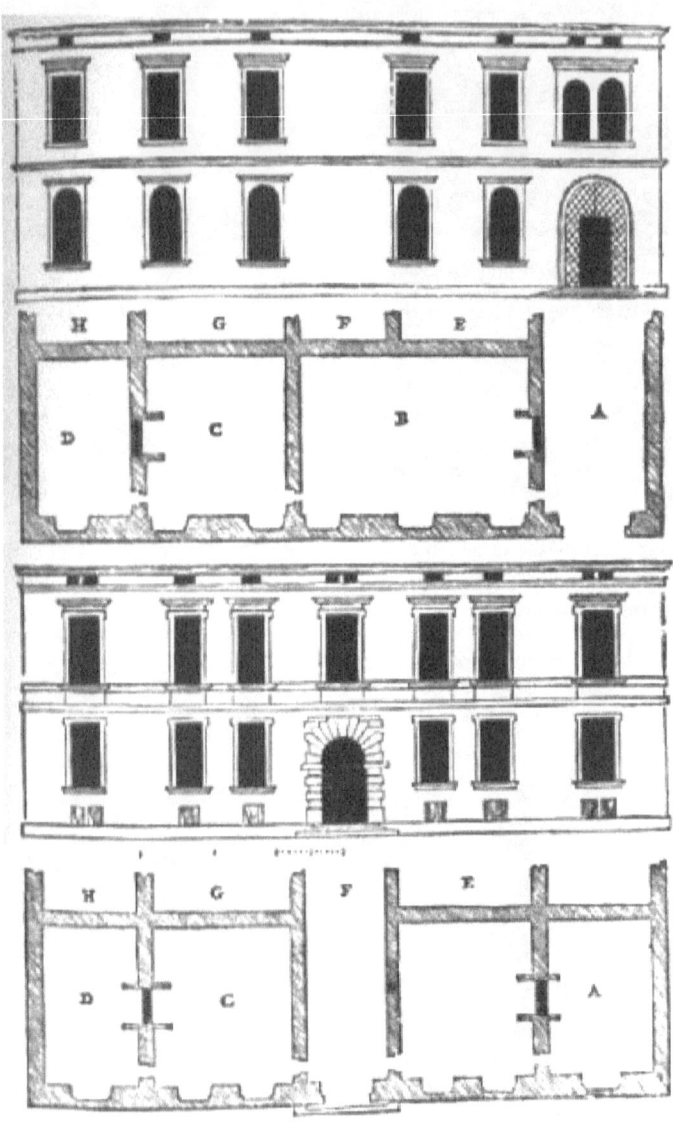

6.17 Serlio: Asymmetric façade "corrected"

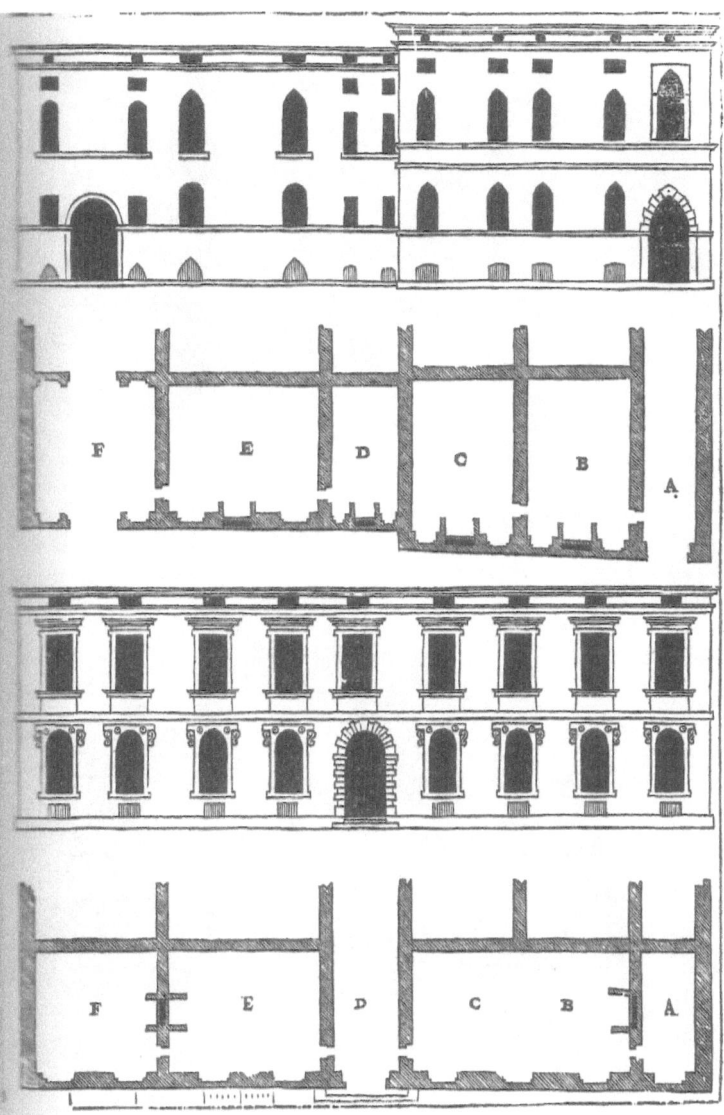

6.18 Serlio: A single symmetric façade replaces
the asymmetric facades of two houses.

6.19 Le Muet, "symmetric" house façade[8]

6.20 Downing's "symmetric irregularity"

7. 2 Stourhead (note the asymmetric bridge)

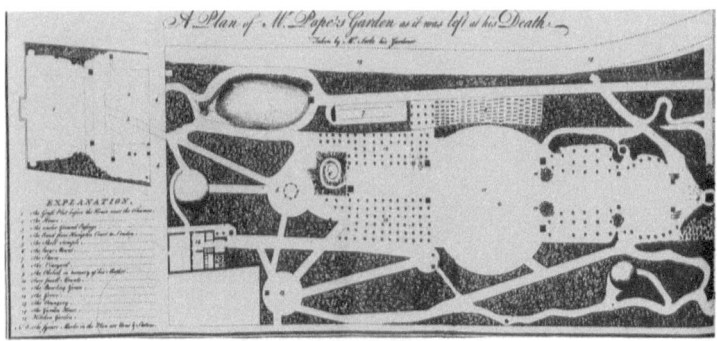

7.3 Plan of Pope's garden in Twickenham

7. 4 London, Chiswick House

7.5 Paris, Rue Castiglione. Bombastic, opaque, joyless: ordered. (The column is surmounted by a statue of Napoleon.)

7. 6 London, MI6 intelligence agency headquarters.

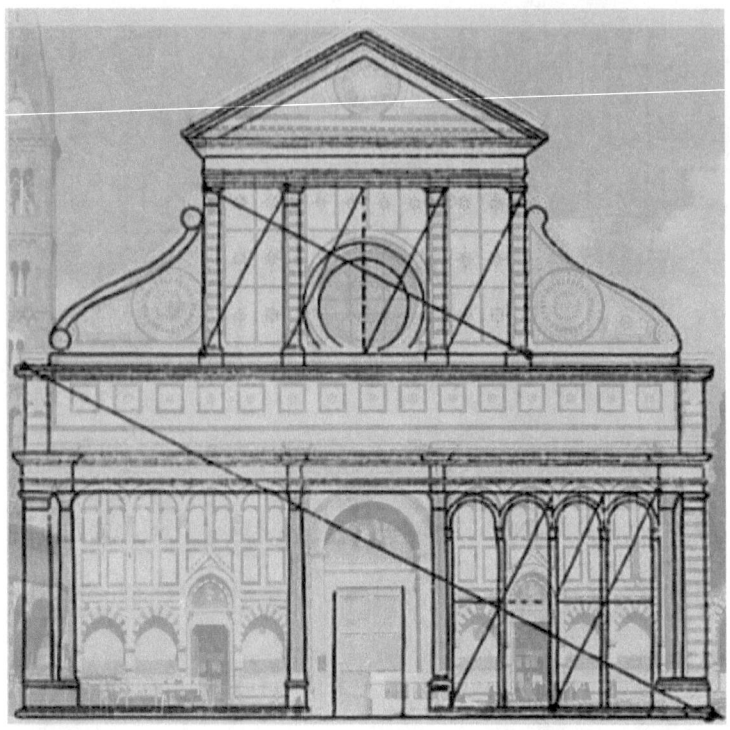

8.2 Wölfflin's drawing of the proportions of the Santa Maria Novella façade superimposed on a photograph of the actual façade.

CHAPTER FIVE
THE ASYMMETRY OF
PRIMITIVE ART

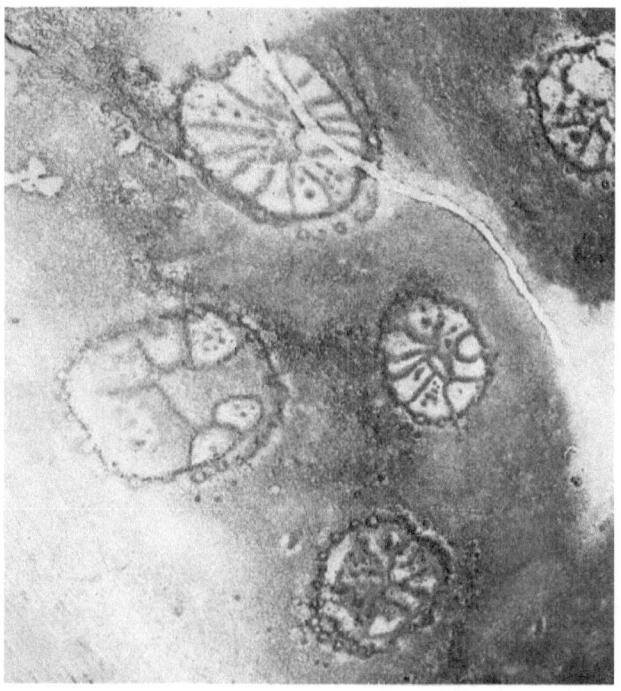

5.1 Aerial view of giant stone structures in Azraq Oasis, Iraq[1]

[1]news.yahoo.com/visible-only-above-mystifying-nazca-lines-discovered-
mideast-0114306-668.html

"Primitive art on the whole is an art of rigid symmetries."
- Gombrich 1966, p. 94.

According to Franz Boas, artisans who have attained "technical perfection" or a "high degree of mechanical skill" are found very widely in primitive cultures. Work done by Indian joiners and carvers in the regions north of the Puget Sound, for example, "rivals that of our very best craftsmen" in the civilized world.[1] In a remarkable passage Boas declared, indeed, that "the appreciation of the aesthetic value of technical perfection is not confined to civilized man. It is manifested in the forms of manufactured objects of all primitive people that are not contaminated by the pernicious effects of our civilization and its machine-made wares... In the households of the natives we do not find slovenly work, except when a rapid makeshift has to be made. Patience and careful execution characterize most of their products. Direct questioning of natives and their criticism of their own work also shows their appreciation of technical perfection... Slovenly work does not occur in an untouched primitive culture."[2]

For Boas bilateral symmetry represented "perfection of form".[3] According to him, craftsmen and craftswomen in primitive cultures generally possess the skill needed to give their artifacts symmetric form. Symmetric design, he went on to say, is

---

[1] Boas 1927, p. 2.

[2] *ibid*, pp. 19 - 20; 352.

[3] In this passage he referred, rather obscurely, to bilateral symmetry as the only "true" symmetry.

one of the "characteristic features" of "the art of all times and all peoples". It is "one of the most ancient and most fundamental characteristics of all art ... a common characteristic of art the world over".[4]   Asymmetric designs, on the other hand, are regarded by Boas as aberrations. They are found only quite rarely, he claimed, either when primitive people have not acquired the skill needed to make things symmetric, or when – as in the case of the unequal halves of the image of a killer-whale's tail on a Tlingit blanket – the artisan happens to have made "a mistake".[5]

Although artisans in primitive cultures are led by their appreciation of the aesthetic value of technical perfection to give symmetric shape to their work, Boas did not believe that their preference for symmetry has a doctrinal basis.   Rather, he supposed that there is something inherent in the way things are made that causes them to acquire symmetric form. "Symmetry" he stated, "results from the process of manufacture" of such objects as coiled pottery and coiled baskets.[6]   Another, more important, source of symmetric design however, indeed one of its "fundamental determinants" is, Boas declared, the symmetry of animal forms. In particular "the symmetry of the human body"- one consequence of which is the symmetric motions of our arms and legs[7] – engenders in all humans "the feeling of symmetry", and this leads us naturally to create artifacts that are symmetric.[8]

---

[4] Boas 1955, pp. 32, 49. Fifty years before the publication of *Primitive Art* Allen (1879) had claimed that "A savage ... makes his arrowheads and his club bilaterally symmetrical with an amount of care that puts to the blush his civilised companions".   Thus the myth of the primitive craftsman's high standard of precision and his preference for symmetry were evidently not invented by Boas (and possibly not by Allen, either, though I have not been able to trace it back further).

[5] *ibid*, p. 230, *fig*. 234. Comp. Reichard (1922) who declared that the asymmetry of a beaded design was "doubtless due to the fact that the maker ... misjudged her distance".

[6] *ibid*, p. 34. Boas also claimed that the use of two-handed implements such as the bowdrill resulted in symmetric forms.

[7] "Symmetrical motions of the arms and hands are physiologically determined ... I am inclined to consider this condition as one of the fundamental determinants, in importance equal to the view of the symmetry of the human body and that of animals; not that the designs are made by the right and left hand, rather that the sensation of the motions of right and left leads to the feeling of symmetry" *ibid*, pp. 33 - 34.

[8] *ibid*, p. 33.

I am not aware that Boas' ideas regarding the alleged preference for symmetric design in primitive cultures have been challenged. We find them echoed (as we shall presently see) not only by anthropologists but also by art historians, such as Gombrich whom I have quoted in the motto of this Chapter. There are, however, certain fundamental objections to these views that we ought to consider. Most obviously, as we saw in Chapter Two, our bodies and Nature's forms altogether are *not* symmetric. They therefore could not be the inspiration that leads people to make symmetric objects.

How, one also wants to know, is the "feeling of symmetry" consistent with our pleasure in making and looking at asymmetric forms – at landscapes, for example, or paintings of them? Moreover, if "the feeling of symmetry" arises naturally it is surely a paradox that we must first acquire highly specialized skills before we can make things that are symmetric. Boas meets this latter objection by arguing that "a feeling for symmetry may exist without the ability of perfect execution". Both the Bushmen and the inhabitants of Tierra del Fuego, he claimed, have "the intent to give [symmetric] form" to their artifacts but lack the ability to do so.[9] Boas did not provide any evidence for these bold statements, and so one wonders how he could have learned of that "intent", and of that "feeling for symmetry" on the part people who lack the ability to make (and probably even to describe) artifacts that are symmetric. Might it not be more plausible to see in their artifacts evidence, if anything, of an innate feeling for *asymmetry* instead? Indeed, why do we assume that the creators of those artifacts gave any thought at all to whether they should make things that are either symmetric or asymmetric? In all likelihood indeed, this question, this choice, did not occur to them! It seems significant that neither Boas nor any other anthropologist has ever reported finding words or expressions that denote "symmetry" or "asymmetry" in the primitive cultures they studied.

One of Boas' claims was that "the tribes of Tierra del Fuego decorate their faces and bodies with designs, many of which are symmetrical".[10] This is a puzzling statement, in part because we have already seen him declare that the Fuegans

---

[9] *ibid*, p. 24.

[10] *ibid*, p. 23.

lacked the technical skills needed to create symmetric designs. If, as it now appears, they *did* in fact possess those skills, we must wonder why they did not employ them in *all* of their artifacts rather than only in "many" of their bodily decorations. Setting this concern aside for the moment, Boas' association of the impulse to create symmetric shapes with the purported symmetry of the human body and its movements makes instances of body ornamentation of especial interest for our purposes here.

For when we look at the drawing that Boas used to illustrate the symmetry of Fuegan facial painting – he referred to it as a "series of symmetrically arranged dots running from ear to ear across the nose" - we can at once see that the dots are in fact arranged *asymmetrically* (fig. 5.2).[11] (The painted board from Tierra del Fuego that he described as "symmetrically decorated" is also *not* symmetrically decorated.[12]) The Andaman Islanders too, Boas wrote, "like to decorate their bodies with symmetrical patterns", but the photograph with which he illustrated this statement (fig. 5.3) also shows decorations that are asymmetrically arranged. The striations on both arms for example do not converge on a central axis but run in the same direction and are therefore asymmetric. Further, the more or less central dark line is flanked on one but not on the other side by a broad, light-colored band, and the band that runs vertically down from the woman's right shoulder to her waist is matched only very imperfectly on the left side of her body. (Indeed, one would do well to note the markedly asymmetric arrangement of her body itself, her left shoulder and nipple, for example, being lower than those on the right). These pronounced irregularities probably reflect conscious aesthetic decisions; and we have no grounds for supposing that they manifest a lack of technical skill.[13]

In *Primitive Art* Boas documented his claims about symmetry with numerous illustrations of purportedly symmetric designs, yet in every instance these illustrations show objects that are asymmetric. Boas' statements, for example, that

---

[11] *ibid,* p. 32.

[12] *ibid,* p. 3 and *fig.* 7, p. 23.

[13] *ibid,* p. 32 and pl. II. Comp. the remark of Alsop (1982, pp. 30-31) about the "*brilliantly skillful* [my emphasis] and idiosyncratic use of asymmetry" that characterizes the body-painting of the southeastern Nuba.

"many of the designs of the Australian aborigines are symmetrical" and that "in paleolithic painting geometrical forms occur that exhibit bilateral symmetry" are contradicted by the objects with which he illustrated these claims, for they are all unambiguously asymmetric.[14] The same mischaracterization occurs in the description of a slate Haida dish carved with the representation of a sea-monster (*fig.* 5.4). Boas rather strangely remarks of this object that although its design "appears asymmetrical", in fact it "is perfectly symmetrical".[15] As even a casual observer can see, however, the design is an asymmetric one. Note for example the differences between the two eyes, or between the two nostrils. Boas also specifically identified as symmetric decorated boxes from British Columbia; Kaffir neck rests; Melanesian shields and paddles; painted rawhides of the Sauk and Fox Indians; and prehistoric Peruvian heraldic devices: yet it is easy to see from his illustrations of these artifacts in *Primitive Art* that all of them are asymmetric. Other objects illustrated in the book, though not directly identified by Boas as symmetric, are also appropriate as a test of his contention that symmetry is a characteristic of primitive art. They include the painting on a Haida box; the decorations of Kwakiutl house fronts; a Haida painting representing a dog-fish;[16] an Arapaho pouch-painting; and the woven pouches from British Columbia.[17] All of these objects are asymmetric. So is the arrangement of beaded thongs on Thompson Indian leggings, that do not follow the symmetric abcba | abcba progression Boas attributed to them, but an asymmetric 6.6.7.6 | 6 | 6.6.6.8 pattern.[18]

Another of Boas' illustrations is reproduced in *fig.* 5.5.[19] It shows a Congolese mask, intricately decorated with geometric patterns, not one of which is symmetric in itself or with its

---

[14] ibid, p. 32. That this error continues to be made by modern anthropologists is attested by Sutton, 1988, in whose important study of Aboriginal art the Lake Eyre toas (p. 62; *fig.* 89); the Lumarluma bark painting from Arnhem Land (p. 66; *fig.* 99); and another untitled bark painting, also from Arnhem Land (p. 69; *fig.* 96), are described as symmetric but are in fact quite obviously asymmetric.

[15] *ibid*, p. 246, *fig.* 258.

[16] This is the same painting that Adam (1936) described as having "two symmetrical profiles".

[17] Boas 1955 *figs.* 246-248; 232; 151a; Pl. VII.

[18] *ibid, fig.* 16, p. 29.

[19] *ibid, fig.* 64, p. 70.

nominal opposite. The bottom corners for example each have a triangle. That on the left is much larger than the one on the right, but the larger one contains 4 horizontal lines within it whereas the smaller one has 6. Note too the completely different set, shape and size of the two eyes. How Boas could have thought that this remarkable artifact upheld his claim that primitive art is symmetric is a mystery.

An even more curious instance of Boas' misperception is the painting on the edge of a Tlingit blanket representing a killer-whale. According to Boas the asymmetry of the two tail-halves on one side of the blanket was the result of a mistake made by the artisan.[20] In fact, the mistake was by Boas, who declared that the design he reproduced in the book was "repeated" on the other side of the blanket "but with symmetrical tails". A photograph of that other side (*figs. 5.6 a and b*), kindly taken for me by a member of the Amerian Museum of Natural History staff, shows that both halves of the blanket are asymmetric - in themselves as well as in relation to each other!

In fact, not one of the 15 plates and 308 text figures that Boas included in his book, Primitive Art, *shows an object that is symmetric. Contrary to Boas' descriptions, these illustrations document the asymmetry of primitive art, not its symmetry.*

In other publications Boas wrote at length about the designs of blankets made by Chilkat women on the northwest coast of the United States. The production of these blankets began with the weaver's husband, who sketched on a wooden board the general concept and principal features of the blanket's design. The weaver transferred to her loom some, though not necessarily all, elements of the pattern drawn by her husband, and then filled in (evidently as she saw fit) the areas for which no designs had been indicated on the board. Remarkably, a number of these pattern boards have survived, and in some instances it is even possible to match a board to the blanket that was derived from it.

Boas misunderstood the nature and function of the pattern boards. In the note that he contributed to Emmon's exhaustive study of Chilkat blankets he declared that the boards show "only one part of the whole middle pattern and one wing … the other side being symmetrical with the one shown on the

---

[20] *ibid, fig.* 234, p. 230.

pattern board".[21] Later anthropologists echoed this claim.[22] However, the design on pattern boards is not usually limited to either the right or left half but typically extends from one side to between one-eighth and one-half of the other side: *and the extension is not the mirror image of the equivalent area on the other side of the central axis.* We do not know why the design extends across the center – no researcher appears to have investigated this question – but that the extension does not reflect the other side refutes the notion that the design of Chilkat blankets was intended to be symmetric. Nor indeed are any symmetrically-woven Chilkat blankets known. *Fig.5.7a* shows the pattern board that appears in Emmons' study as *fig. 576.* The design extends beyond the central axis into the left half of the board. It can readily be seen that the details on the left differ in size, shape and spacing from their equivalents on the right side. Thus, the eyebrow in each pair differs asymmetrically from the other, as does the eye in each pair; and the nostrils are not centered on the mouth and teeth below them. *A fortiori* this also holds true of the blanket – Emmons' *fig. 575* – whose design was apparently derived from this pattern board (*fig. 5.7b*). It should be noted that the asymmetries of the blanket itself are not copies of those on the pattern board. Different as the uppermost pair of eyes and eyebrows on the blanket are from each other, for example, they are different from their equivalents on the pattern board. Indeed, the layers above and below the principal face are rather more closely aligned on the blanket than they are on the pattern board.

---

[21] Boas 1907, pp. 188-189. He claimed that the symmetry is achieved by using a reversible bark stencil, and that this technique is also used to obtain symmetry in painted designs.   No such stencils exist in ethnographic collections, as far as I have been able to determine, and no other reference to them exists in the literature.   The point of course is moot, for the blanket designs are asymmetric and so the technique described by Boas could not have been used on them.

[22] Crawford (1978) writes: "Since the design is bilaterally symmetrical, the pattern board need show only half of the design". Comp. Holm (1965, pp. 84-85): "The principles of splitting and of representing the whole animal naturally leads to bilateral symmetry". But what if the sides of the animal that has been "split" are not symmetric? Holm concedes that "there are to be sure many examples of asymmetric design" but then states that the principle of bilateral symmetry holds for many of the individual design units such as the ovoid eyes that "are essentially symmetrical within themselves".   However, as we established in Chapter One, terms like "essentially symmetrical" are a circumlocution for "asymmetric"!

Moreover, details (such as the peaks in the white spaces above the eyebrows of the lower face) that are original to the blanket, and were not derived from the pattern board, are also asymmetric.

Some of the asymmetries of the blanket accordingly are not attributable to the designer of the pattern board but are the invention of the weaver herself. We conclude that the pattern board and the blanket, while strongly bilateral (or, in the case of the board, strongly suggestive of bilaterality) are – each in its idiosyncratic way – *asymmetric*.

One anthropologist has made a point of emphasizing the avoidance of symmetry in the work of potters in the culture she studied. The decorative designs on Zuni pots, according to Ruth Bunzel, are characterized by their "very marked lack" of bilateral –   or as she preferred to call it, "duplicating" – symmetry:

> To paint a bird or animal with two heads in order to preserve the symmetry of the design would be utterly foreign to Zuni taste. This applies to the geometrical patterns as well as the representative. The "steps" designs which could so easily be made symmetrical without destroying their character are never so drawn. There seems to be a careful avoidance of this particular type of symmetry, in spite of the fact that the parts of the design are carefully balanced. This lack of duplicating symmetry extends also to the arrangement of the motives. There is no feeling that the designs on any field must be arranged with reference to an imaginary center line. The decorative importance of this principle is most apparent in the treatment of deer and sunflower designs. An artist trained in our traditions would certainly treat these motives differently. He would turn the two deer in each horizontal field either towards one another or away from one another, dividing the field into two halves, each of which mirrors the other, and bringing the two deer on each side of the sunflower into the same relationship with it... However the Zuni artist... turns all his deer with their heads to the right... The lack in Zuni design of the particular kind of symmetry which we expect in our own decorative art, but which is by no means common to other styles, does

not mean that Zuni designs are not constructed without a careful balancing of the various units.[23]

Bunzel stands alone, however, in her recognition of asymmetric design in primitive art. Most anthropologists seem instead to follow Boas in his view of the near-universal occurrence of symmetric design, and to account for this by alleged universal characteristics of the human and other natural forms. Anna O. Shepard, for example, attributed what she saw as the prevalence of symmetric design in primitive art to a preposterous claim that "bilateral symmetry is most conspicuous in nature and, above all, it is expressed in the human body".[24] Dorothy K. Washburn, who has written extensively on the subject, agreed with Shepard. "Just as symmetry has been found to underlie the structure of the natural world"[sic!], she wrote, "so too, by extension, do humans use the property of symmetry in their perception of the world".[25] It is these influences, she explained, that established symmetry as one of the "universal properties of form" and account for the fact that "most designs produced by most societies are symmetrical".[26] A similar nonsensical claim was made by the ethno-mathematician Slavik Jablan, who alleged that symmetry "has been present from the earliest time in all that has been done by man", a circumstance that he explained as "the reflection in human artifacts" of "the

---

[23] Bunzel 1929, pp. 28-29. Bunzel adds that in Zuni pottery "the whole jar is symmetrically laid out" but seems to imply by this merely that it is shaped symmetrically on a potter's wheel. Regrettably, Bunzel never discussed *why* the Zuni eschew symmetric decorations. When Dillingham (1992, pp. 85-86) asked an Acoma or Laguna potter "what makes a pot beautiful?" she replied, "the overall shape of the jar ... symmetry in motion". Unfortunately he did not ask her to elaborate on her statement which, in truth, evokes the college seminar room more than the native pueblo. Perhaps worth noting here is the observation of the writer Zora Neal Thurston, (Thurston, 1983, p. 54) that "Asymmetry is a definite feature of Negro art... the sculpture and the carvings are full of this ... lack of symmetry". It is unclear whether she was referring to African art or to the art of African-Americans.

[24] Shepard 1948, pp. 221, 231. Shepard quoted approvingly the opinion of Puffer (1905, p. 10) that an asymmetric arrangement can only be pleasing if it has a "hidden symmetry".

[25] Washburn 1995, p. 525.

[26] Washburn 1999; Washburn and Crowe 1988, p. 33.

symmetry existing in Nature".[27]

It is perhaps worth emphasizing again that, as we established in Chapter Two, Nature's forms – among them the human body - are of course *not* symmetric.

Views like these about the universality of symmetric design are very often reflected in anthropologists' descriptions of the artifacts of specific cultures they study. Examples include Fewkes, in an important study of prehistoric Hopi pottery, who says of a pot in one of his photographs that "the vessel is symmetrical"; and of another,that its "form is regular and symmetrical".[28] Adam similarly tells us that the Haida painting of a dog-fish is "symmetrical".[29] Mainzer, pointing to a sand painting, remarks that he is "astonished" by the symmetry of Navajo artifacts.[30] Mokhopadhyay refers to the symmetry of the decorations on Tlingit baskets, and Patkau declares that "symmetry of design is very important" to the Nlaka'pamux as they decorate their baskets.[31] Crawford states that the design of Chilkat blankets "is bilaterally symmetrical"; Holm, that asymmetric designs in Northwest Coast nevertheless uphold "the principle of bilateral symmetry" because their component parts are "essentially symmetrical within themselves".[32] Reichard states that in the work of artists in the Admiralty Islands "the feeling for symmetry [a Boasian phrase] … is very pronounced" and attributes the few designs she recognizes as asymmetric "to poor technique rather than to the artist's taste".[33] She finds that the Solomon Islanders too show a "liking for symmetry" in all their work.[34]

Yet, none of these descriptions is upheld by the illustrations that accompany them. One can readily see that Fewkes' two pots are asymmetric; that Adam's Haida painting of a dog-fish (*fig.* 5.8) is asymmetric; that Mainzer's sand

---

[27] Jablan 1955, p. 4. Michelis (1955) states that humans have an "innate sense of symmetry".

[28] Fewkes 1895, p. 651; plate CXIX, a and c.

[29] Adam 1936.

[30] Mainzer 1996, p. 16.

[31] Mokhopadhyay,2009; Petkau n. . d. .

[32] Crawford 1978; Holm, 1965, pp. 84-5.

[33] Reichard 1933, p. 148.

[34] *ibid,* p. 118.

painting (an example as he thought of the "astonishing" symmetry of Navajo work) is asymmetric; and the same is true of the baskets described by Mokhopadhyay and by Patkau; of the Chilkat blankets to which Crawford refers; and of *all* the artifacts that Holm illustrates in his *Northwest Coast Indian Art*. Similarly, the photographs published by Reichard clearly show that all the works she described as symmetric are asymmetric.[35]

Asymmetry, it would appear, is indeed one of the most pervasive characteristics of primitive art. Any collection of primitive art confirms this. One such collection is Torbrügge's *Prehistoric European Art*.[36] This work contains 350 illustrations of artifacts; my selection here is made at random from among them except insofar as certain photographs do not allow us (because of the camera angle) to determine whether or not the objects they show are symmetric:

Fig. 5.9

A terracotta bowl of the 6th or 7th century BCE from Dietldorf, Germany has an erratic figure, like a six-pointed star, at its center.[37] The twelve sets of lines that describe the outline of the star vary markedly in length, and range from 8 lines on one side to 5 on another; the paired limbs of two of the tree-like figures branching out from the center are not aligned with each other, and vary erratically in the numbers of roundels along their lengths (in one pair the arms have 9 and 12 such figures); the triangular patterns on the circumference are of different sizes; the abstract human figures are not the same size or placed at equal distances from each other (the one near the top has two thumbs on its right hand).

---

[35] eg pl. cxxvi, nos. 492, 493; pl. xxiv, no. 69.   Remarkably, Reichard *ibid*, p. 149 acknowledges that among the Tami and the Massim designs are characterized by "the avoidance of symmetry... The better a composition is conceived and the more carefully it is carried out the more likely it is to be asymmetrical, and this always with the preservation of perfect balance".   It would have been hoping for too much to expect that this observation would lead anthropologists - including Boas himself, to whom Reichard dedicated her work - to reconsider their assumptions about symmetry and asymmetry in primitive art.

[36] Torbrügge, 1968

[37] *ibid*. fig 10, p. 11.

Not a single feature of this complex design appears to be part of a symmetric configuration.

Fig. 5.10
The voluptuous "Venus of Vestonice" has vast pendulous breasts of unequal size that are not centered on her body; a head and perhaps shoulders that are asymmetric; eyes that are set at varying angles, differ from each other in length, and are not equidistant from the center of the face. The shapes and sizes of her two hips do not match each other; and the oval of her navel is irregular and is not at the center of her stomach.[38] .

Fig. 5.11
Two highly-stylized female ivory idols from Dolni (*fig.* 5.11) are both asymmetric. The legs of the figure on the left are not equally thick, and the torso is on a different axis from the legs. The lower torso of the idol on the right is quite irregularly shaped; her breasts differ notably from each other in size and shape.[39]

Fig. 5.12
The rock drawings in Skavberg, Norway show two human figures. The outlines of their bodies and heads as well as the shapes and lengths of their arms and legs are markedly asymmetric.[40]

Fig. 5.13
The decorations of a female idol from Cucuteni-Baiceni in Romania are asymmetric; so is the shape of her body. Note, for example the discrepancy in shape and size of her two shoulders; the marked difference between the outline of the two sides of her head; and the un-matching positions of her eyes.[41]

Fig. 5.14
The design of a bronze belt-hook from

---

[38] *ibid,* p. 15

[39] *ibid,* p. 26.

[40] *ibid,* p. 56.

[41] *ibid,* p. 65.

Hoelzelsau, Austria, is strongly bilateral in emphasis but no less strongly asymmetric: in the two bird-like heads flanking the human figure, for example, the eye of that on the left is placed near the bottom of the head and of the other near the top. The bird on the right has a longer beak than the other and its neck protrudes much further beyond the curve of the buckle than that of the bird on the right. The left-hand portion of the cross-bar at the top of the hook is longer and appears to be broader than its opposite on the right.[42]

Fig. 5.15
The exquisite sinuous forms on a bronze mirror from Desborough, England are asymmetric in themselves and do not mirror each other on either side of the central axis of the design; both the handle and the frame are asymmetric.[43]

A random selection from among the 570 photographs of artifacts in Sydow's *Kunst der Naturvölker und der Vorzeit* also shows the prevalence of asymmetric design.[44]

Fig. 5.16
The remarkable tribal house in Togoland is made up of a cluster of cylindrical towers, each evidently of different height, circumference and shape; the oval entrance to the structure is irregularly-shaped and not placed on the central axis of the cylinder in which it is set.[45]

Fig. 5. 17
A bronze statue of a king from Benin is flanked by two attendants, one of whom is notably shorter than the other. The object that the king holds in one hand is different from that in the other. His nose and mouth are not centered on the face, though the eyes are. However,

---

[42] *ibid*, p. 191.

[43] *ibid*, p. 222.

[44] Sydow, 1932.

[45] *ibid*, p. 131. Cf. also the complex of buildings from the northern Cameroons, p. 132.

the left eye is smaller than, but has a pupil that is both larger and lower than, his right eye. The left nostril is smaller than the right. On the pedestal the arms holding what appears to be a bush are of different sizes.[46]

Fig. 5.18

The decorative edge of the "dance-cap" from the Cameroons alternates three rows of black and white beads (they are described as pearls) that slope in the same direction along the entire perimeter without regard to a central axis. A triangle with similar bands of beads reaches up from the front of the cap; its left and right sides are of unequal lengths. The bird's head at the apex of this triangle points asymmetrically to the left. The three other pairs of birds on the cap either point toward or away from each other but they are asymmetric in their size, shape and location on the cap.[47].

Fig. 5.19

The prominent geometric designs of a dance mask from the Bena Lulua of the Congo are asymmetric. The left-half of the mouth and the left eye are lower than their opposites; the axis implied by the bottoms of the two "V"- shaped designs on the forehead veers sharply away from the middle of the mask; the three black-and-brown stripes reaching down from the left eye extend much lower than those on the right side of the face. The arrangement of black triangles above each eyebrow is different, as is the arrangement of triangles on either side of the upper lip. Thus, the row of triangles on the upper right lip contains four triangles and ends at the corner of the mouth, whereas the row on the upper left lip contains at least six triangles and extends to the edge of the jaw.[48]

Fig. 5.20

A pavilion-like structure housing the graves of the family of a Borneo rajah has elaborate bow-like

---

[46] *ibid*, p. 139.

[47] *ibid*, p. 155.

[48] *ibid*, plate VI, opp. p. 176

projections on either side of the roof that are different in size and shape from each other; asymmetrically-placed carved birds on the ridge of the roof; and two large panels contain mythical, dragon-like beasts that both face away from the center but are unlike each other.[49]

Fig. **5.21**
Reliefs in the entrance chamber of tombs in Croizard, in the Petit Morin valley of the Marne (France) are exquisitely stylized images of a woman and a man The heads are not centered on the bodies of these figures, nor the noses on their faces; the curved lines that possibly indicate the man's beard extend much further on his left cheek than they do on the other side. The woman's breasts are not centered on the same axis as her face.[50]

Fig. **5.22**
The elaborate decorations of a 7th-century silver buckle from Fonnaas in Norway have a strong bilateral quality but are asymmetric throughout. To take just one area of the design: the face at the bottom of the object is set well to the left of the central axis but the nose on this face is far to the right; the eyes and the curves above them are also asymmetric. The motifs that border the buckle immediately above the face are different in both size and scale.

The examples given here do not preclude the possibility that symmetric designs are to be found in some works of primitive art. Indeed, I have found a few myself. One is a headhunter's wooden shield from northwest Borneo that is particularly noteworthy for having an outline that is symmetric while the interior decorations are asymmetric, including a floral pattern that straddles the clearly-delineated, central entirely different, axis.[51]

Also paradoxical is the fact that symmetric designs recur with some frequency in the woven straps, bands and pouches of the Huichol Indians of northwest Mexico, but not in other objects

---

[49] *ibid,* p. 293.

[50] *ibid,* p. 475.

[51] Sydow, *op. cit.* , p. 308.

made by Huichol artisans.[52] It is curious, too, that Navajo blanket designs seem to be symmetric as often as they are asymmetric, even when they are from the same period. A mid-19th century serape blanket in the Arizona State Museum has three rows of rectangles in its middle that are asymmetrically aligned with each other on both vertical and horizontal axes; the rectangles within any of these rows vary asymmetrically in shape and size.[53] By contrast another serape blanket from the same period in which the basic diamond shapes of the design are linked together to create a web-like overlay, is symmetric. Scholars have not accounted for, and indeed appear not to have recognized, the fact that both symmetric and asymmetric designs are commonly found in these blankets.[54] We should not infer from such designs, however, that the artisans who made these objects possessed the concept of symmetry. On the other hand, in view of the fact that symmetric forms are almost never created inadvertently, it seems most unlikely that the creators of these artifacts did not intend to make them (as we would now say) symmetric. Symmetry then, is evidently not completely unknown in primitive art, though it is very rare and perhaps reflects Western influences. Its occurrence poses a very interesting problem. Unfortunately, this is a problem that remains unrecognized by anthropologists, who are trained to believe that primitive art is naturally and almost invariably symmetric, and can be accounted for by glib references to the shape of the human body and other natural artifacts.

In conclusion it perhaps is worth speculating briefly about why Boas insisted, despite clear evidence to the contrary, that almost all works of primitive art are symmetric. I start with the premise that his opinion was not based on scholarly, objective, considerations.

Boas was born to a cultured, liberal, Jewish family in Minden, Germany. The pervasive anti-Semitism of German academic life, and his distaste for the rising tide of Prussian nationalism, prompted him to leave his native land soon after he

---

[52] Powell 2010 fig 4. 19, p. 62. Although the design of the fabric in 4. 23, p. 63, is symmetric, it is used asymmetrically on the pouch. Most non-woven Huichol artifacts have the marked asymmetry of e. g. the decorated votive bowls shown in *fig.* 3. 9, p. 39.

[53] Kallenberg 1972, pl. 16, p. 38. Comp. also pl. 37, p. 62; pl. 39, p. 64; pl. 47, p. 72; pl. 49, p. 74; pl. 61, p. 87.

[54] *ibid,* pl. 17, p. 39.

completed his doctoral studies. He settled in the United States, and before long was appointed assistant curator of ethnology at the American Museum of Natural History in New York City, as well as head of the new anthropology department at Columbia University.[55] In Germany Boas had worked under Adolf Bastian, the ethnographer – and a specialist in primitive art – who was an early advocate of the belief that people in all human cultures have essentially the same intellectual capacity. This was a radical idea at a time when the theory of evolution readily lent itself to concepts of racial superiority and inferiority. It reinforced the liberal and idealist values with which Boas had grown up at home, and would play a profound role in shaping his work as America's leading anthropologist. Boas' claim that the precision routinely achieved by artisans in primitive tribes was no less than that of "our very best craftsmen" in modern advanced societies is not borne out by any facts adduced by Boas himself – indeed it is on the face of it a preposterous claim! – but it reflects his *desire* to affirm the equality of all branches of the human family. However, his pursuit of this objective caused him, more often than one would have expected, to stray from the confines of intellectual probity.[56] From a purely humanitarian standpoint that desire could perhaps be thought of as commendable; but it led Boas, the scholar, to the cognitive errors – the misperceptions, the spurious reasoning – noted here. In Chapter Two I had suggested that the misperception that snowflakes and bees' cells are symmetric may have reflected a belief – arrived at on *a priori* grounds that are scientifically invalid: yet none the less influential down the generations for that – in the orderliness and rationality of Nature and its Creator. A similarly *a priori* conviction, now in the equal capacity of all cultures, seems to have shaped Boas' work.[57]

---

[55] Boas' influence would be spread by a number of his former students, who themselves went on to establish anthropology programs at American universities. Among his most prominent students were Alfred L. Kroeber, Margaret Mead and Ruth Benedict.

[56] " ... no trace of a lower mental organization is found in any of the extant races of man... the mental processes of man are the same everywhere, regardless of race and culture, and regardless of the apparent absurdity of beliefs and customs" - Boas 1955, p. 1. The unequal distribution of intelligence in different parts of the human family is now widely documented, even if still not always acknowledged.

[57] See the interesting comments on Boas by Gombrich (2002, p. 269*ff*). Gombrich, who as we saw in this Chapter's motto regarded primitive art "on the whole" (a qualification he unfortunately did not explain) as "an

The task of scholars, of course, is *not* to defend *a priori* convictions. When that is what they do, scholarship, truth and honor are betrayed – *la trahison des clercs* – and we are left with mere propaganda and ideology: that is to say, with deceit, which can only be protected by intimidation. The *trahison* of the *clercs* is a decidedly social activity, and from the outset – under Boas – it was a distinguishing mark of the anthropology fraternity. It is no coincidence, therefore, that it was this fraternity that was among the first in the academic world to establish a virtually inflexible insistence on intellectual conformism – a value that is *absolutely* incompatible with the search for truth and understanding, but that is an indispensable tool for the entrenchment of intolerance and authoritarianism.[58] The near-universal but unequivocally mistaken agreement of anthropologists that artifacts of primitive cultures are almost always symmetric is one important manifestation of that sorry phenomenon. Ultimate responsibility for it can be laid at the feet only of Franz Boas himself.

Boas acknowledged that he was still "undecided" about whether his analysis sufficiently accounted for the symmetry (as he saw it) of primitive art around the world.[59] He did not indicate however which were the aspects of his work that he thought were open to doubt, and there is no suggestion in his writings that he ever retreated from his claim that primitive art is symmetric.

Boas' contributions to the study of symmetry and asymmetry, it must be said, are far from unimportant. His focus on the technical skills needed to create symmetric design was a valuable original insight that, unfortunately, neither anthropologists nor art historians have explored. Nor should we dismiss Boas' notion that the forms we make are unconsciously inspired by Nature's own forms, including the forms of the human body. To the extent that that is so, however, it is important that we not lose sight of the fact that natural forms, including our

---

art of rigid symmetries", joined Boas in questioning the notion of "primitive", and the idea that there is a continuum of artistic development from primitive to advanced. Whatever the merits of Gombrich's position, it is indisputable that Boas failed to offer any evidence of the concept of symmetry or of symmetric design in the cultures about which he wrote in *Primitive Art*. Nor do we find any such evidence in Gombrich's discussions of primitive cultures.

[58] It has since spread, of course, through most of academia.

[59] Boas 1955, p. 34.

bodies, are not symmetric but asymmetric. As such, it is possible that they are part of the reason why the things made in primitive cultures (and until not all that long ago, in our own civilization, too) were almost invariably asymmetric.

CHAPTER SIX
THE HISTORY OF
THE CONCEPT OF SYMMETRY[1]

6.1 Serlio: The "miser's" house renovated[2]

---

[1] [We distinguished, earlier, between the idea of symmetry and the concept of symmetry. We saw that the idea of symmetry stands by itself, but that the concept of symmetry subsumes the idea of it and would not be possible without it. It is likely, therefore, that in the West the idea of symmetry must have been formulated *before* – but possibly only shortly before – the appearance of the concept of symmetry. As will be evident in this Chapter, we know almost nothing of the history of the idea of symmetry until it became largely inextricable from the history of the concept of symmetry. (In Byzantium, on the other hand, the appearance of a descriptive idea of symmetry did not lead to concept of symmetry: see Selzer 2021, chapters 5 and 6.)]

[2] Serlio 2001, Bk. VII, Cap, LXII

> *"[Symmetry is] an invention of the Italian architects*
> *in the worst age of an attempted revival of Classical art".*
> – James Fergusson (1849, p. 399)

As we have seen, there are those who claim that the concept of symmetry has been known in all cultures since the earliest times – even that it is an inherent part of human consciousness. As we have also seen, however, such opinions are certainly mistaken. *There can be little doubt that the concept of symmetry originated in Italy in the fifteenth century, at the outset of the Renaissance.*

On the other hand, we do *not* know who first discovered this concept, or what the train of ideas and circumstances was that led to its discovery.[3] Our ignorance in this regard is particularly regrettable because the real nature of the concept, and of the

---

[3] Fergusson and Viollet-le-Duc, as quoted in the mottoes to this chapter and Chapter Three, respectively, and Sitte (2006, p. 55, fn. 4; 284), are among the few writers who have gone some way toward recognizing that the concept of symmetry had specific and identifiable origins. The German poet Goethe (see p. 322 *ff, below*), thought it been known from the earliest times in Byzantium and was transmitted from there to the Latin West. Berenson (1953, p.163), elaborating an idea broached earlier by Burkhardt (1985, p.163), suggested that during the early years of the Renaissance, a "tropism of pattern" led "by automatic reaction toward ... symmetrical design..." The discovery of Antiquity, Berenson added, "helped to accelerate this process by lighting the way and cheering with examples of successful effort". Another suggestion about the concept's origins is that by the historian Richard Goldthwaite (1993, p. 180), who attributed the emergence of symmetric design in architecture to a desire at the beginning of the fourteenth century "to heighten and refine the physical presence of the city". This desire, he claims, brought "a genuine urban aesthetic into focus" that "consisted in [sic] the organizing principles of spaciousness, regularity, orthogonality, symmetry and centrality".

Berenson's "automatic reaction" is an empty concept, as is

persistent fallacies that, as we have seen, are tied to it (such as the belief that Nature's forms are symmetric), probably can only be fully understood by reference to its origins.

Lacking solid evidence, accordingly, we must either ignore the very important question of how the concept of symmetry came into being, or else give ourselves permission to speculate about its origins.

I opt here for the latter.

As we will presently see, the environment in which the concept of symmetry made its first appearance was one of immense and prolonged danger. In seeking to understand the concept's origins, I will be guided by the dictum of Bloch, the great French historian, that "historical facts are in essence psychological facts"[4], which I buttress with the extraordinary insight of psychologist Karen Machover that symmetry offers "protection against a menacing environment".[5]

I shall propose, accordingly, that the concept of symmetry emerged in response to an urgent need to cope with the very menacing environment in which it made its first appearance.

I must acknowledge that no data attest in so many words to a connection between that concept and that environment; and I do not pretend that my hypothesis linking them is anything but speculative. Nevertheless, I hope that readers will find it suggestive, and will think that it may have some merit.

We can begin by noting that the concept of symmetry did not emerge gradually as the outcome of a prolonged evolution. It appeared, rather, quite suddenly and unexpectedly, and as a drastic and discontinuous shift in taste. This shift occurred during the transition from the Middle Ages to the Renaissance, broadly speaking, and it is in the contrast between the aesthetics of the two periods that its character can most readily be discerned.

The aesthetic of the Middle Ages mirrored the almost

Goldthwaite's "heightening and refining the physical presence of a city". Neither explains why, out of almost endless possibilities, that of symmetry was chosen.

[4] Bloch 1953, p. 194. In tribute, I will mention that Bloch was a member of the French Resistance, and that in 1944 he was caught, tortured, and executed by the Germans.

[5] Machover 1949, pp. 87 - 8. Machover pioneered the widely-used figure drawing method of personality assessment.

endless visual variety and complexity of Nature's own, invariably irregular and asymmetric, forms. It is exemplified by the richly and erratically elaborated textures of the facades of mediaeval cathedrals (*figs.* 4.1–4.4), and by the "picturesque confusion of roses, hawthorns and honey-suckle mixed with fruit-trees and shrubs, all growing in wild profusion" found in many medieval gardens (*fig.* 6.2).[6]

By contrast, the aesthetic of the Renaissance was an *anti-naturalistic* one of symmetry and visual simplicity of form.[7] It is exemplified by the vacuous appearance of such structures as Palladio's Saraceno and Godi villas (*figs.* 6.3a and b), and the Palazzo Farnese, Rome, which is partly the work of Michelangelo (*fig.* 6.4). It is also manifested in the formal gardens of the Renaissance (*figs.* 6.5 - 6.7). Laid out on a vast scale and in precise, spare, symmetric and immediately-comprehended geometric forms, these gardens were characterized by "extreme simplicity" and "sameness".[8] No pleasing intricacy or artful wildness ever perplexed the scene in the Renaissance garden; no chaos was ever permitted to shimmer through its veil of order.

---

[6] Crisp 1924, p. 27. For Dante's view of the irregularity of Nature's forms cf. *Paradiso* XIII, 76 - 79 : "*ma la natura la dà sempre scema, similemente operando a l'artista ch'a l'abito de l'arte ha man che trema*" (i.e. Nature resembles the artist who knows his art but has a hand that trembles). Cf. remarks by Panofsky (1955, p. 182) concerning Renaissance theorists who regarded Medieval architecture as a "naturalistic" style that originated in the imitation of living trees (a notion vigorously disputed by Ruskin [1903-1912, X. 237-8]).

[7] The distinction between Medieval and Renaissance is not an absolute one, of course. Cf. Panofsky's famous essay (1955) "The First Page of Giorgio Vasari's' *Libro'*", in which he discusses instances of "Gothic" elements in Renaissance designs; and more broadly, the brilliant essay by E. F. Jacob, (1953, pp. 170 – 184), '"Middle Ages" and "Renaissance"'. The survival of Medieval idioms well into the Renaissance and beyond is particularly evident in England (Clark 1974, chap. 1). In Oxford, the fan-vaulting over the staircase leading to the Christ Church dining hall dates from 1638; the Codrington Library at All Souls College, with its seemingly Medieval exterior, is about a century later. Tom Tower (Oxford, Christ Church, 1682), shows that even the neoclassical paragon Christopher Wren was not always averse to designing in a "Gothic" style. Post-Medieval English architecture in a Medieval idiom is however often (and inauthentically) symmetric.

[8] Walpole 1995, pp. 25 - 29. See Chapter 7, "The Natural Garden in England", *below*.

The change from the medieval aesthetic to that of the Renaissance is cast into vivid relief by the contrast between a passage in *The Divine Comedy* and a drawing made by Botticelli about one hundred and fifty years later to illustrate that very passage. In a brief but evocative phrase Dante had described a forest as "*spessa e viva*" – dense and living.[9] Botticelli's illustration of it however shows merely a sparse and barren arrangement of a few forlorn and largely leafless trees (*fig.* 6.8). It is a forest with nothing of the "*spessa e viva*" about it. Indeed, if it resembles a forest at all, it is one in which only a small number of denuded trees are still standing after it has been ravaged by a devastating fire or blight. Ruskin might well have had this very drawing in mind when he commented on "the expiring naturalism of the Gothic school" with these words: "Autumn came - the leaves were shed, - and the eye was directed to the extremities of the delicate branches. *The Renaissance frosts came, and all perished!*"[10]

The intellectual foundation of the Renaissance's new aesthetic was laid by theorists who deplored the rich complexities of medieval design as "barbaric", "confusing" and "irrational".[11] Instead, they equated beauty with forms that are simple enough to be comprehended in the first moment that one sees them.

"Anything that impedes the sight in any way", declared Filarete (c. 1400 - c. 1469), "is not as beautiful as that which leads

---

[9] *Purgatorio*, cant. XXVIII, 2.

[10] *Stones of Venice*, Ruskin 1903 - 12, v. XI, p. 22. (The italics are Ruskin's). Ruskin's comments on the nature of Gothic architecture and its contrast to the architecture of ancient Greece and of the Renaissance are among the finest passages in the literature of art. See the appendix, "Ruskin on Gothic Architecture and Art", to Chapter Four, *above*.

[11] For example, the opinion of Vasari (1908, p. 38) that "German" (i.e. Gothic) architecture was "*mostruosi e barbari ....confusione o disordine*"; and Bramante's of it as "beyond all natural reason" - *fuori d'ogni ragione naturale* (quoted Germann 1973, p. 29). Cf. Albertini (1510, p. 10), who regarded the old façade of Florence's cathedral as "*senza ordine o misura*"; and Michelangelo's view (Hollando 2006, pp. 46 - 6) that the densely naturalistic landscapes of Flemish painters were "done without reason...without care in selecting or rejecting". (Michelangelo added that Flemish paintings are - in Hollando's Portuguese text - "*sem simetria nem proporção*" ["with neither symmetry nor proportion"]. On the conjunction of the two terms cf. the title of Dietterlin, 1968, and text above fn. 113, p.321, *below*. The conjunction of the two terms occur together as late as the 18th century [D'Alembert in Gerard 1759, p. 231; Reynolds 1997, p. 241], though its exact meaning remains uncertain.)

the eye and does not restrain it"[12]. He explained that this was what made him prefer the circle to other forms, for when one looks at a circular shape "the eye, or, better the sight, quickly encompasses the circumference at first glance". By contrast, Filarete continued, the pointed arch "departs from perfection" because "the eye does not run along it as it does on the circle" but "must pause a little at the pointed part".

Leon Battista Alberti (1406-1472), Filarete's slightly younger contemporary, expressed his preference for simplicity by stating that it is most disagreeable when variety leads to "discord and difference". Indeed, Alberti found the sight of varied objects acceptable only when they were seen from so far away that they came to "conform and agree with each other" – which is to say, when their differences were no longer apparent![13]

Two hundred years later we find the preference for simple forms implied in a statement by Bernini (1598 - 1680). "When at the first instant the eye meets a form that satisfies by its contour and fills the beholder with admiration", he wrote, "then the aim of art has been fulfilled."[14] Bernini's French contemporary Freart seems to have had much the same criterion in mind when he declared that "the more simple the principles of art are, and the

---

[12] Filarete 1965, 59*v*.

[13] Alberti 1966a, I.9. Wölfflin (1952, p. 260), in a section entitled "Simplicity and Lucidity" errs, I think, in assigning to sixteenth (rather than to fifteenth) century Italian architecture "the process of purification and the exclusion of all details which played no part in the total effect, the choice of a few large forms…" He is on much firmer ground when he discusses (*idem*) the parallel development in *sixteenth* century painting: "Pictures were carefully planned so that the great dominant lines should be emphasized. The old method of looking at close range, searching for individual detail and wandering about the picture from part to part, is abandoned, and the composition has to make its effect as a whole, clearly intelligible even from a distance: sixteenth-century pictures have a higher degree of 'readableness', and perception is made extremely easy for the spectator. The essentials emerge at once". Comp. the insistence of Wotton (1968, p. 86, mislabeled as p. 78) "that we bee not distracted by too many things at once" in a painting.

[14] Quoted Wittkower 1974, p. 54. The form of Bernini's Vatican colonnade is indeed evident "at the first instant", though I would like to believe that not everyone accepts that it "fulfills the aim of art".

fewer in number", the more are they "worthy of our admiration".[15]

Underpinning these assertions was the even bolder one that Nature's forms are simple, and that it is because they are our templates of beauty, that no shape can be beautiful unless it is a simple one.[16] Thus Alberti characterized Nature's forms as "modest" (almost certainly to be understood as "simple"[17]) when

---

[15] Freart, etc., 2017, p. 38. The idea that simplicity is a necessary element of beauty, in part because it enables forms to be understood quickly – bizarre as this idea surely is - would remain a staple of much aesthetic theory. Thus, in the 18th century, we find the declarations of Hemsterhuis (1769, p. 5; cf. Scholten 2011) that "the mind judges as most beautiful that of which it can form an idea in the smallest space of time"; and of Lord Kames (1845, pp. 159-60) that "a picture ... like a building ought to be so simple as to be comprehended in one view". Similarly, a late-19th century garden historian (Blomfield 1892, p. 54) argued in favor of the "extreme simplicity" of the formal garden that "There is no difficulty in grasping the principles of a garden laid out in an equal number of rectangular plots. Everything is straightforward and logical; you are not bored with hopeless attempts to master the bearings of the garden". More recent instances are to be found in Mies van der Rohe's influential – if plagiarized (from Browning's *Andrea del Sarto*) - dictum that "less is more"; and the assertion of Corbusier (1986, p. 23) that simple primary forms such as the cylinder "are beautiful forms because they can be clearly appreciated." See also Hall, (2008, p.3; above fn.9) on Clement Greenberg's criterion of "at-onceness". I am unconvinced by Hall's attribution of this notion ultimately to Leonardo. For Leonardo, the immediate intelligibility of painting was not a criterion of excellence but a characteristic that matches our perception of natural sights, and differs from the lengthier process of digesting a verbal description of such sights. "See what difference there is between hearing an extended account of something that pleases the eye and seeing it instantaneously, just as natural things are seen" (Kemp, 1989, pp.22-32). Leonardo's point is overstated, of course, because the longer and more closely we look at something, the more we are often likely to see in it.

[16] Dissenting just a little from this view, Wotton (1968, p. 32), in dismissing *entasis* as contrary to the "Natural Type" which was imitated (as he thought) in pillars, conceded however that Nature "(though otherwise the comliest Mistresse) has now and then her deformities and Irregularities". He argued too (*ibid*, p. 21) that, "in the great Paterne of Nature there can be a good reconcilement" of "Uniformitie and Varietie", citing as an example the (alleged) symmetry of the human body and the variety of its parts.

[17] Portoghesi renders Alberti's *modestiam naturae* as "la semplicità della natura".

he urged builders to imitate them – *modestiam naturae imitari.*[18] According to him Nature preferred the circle (which is the simplest of shapes) to all the other forms "created under her influence" – *rotundis naturam in primis delectari.*[19] In a similar vein Palladio, a century later, spoke against the way of building that "departs from the simplicity which appears in things created by Nature", and urged that churches be built on a circular plan, "as that alone … is simple, uniform, equal".[20]

---

[18] Alberti I. 9.

[19] Alberti VII. 4. Alberti illustrated his point by declaring that the earth, the stars, the animals, their nests, and so on are all circular in shape (*omnia esse rotunda*) – only to add, then, that Nature also delights in the hexagon, as we can see from the cells of bees, etc.

[20] Palladio 1570, Bk. I, 20: "*quella semplicità, che nella cose da lei* [la Natura] *create*". (Ackerman [1991, p. 160] seeks to rescue Palladio from this absurdity by claiming that, for him, "the imitation of Nature was quite the opposite of copying what one sees around; it was a search for abstract principles", but I do not believe that this gloss reflects what Palladio actually wrote. His reference was to the things Nature creates, not to the principles by which she designs them.) An early-medieval expression of the view that Nature's creations are simple in form may be found in the *De Lineis, Angulis et Figuris* of Robert Grosseteste (c. 1175-1253), who writes there that the form of the entire universe and of all its parts can be reduced to lines, angles and regular figures. (*Utilitas considerationis linearum, angulorum et figurarum est maxima, quoniam impossibile est sciri naturalem philosophiam sine illis. Valent autem in toto universo et partibus eius absolute.*) See Sparavigna (n. d.); also Simson (1962), p. 198. Concepts like these would become a staple of aesthetic theory. An Enlightenment echo are Voltaire's references to "the beautiful simplicity of Nature" (Voltaire 1759, pp. 215, 220). "*Tout dans le nature*", Cezanne declared (Bernard 1912, p. 35), "*se modèle selon la sphère, le cône et le cylindre*": everything in Nature is modeled after the sphere, the cone and the cylinder". Corbusier (1986, p. 7) showed himself a child of the Renaissance in this regard when he extolled "the great primary forms" of Nature that are "distinct… without ambiguity": adding that "It is for this reason that these are beautiful forms, the most beautiful forms."

Such ideas are not original to aesthetics. Their antecedents indeed reach back to Aristotle himself, who declared (*Posterior Analytics*) that Nature does nothing that is superfluous and that our theories about her must accordingly be as simple as possible. From this emerged the axiom – most famously in the form of Occam's Razor – that, all else being equal, simple and supposedly therefore "elegant" or "beautiful" scientific ideas are the most likely to be correct. More recently we find Einstein himself declaring (Norton 2000) that "Our experience justifies us in be-

The Renaissance requirement that forms be simple was accompanied by the requirement that they be symmetric, too.

Here again Nature's creations were alleged to be the prototype that all good design must emulate. "It is of the essence of Nature (*tam ex natura est*)", Alberti declared in *De re aedificatoria*, (1452) "that things on the right should correspond in every respect to those on the left (*ut dextra sinistris omni parilitate correspondeant*)": and he then went on to claim that nothing can be beautiful until its lateral halves, too, correspond in every respect.[21]

The connection between the two new standards of simplicity and symmetry, their more or less simultaneous appearance, was not an accident. For symmetry, by limiting a form's

---

lieving that nature is the realization of the simplest conceivable mathematical ideas". Belief in the ultimate simplicity of truth is also found in the Christian concept of Divine Simplicity, parallels to which can be found in some Jewish and Muslim theological writings.

Nevertheless, the criterion of simplicity was not without its detractors. For Pico della Mirandola and others in his circle, the truth can only be couched in obscure terms, a view that Wind (1968, p. 16) regarded as "pernicious". In modern times Frances Crick, referring to the patent complexity of the biological world, declared that Occam's Razor "can be a very dangerous implement" (Crick 1988, p. 138); while the Cambridge cosmologist John Barrow (1990, pp. 342 – 343) has declared that "a universe simple enough to be understood is too simple to produce a mind capable of understanding it". Worth considering too is the point made by Fitzpatrick (2013), in his critique of the philosophical foundations of the concept, that there is no real understanding of what makes one theory simpler than another. More important, he also points out that no one has yet come up with persuasive reasons for supposing that simpler explanations of natural phenomena are likely to be more correct than complex ones. Some persons have responded to these difficulties by questioning the juxtaposition of complexity and simplicity. Thus Delacroix (cited Johnson 1991, p. 23) claimed that his paintings amalgamated "all those qualities which enhance the impression by creating a final simplicity". And John Cage (Nyman, 1980) stated that the repetition of a single sound, or the continued performance of a single sound over a period of twenty minutes in the music of La Monte Young, led him to discover "that what I have all along been thinking was the same thing is not the same thing at all, but full of variety".

[21] Alberti 1966, IX, 7. An early statement implicitly combining the criteria of symmetry and visual simplicity is Alberti's dictum (*ibid*, IX, 4) that trees "ought to be planted in rows exactly even and answering to one another exactly upon straight lines". Cf. Hogarth's (2015, pp. 33-38), equation of "symmetry" with "uniformity" and "regularity".

complexity to that of one of its halves, is itself a simplifying device. It was this attribute, indeed, that provided Montesquieu with one of his principal arguments in favor of symmetric design which, he said, "pleases the mind by the ease with which it allows it to embrace the whole object immediately".[22]

We should be clear that these ideas were not mere intellectual playthings. They were, rather, basic elements of a comprehensive, radical and highly successful program for reshaping the physical appearance of civilized Europe.

The triumph of this program is nowhere more apparent than in the design of the Renaissance garden. There, the gardener's traditional task of sympathetically facilitating Nature's beautiful complexities was subordinated to – in fact, largely abandoned in favor of - the revolutionary agenda of symmetry and simplicity.

Nature's forms of course are not symmetric and they are not simple. They encompass, rather, an almost unlimited range of shapes, colors and textures: and they do so, as we saw in Chapter Two, not incidentally or accidentally, but out of their very essence.

It was these facts – irrefutable and self-evident though they are - that Renaissance horticulturists set out to negate. They did so in their gardens by forcing Nature into forms that are intrinsically alien to her, concealing her variety, complexity and asymmetry beneath a *burqa* of inauthentic, man-made symmetric simplification. It was not that beauty was excluded from the Renaissance garden but that its beauty was regarded as the creation of Man, not of Nature, and was subject to his dictates. Thus Charles Cotton, referring to Chatsworth's gardens, said that their loveliness was achieved "[de]spite Nature", and not with her help.[23] For all the rationalizations about Nature's simplicity and symmetry that we noted on earlier pages, the severing of the age-

---

[22] ..."*qui plait a l'âme par la facilitité qu'elle lui donne d'embrasser d'abord tout l'objet*" (Montesquieu 1825, p. 616). Montesquieu continued: "The reason that symmetry pleases the mind is that it saves it trouble, that it gives it ease, that it cuts its work, so to speak, in half" ("*la raison que la symetrié plaît à l'âme, c'est qu'elle lui épargne de la peine, qu'elle la soulage, et qu'elle coupe pour ainsi dire l'ouvrage par la moitié*".)

[23] Cotton 1683, p. 78. Contrast Cotton's contemporary Marvell (in "Appleton House"): "Nature here hath been so free/As if she said, 'Leave this to me'./Art would more neatly have defaced/What she had laid so sweetly waste".

old connection between Nature and Beauty was a recurring theme of Renaissance aesthetics.

The attitude of Renaissance gardeners toward Nature was indeed one of hostility; their goal, that of domination. Far from being Nature's helpmeets or partners, they made themselves her taskmasters. "They presume to do ... what they list with nature, and moderate her course in things as if they were her superiors", William Harrison wrote in 1577.[24] One such gardener was the clergyman John Laurence (1668 - 1732), a martinet whose garden was a parade ground:

> .... And Hedges all streight and regular.
> And Trees in rank and file, with Order stand,
> Improv'd by Discipline, like a Martial Band. [25]

Another who presumed to do what he chose with Nature was Cromwell's general, Fairfax, who figured in a series of poems by Andrew Marvell (1621-1678).[26] In Marvell's portrayal, gardening for Fairfax was not a respite from the "warlike studies" he had pursued during a long and distinguished military career, but an extension of them. His garden was square and walled, laid out "in the just figure of a fort"; and within it he "stupif'd" and "enforced" Nature's creations:

> See how the Flow'rs as at Parade
> Under their Colours stand displaid

---

[24] Bk. II, chap. 20. Harrison's *Description of England* appeared as part of the second edition (1587) of Holinshed's *Chronicle*.

[25] Laurence, *Paradice Regain'd, or the art of gardening* (1728), p. 16. Such attitudes persisted, at least in France, well into the 18th century, when we find Montesquieu declaring in his essay on taste, "*Nous admirons le soin que l'on a de combattre sans cesse la nature, qui, par des productions qu'on ne lui demande pas, cherche à tout confondre*". An English translation (Montesquieu 1759, p. 290) extends this thought so enthusiastically that I may be forgiven for including, it, too, here: In it Montesquieu expresses his admiration for the zeal with which gardeners "combat and correct perpetually the irregular fecundity of Nature"!

[26] It is worth remarking that according to Vita Sackville-West (1929, p. 42), who was herself an accomplished poet of the garden, Marvell's appreciation of uncultivated Nature, reflected in his bitter aspersions on Fairfax, "was not at all proper to the seventeenth century".

> Each Regiment in order grows,
> That of the Tulip, Pinke and Rose.[27]

And then there was Jacob Bobart (1641 - 1719), Keeper of Oxford University's relentlessly ordered and symmetric Physick Garden. Bobart's "despotick" command of his plants was celebrated in the poem *Vertumnus* by his friend Abel Evans (1675 - 1737):

> Full in the centre, there he stands
> Encircl'd with his verdant bands;
> Who all around obsequious wait,
> To know his pleasure and their fate:
> His royal orders to receive,
> To grow, decay, to die or live:
> That, not the proudest Kings can boast,
> A greater or more duteous host.
> Thou, all that pow'r dost truly know,
> Which they but dream of here below;
> Thy absolute despotick reign,
> Inviolably dost maintain:
> Now, with ill - govern'd wrath, affright
> Thy people, or insult their right ...
> ... Thou, on thy botanic throne,
> Sits't fearless, uncontrol'd, alone:
> Thy realm in tumult ne'er involve'd,
> Or, rising, are soon dissolv'd:
> Free from the mischief and the strife
> Of a false friend or fury wife:
> And if a rebel slave, or son,
> Audacious by indulgence grown,
> Presumes above his mates to rise,
> And their dull loyalty despise:
> Thou, awful Sultan! with a look
> Cans't all his arrogance rebuke;
> And, darting one imperial frown,
> Hurl the bold traitor headlong down:
> His brethren, trembling at his fate,
> Thy dread command with reverence wait...[28]

---

[27] "Upon Appleton House", xliii (Marvell 1927, v. 1, p. 69).

[28] Evans, 1713.

To Horace Walpole in the eighteenth century it was clear that the purpose of the formal garden was not to celebrate Nature but actually to "oppose" her. Saint-Simon, the political philosopher, regarded the gardens of Versailles as the product of Louis XIV's *"Plaisir superbe de forcer la nature"* – his haughty pleasure in forcing Nature". Modern scholars too have commented on Renaissance horticulture's palpable hostility to Nature. Thacker declared that the gardens at Vaux-le-Vicomte "appear as an immense gesture of power, where the elements of Nature have been tamed, disciplined, brought together to serve as parts of a human scheme";[29] while Comito, in a similar vein, wrote that Nature was "overcome" in the gardens of the Belvedere.[30] Hyams described the Villa d'Este's gardens (*fig.* 6.6) as "regular, symmetrical, ordered, *anti-natural*" (my italics).[31] And Strong wrote of the same gardens that they were "a *coup d'oeil* aimed at establishing immediately in the mind man's total control over the forces of nature."[32] These remarks can be applied generally. The gardens of the Renaissance were not intended to be Nature's apotheosis but her nemesis.

The Renaissance garden thus asserted Man's dominion over Nature. The garden had become *his* domain and in it Nature was *his* creature, existing meekly within the tight restraints that he placed upon her. Curbed and enfeebled, she appeared here as a gaunt caricature of herself; and her organic fabric was often impossible to discern in the spare, tidy and closely-cropped symmetries into which she had been forced.[33]

It would be mistaken to regard the radical innovations brought about by Renaissance horticulturists as a mere random shift in taste. They should be seen, rather, as arising from a startling new view of Nature as a force of immense, and lethal, malevolence that must be battled, and overpowered.

---

[29] Thacker 1979, pp. 143-144.

[30] Comito 1978, p. 153.

[31] Hyams 1971, pp. 132, 134.

[32] Strong 1998, p. 20.

[33] John Evelyn thought the gardens of his day looked as if they were made of pasteboard and pine planks and that they smelled "more of paynt then of flowers and verdure" (in a letter to Sir Thomas Browne, quoted Strong 1998, p. 221). At the very outset of the Renaissance the suggestion was already being made (Alberti 1966, IX: iv) that gardens should be laid out in the "geometric shapes that are favored in the plans of buildings".

This was not how Nature had been experienced previously. To be sure, the medieval mind, as Kenneth Clark has written, found Nature to be "disturbing, vast and fearful".[34] Yet the dangers inherent in Nature had been mitigated in the Middle Ages by an awareness of her benevolent, or at least harmless, aspects. Thus Dante, immediately after the opening lines of the *Divine Comedy* in which he recalled the terrors "almost as bitter as death" that he encountered in the tenebrous forest, went on to refer to "the good I discovered there" – *del ben ch'I vi trovai*. Later, as we have already noted, he wrote appreciatively of the "dense and living forest" that he chanced upon.

For his part, Petrarch did not experience anything to disconcert him as he passed through the forests, "inhospitable and wild" (*boschi inospiti e selvaggi*) though they were, of the Ardennes (#176). It was outlaws and not Nature that he felt posed the greater threat, but not even they could mar his sense of well-being, and he writes that, for him, it was "sweet to be alone and unarmed there where Mars takes up arms without warning" - *Dolce m' è sol senz' arme esser stato ivi, dove armato fier Marte, et non acenna*, (#177). And it should be recalled that Boccaccio's protagonists, when they wished to escape the plague, fled from Florence to the country, where among much else they reveled in the beautiful garden to which we have already alluded. For Boccaccio, then, Nature was not the source of the plague that it would presently become, but a delight-filled refuge from it.

Clark himself, having written of the fearsome aspect of Nature, then went on to say that "in this wild country man may enclose a garden".[35] And it is clear that the medieval garden was not inspired by anything like the Renaissance's determination to oppose and subdue Nature. If the shelter it provided was from the harshness of Nature, it was also an evocation of the Garden of Eden's delights, a place in which Nature could show her benevolent aspect, her loveliness. Nature's harshness, moreover, was a perennial and necessary consequence of the Fall, very different from the dangers of unprecedented lethality that arose suddenly at the end of the Middle Ages and called into question the very survival of the human race.

The Renaissance garden on the other hand had none of the sweet and gentle charms of its medieval predecessor. It was, rather, a triumphant advertisement of man's (symbolic) success

---

[34] Clark 1949, p. 8.

[35] Clark *idem*.

in subjugating and eliminating the terrible dangers that Nature posed. Where the medieval garden was a loving, artistic, affirmation of Nature's beauty, the Renaissance garden, as we have seen Cotton write of Chatsworth, was created "[de]spite Nature".

The new perception of Nature as a malevolent, lethal force can be discovered in a number of early Renaissance paintings. Uccello's "The Hunt" (c. 1470) shows twenty or more well - armed men, supported by twice as many hunting dogs, at the edge of a forest (*figs.* 6.9a – c). Despite their numbers, the men seem fearfully hesitant about entering the dense, dark wood. The horseman in the enlarged detail actually seems terrified by what he thinks may lie ahead of him. It is as if Death awaits the hunters in the forest, as well as their prey.

Piero di Cosmo's *The Hunt* (c. 1480) suggests an extraordinarily brutal view of Nature with a powerful fire raging in the dark, dense forest; men and satyrs engaged in violent struggles with a variety of fearsome beasts; and the pallid corpse of a naked man, his left arm apparently severed, lying unnoticed on the ground (*fig.* 6.10).

The relation of Death and Nature is also apparent in Antonio Pollaiuolo's great print, "Combat" (c. 1470) as it is sometimes called (*fig.* 6.11), that depicts two groups of naked men engaged in what is evidently a struggle to the death. The impenetrably dense background of denuded trees and ripe crops against which the struggle takes place is mysterious, but one senses that it is somehow associated with the fatal outcome that is in the offing.[36]

In Giovanni di Paolo's "Death on Horseback" (c. 1440?) the association of Nature with death is explicit (*fig.* 6.12). The horrifically sinister figure of Death rides out of the massive, dark and impenetrable forest that is his home to snare his next victim, who stands frozen with fear.[37]

---

[36] Hall (2005, p. 74) interprets that link as follows: "This affinity between leaves and muscles suggests that it is the autumn of all their lives. The grim reaper will cut down both crops and men indiscriminately". But the plants are not engaged in combat with each other, as the men are, and I tend to think that it is they – the plants – that somehow *are* the grim reaper (or perhaps he is hiding in them)!

[37] In the "Triumph of Death" (1446) in Palermo the figure of Death also appears to have emerged from the impenetrable gloom of the surrounding forest.

In these works Death seems associated with, is almost identical to, Nature herself. Nature is not a source of loveliness, or a comfort. Instead, she carries within herself – or perhaps more correctly, she *is* - a mortal menace that may at any moment leap out of concealment and snatch another hapless human to his or her doom.[38] This was a very different view of Nature from the one that had prevailed earlier. It is not how Dante saw Nature, nor yet Petrarch or Boccaccio.

What then caused the change? What brought about this drastic, disturbing new view of Nature?

I speculate that it was – the plague.

The plague of 1348 - 1350, known as the Black Death, is by any measure one of the greatest calamities in recorded history. At least one-third (and by some estimates more than one-half) of western Europe's population succumbed to it, and it severely dis-rupted the social and economic fabric of European life.[39] One can scarcely exaggerate the sense of hopelessness – "the general sense of malignancy"[40] - that it instilled in those who had not, or had not yet, been infected. "There was such a fear", a Florentine chronicler wrote, "that no one knew what to do"; while another chronicler feared that it would bring about *"la sterminio della generazzione umana"*- the extinction of the human race.[41] "Everywhere sorrow, everywhere terror", Petr -

---

[38] Clark (1949, pp. 46, 177) writes of the "consciousness of the infinite, unknown destructive powers of nature" possessed by Leonardo da Vinci (1452 – 1519). Leonardo's studies of rocks, water and landscapes, Clark remarks, show "the forces of nature rising in revolt against man with his absurd pretence to ignore them, or use them for his advantage", an inter-pretation (with its sense of Nature violated by human exploitation) that evokes the 20th century hippie or political progressive too strongly to be plausible. It seems likelier that those studies reflect Leonardo's percep-tion of the lethal dangers that Nature posed to defenseless Man in the era of the plagues. Leonardo's awareness of those dangers is evident from the report of Kruft (1994, p. 59) that Leonardo, "under the impact of the great plague of 1484/85" in which about one-third of the population of the Duchy of Milan died, conceived a radically new design for a plague-resistant town based on the principles of decentralization and hygiene.

[39] For example, "Because of the chaos of the present age the judges have deserted the courts [and] the laws of God and of man are in abeyance" – *Decameron* (conclusion of the sixth day.)

[40] Southern 1995, p. 52.

[41] Villani Bk. I, end of chap. 1, quoted Meiss 1951, p. 66

arch wrote to his brother, who was the sole survivor of 35 monks in his monas-tery.[42] His letter continues: "When has any such thing ever been heard or seen; in what annals has it ever been read that houses were left vacant, cities deserted, the country neglected, the fields too small for the dead and a fearful and universal solitude over the whole earth?"[43]

---

[42] "*undique dolor, terror undique*". Watkins, 1972; Deaux 1967, p. 92. It was the plague that brought about the greatest calamity of Petrarch's life, the death of his beloved Laura.

[43] Despite the magnitude of this catastrophe there are some historians who believe that the Black Death had little if any lasting impact on European society or culture. Huizinga for example, in his *Waning of the Middle Ages* (1954), attributes to causes other than the plague itself phenomena that common sense tells us must surely have been consequences of it. "At the closing of the Middle Ages" he declared, "a somber melancholy weighs on people's souls. Whether we read a chronicle, a poem, a sermon, a legal document, even, the same impression of immense sadness is produced by them all" (p. 31). Huizinga does not connect this "immense sadness" to the experience of the plague, however, but rather oddly ascribes it to the "asceticism of the blasé, born of disillusion and satiety" (p. 37). Similarly, he attributes the terror-filled view of death that came to the fore in the late fourteenth century, not at all to the plague but to "a kind of spasmodic reaction against an excessive sensuality" (p. 141). The morbid tone of French poets of the period, he also says, arose from "a sentimental need of enrobing their souls with the garb of woe", and evidently it too was not a consequence of the plague (p. 34). Another historian, Millard Meiss, in his *Painting in Florence and Siena after the Black Death,* also minimizes – despite his book's title - the effects of the plague. He reports that in the second half of the fourteenth century artists turned away from the naturalism of Giotto and his circle, while Boccaccio and other humanists repudiated their enthusiasm for the literature of Classical paganism. Meiss argues that these reactions would "inevitably" have occurred anyway, though he concedes that they were made "far more acute" by the experience of the plague. G. G. Coulton's view of the Black Death's impact is also ambiguous (Coulton 1930, p. 101.) "This catastrophe did deeply affect the later course of European civilisation" he writes, on the one hand: while on the other declaring that by the time of the Renaissance "most men had forgotten even this notorious calamity"! Coulton overlooks the fact that people in the Renaissance were themselves subjected to recurring assaults of the plague. A historian with a lively sense of the impact of the plagues is Herlihy, who declared that European civilization "has still not outlived all the *sequelae* of the great epidemics" (Herlihy 1977, pp. 69*ff*; 81; but cf. the critical assessment of Cohn in his introduction to this work). The cultural and psychological consequences of the plagues should not be confused with the *economic* impact of the

The plagues did not end with the Black Death (figs. 6.13, 6.14). For the next three hundred years Europe continued to be assaulted by wave after wave of further outbreaks. Although the severity of these outbreaks gradually diminished, they remained so lethal, even three centuries after the Black Death, that about one-third of Venice's population perished in the plague of 1630, and about one-fifth of London's population in the Great Plague of 1665 - 1666.

In the long years during which these outbreaks occurred no useful understanding was reached about what caused the sickness or what could be done to cure those stricken by it. Familiar remedies such as penitential prayer, appeals to saints,[44] or the massacre of defenseless Jews,[45] no matter how zealously resorted to, proved unavailing.[46] Nothing seemed to prevent the plague from coming back, and each – seemingly arbitrary - recurrence further intensified people's feelings of helplessness. No one knew when the plague would strike again, no one knew what to do when it did, and no one understood why it kept on returning. As John Donne said in a sermon he preached while the calamitous outbreak of 1625 was still raging in Britain:

> God can call up ... a plague that shall not onely
> be uncureable, uncontrollable, unexorable, but
> undisputable, unexaminable, unquestionable;

---

plagues, which Bridbury (1973; see also Whittow 1996, pp. 66-68.) suggests may, paradoxically, have been minimal or – for Malthusian reasons – actually positive. Bridbury also makes a good point when he notes that, however devastating the plagues were, they coincided with one of the greatest cultural developments in human history, the Renaissance. For a careful study of the "great social and economic dislocation" that the Black Death caused in Siena see Burowsky, 1964.

[44] Piero di Cosimo, whose "The Hunt" was mentioned on a previous page, also painted a Madonna and Child with the two "plague saints", Lazarus and Sebastian. See more generally, Mormando and Worcester, 2007.

[45] For instance: on February 14, 1349 more than 900 Jews accused of spreading the plague were burned at the stake on the grounds of the Jewish community in Strasbourg (Herlihy 1977, p. 66). This number was surpassed only in the Nazi era, when the Germans murdered 3305 Alsatian Jews.

[46] Not unknown, too, were frenzied mob violence against, and the savage judicial torture of, people accused of spreading the plague: see MacKay 1932, pp. 261 - 265.

a plague that shall not onely not admit a rem-
edy, when it is come, but not give a reason how
it did come.[47]

Thus, in the centuries after the Black Death the plague
continued to haunt the lives and the imagination of people in
western Europe. Had the Black Death been a discrete occurrence
the memory of it may not have played a significant role for very
long. As it was, however, the experience of the Black Death itself
was reworked and channeled into new forms of expression as one
outbreak of the plague was followed by another. Florence, which
holds our interest here as the probable birthplace of the concept
of symmetry, suffered an outbreak of the plague on an average of
every 8 years between 1400 and 1490.[48]

The plague was thus a familiar, recurring fact in the lives
of the great 15th - century artistic and intellectual figures among
whom the symmetry norm originated.[49] The accumulated experience of previous outbreaks, starting with the Black Death, was
part of their inheritance at birth, and was further elaborated
throughout the course of their lives. In its gloom they passed all
their days.[50]

---

[47] Quoted Gilman 2009, p. 203. For long there was no substantial consensus about the source of the pestilence. Shrewsbury (1973) called into
question the extent to which the plague was bubonic; while Herlihy declared that it was "most likely not the bubonic or pneumonic plague" and
did not dismiss Graham Twigg's idea that it may have been anthrax, instead (see Herlihy 1977, pp. 6, 90-91). More recently, however, DNA analysis of a mass grave in rural Lincolnshire (Willmott et al, 2020) establishes, probably conclusively, the source as the flea-borne bacterium *Yer-
sinia Pestis*.

[48] Morrison et al, 1985.

[49] Pius II, whom I have credited (see below) as the first person known to
have used the word "symmetry" in its modern sense, was himself
stricken by the plague when he was at the Council of Basel in 1439. "The
sickness raged so fiercely" there, he recalled in his memoirs (Pius 2003,
vol. 1, pp. 40 - 41), "that more than three hundred people were buried in
a single day".

[50] It must be acknowledged that literary references to the Black Death
and subsequent outbreaks of the plague are few and (with the exception
of Boccaccio's extraordinary eyewitness description of Florence during
the plague) generally rather insubstantial. As Tuchman (1978, p. 109)
points out, Froissart (born *circa* 1337) and Chaucer (born *circa* 1343), say
almost nothing about it. But this should not be taken to mean that people

This then was people's experience of the natural forces that shaped their habitat during the era of the plagues and, in turn, planted in them the hope of gaining mastery over those forces.

And it is in the light of this experience, I believe, that we must consider the radical new aesthetic of the Renaissance.[51]

---

who lived through the Black Death were not deeply affected by the experience. Avoidance or postponement of the memory of traumatic events is after all a well - documented response (see the paper by van der Holk and van der Hart in Caruth, 1995). Defoe, who was about five years old during the Great Plague of the 17th century, did not write his fictionalized *Journal of the Plague Year* until half a century later. We find a more recent parallel in survivors of the Nazi Holocaust, who generally avoided mention of it, and seldom wrote about it, until the catharsis of the Eichmann trial in 1961 gave them permission to do so (Stern, 2000). Sometimes moreover *conscious* memory of the traumatic experience may be set aside permanently. We see this perhaps with Pepys and John Evelyn, both of whom described the Great Plague in numerous vivid entries in their diaries, but never mentioned it again after it passed. An exception is Venice, which to this day celebrates the *Festa della Madonna della Salute* with a solemn procession of thanksgiving for deliverance from the plague; the church of the Salute ("health") was built as an offering for protection against the plague of 1630 - 31. Silence about a traumatic experience, accordingly – whether the silence is temporary or permanent - does not necessarily mean that all memory of it has been erased. It may be buried, rather, in the unconscious mind. There, it enters the storehouse of the affected person's experiences and may help shape (though not always in ways that are apparent) his or her perceptions, expectations and behavior. If the traumatic experience was an historical event, it is shared with others who have also experienced it, and becomes an element of the wider culture that is transmitted to ensuing generations. It is a tricky business, to be sure, to argue *ex silentio*: yet we must ask how likely is it that people who lived through a calamity as great as the Black Death were not deeply scarred by it – or that they would not have implanted their trauma in the generations that followed them? For later generations, indeed, the inherited (if, eventually, largely unconscious) memory of the Black Death itself would have been further entrenched by the accumulated trauma of subsequent outbreaks of the plague, including any that they experienced in their own life-times.

[51] To be sure, this does not pretend to be a *comprehensive* explanation of the origin of the symmetry concept and its attendant fallacies. It does not account for a lag of about one century between the Black Death and the emergence of the concept. Nor does it account for towns like Siena which, although afflicted by the plague, show little evidence of the acceptance of the symmetry concept (a notable exception being the 14th century church of Santo Spirito, rebuilt at the end of the 15th century. The ghastly

Most manifestly in the garden, it led to the creation of an environment in which a person could indulge in the illusion that Nature no longer held sway, but had now been subjugated by Man and deprived of some of her most fundamental (and to Man, most menacing) attributes.[52]

The Renaissance's new aesthetic of simplicity and symmetry, we may say, was rooted in the ambition to neutralize Nature's deadliness, even if only symbolically. Nature lurked in the impenetrable *foresta spessa*, to emerge who knew when. Renaissance Man resolved to deprive her of that hiding place. The simplified – symmetric - forms into which he forced Nature in his gardens created a visual environment that Man could readily monitor; it was too sparse and orderly for concealment. The habitat of impenetrable obscurity in which Nature had formerly hidden as she prepared her next deadly strike – her untrammeled preference for visual confusion - was now transformed by Man's agency into an anti-natural environment that was transparent, predictable, controllable: and answerable to Man.[53] In this garden Man was Nature's master, and here, at least, *she* conformed to *his* every requirement. His fearful uncertainty and helplessness in face of death-dealing Nature could now be replaced by a sense of his dominion over her. He knew and determined what took place in even the smallest corner of his domain.

---

façade of San Francesco was not from that time: it was built in the 20th century). In Siena, the plagues, and the wars with Florence, devastated the local economy (leaving the huge cathedral standing to this day less than half-finished), and perhaps this accounts for the lack of new and renovated buildings – and thus, manifestations of the new enthusiasm for symmetric design - in Siena during the 15th to 17th centuries.

[52] Comp. Ackerman (1991, p. 73): "The Renaissance Italian looked on Nature with suspicion unless it had been tamed by man". Unfortunately, Ackerman did not attempt to account for this view of Nature.

[53] There is a curious echo of this in *News from Nowhere* (Morris 2004, p. 65), where the character Hammond says (evidently with Morris' approval): "Like the medievals we like everything trim and clean, and orderly and bright; as people always do when they have any sense of architectural power; *because then they know that they can have what they want, and they won't stand any nonsense from Nature in their dealings with her*". The aesthetic of "trim and clean, and orderly" is clearly that of the Renaissance and not of the Middle Ages; but it is remarkable to see that Hammond associated it with the desire to control Nature, and not to have to "stand any nonsense" from her.

The plagues, I suggest, thus constituted the "menacing environment" (fn. 5, *above*) in which the concept of symmetry originated. Symmetry, by simplifying the visual habitat and making it predictable to man, was a device for defending himself – symbolically, to be sure – against an environment that tragic experience had taught him to fear and mistrust. It enabled Man to enjoy the illusion that he had gained ascendancy over malevolent Nature: that he had vanquished her and no longer needed to fear her. Had he not, after all, imposed tidiness, predictability, transparency – *symmetry*! – on the face of the earth?

I should perhaps now repeat my statement at the outset of this discussion that we have no data that explicitly link this cause – the plague – to this effect – the emergence of the symmetry concept and fallacies. What I offer here, therefore, is an hypothesis: one that may seem plausible, but that is, of course, speculative and not empirically verifiable. However, I urge readers to withhold their judgment of it until they have read Chapter Seven, in which I attempt to account for the *decline* of the formal garden of the Renaissance, and the return to the so-called "natural" garden, with its complex and unpredictable textures of Nature herself.[54]

One indication of how radical an innovation the concept of symmetry was in the middle of the fifteenth century can be found in the difficulty people had in settling on a name for it. Leon Battista Alberti, who died in 1472, is the earliest theoretician of the idea known to us, though we have no reason to suppose that the concept originated with him.[55] He used the term *collocatio* for it, but no one else is known to have followed his example.[56]

---

[54] It is worth noting that the dense malevolence of Giovanni di Paolo's forest, from which we have seen Death emerging on horseback (*fig.* 6.12), gives way in one of his illustrations for the *Paradiso* (f. 186r., cant. XXXI) to a very different view of Nature in the welcoming and variegated loveliness of a typically Medieval *hortus conclusus*. This may be compared with the lovely and visually-complex garden in a small predella panel entitled "*Paradiso*", dated c. 1445, possibly also by di Paolo, in the Metropolitan Museum of Art. In Paradise, of course, Nature's malevolence does not hold sway and there is therefore no need for Man to restrain her exuberance.

[55] Alberti 1966, VI:3; IX:7; comp. I:9, and I:12.

[56] Alberti borrowed the term *collocatio* from Cicero, who used it to denote the component of rhetorical structure that deals with organization or ar-

Instead, some of Alberti's contemporaries, and occasionally Alberti himself, used a variety of ponderous circumlocutions such as "the mutual correspondence of parts" or "this side answering that" to refer to the concept, and these would continue to be employed during the next three hundred years.[57] In addition the concept was sometimes referred to obliquely, not with a specific term or phrase but by alluding to the (purportedly mirrored) two-sidedness of the human body.[58]

---

rangement: cf. Kemp 1977, p. 356, and Payne 1999, p. 75. (Alberti's *collocatio* is not to be confused with Vitruvius' *conlocatio,* for which see Scranton 1974.) I discuss the substance and the history of Alberti's concept in the appendix to the present Chapter.

[57] e.g. Alberti, 1966, IX:7, "... *ut mutuo dextera sinistris*"; Palladio (1570, p.6): "*l'un membro all' altro convenga*". The circumlocution of parts "answerable to each other" occurs in English as early as 1579 in a letter from Hatton to Burleigh: see Nicolas 1847, pp. 125 - 126. Temple ("Garden of Epicurus") in the 17th century and Burke ("Essay on the Sublime ...") in the 18th also employed "answering" to indicate bilateral symmetry.

[58] e.g. Alberti, *op. cit.,* I:9, "*ac veluti in animante membra membris ita in aedificio partes partibus respondeant condecet.*" In the second half of the 15th century the notion that the human body is bilaterally symmetric was however still open to question. Thus Filarete (1965, Bk. 1, *ff.* iii' – iv), after referring to Vitruvius' claim that the navel is centered on a man's body, demurred, "*Ma a me' nonpare' pero che sia totalmente' imezzo* – "but to me it does not seem that it is totally centered": his remark (Bk. 1, f. 2*v, ff*) that the proportions of buildings should be those of man could therefore perhaps take a different meaning from that generally attributed to it. Dürer's *Vier Buecher von menschilicher Proportion* (1528) contains some asymmetric figures but we do not know whether their asymmetry was intended by Dürer or resulted from inaccurate copying by the engraver. As we saw in Chapter One, Leonardo da Vinci - superb empirical observer that he was - knew that natural forms are *not* bilaterally symmetric. Not just the face of his "Vitruvian Man"(*figs.* 1.3 – 1.5) but its entire body are asymmetric. Its arms, for example, as well as the portions subdivided by vertical lines, are unequal in length, and its navel is not equidistant from the two sides of the body but closer to the right side of the figure. It is not usually recognized, moreover, that the body's proportions are not consistent with those specified by Vitruvius, for the distance from the chin to the top of the forehead is not one - tenth the body's length, as specified by Vitruvius; and the length of the foot is not one-sixth that of the body. Leonardo's seeming disregard of Vitruvius' criteria – along with his claim to be illustrating them (Leonardo 1958, v. 1, pp. 213 - 4) - suggests that "Vitruvian Man" may perhaps be another Da Vinci code waiting to be cracked! It should be noted here that, as Hon and Goldstein (2008, p. 117) correctly observe, Vitruvius "thought of the body in terms of proportions. He did not even call attention to the correspondence between

its limbs". Among those to have mistakenly believed that Vitruvius thought in terms of bilateral symmetry is Tavernor (1998, p. 43), who not uncharacteristically claimed, "By *symmetria* Vitruvius meant ... the mirroring of form across an axis, the modern reading of 'symmetry'": similarly Lowic (1983) in his discussion of di Giorgio's use of Vitruvius. It is not unusual to find modern art historians claiming that our bodies are symmetric: for instance, Wittkower (1978, p. 128): "Bilateral symmetry is the symmetry of the human body and for that reason of towering importance to mankind"; or, Hommel (1987, p. 20): "*Frontal beobachtet scheint der menschlicher Koerper vollendet symmetrisch gebaut*"; or Ackerman (1991, p. 163), "The human body also is symmetrical". Wölfflin (2017 p. 26), believed that awareness of our own bodies leads us to empathize with the emotions and moods of buildings, and declared that "the demand for symmetry arises from the arrangement of our bodies. Because we are built symmetrically..." (*die Forderung des Symmetrie ist abgeleitet von der Anlage unseres Koerper. Weil wir symmetrisch aufgebaut sind...*")

The view that because the human body is (allegedly) symmetric buildings, too, must be symmetric, first voiced by Alberti (see first lines of the present footnote) did not become commonplace before the middle of the sixteenth century. In a letter of December 1550 to an unidentified cardinal, Michelangelo argued that "*i mezzi sempre sono liberi come vogliono, siccome il naso che è nel mezzo del viso non è obligato nè all'uno nè all'altro occhio, ma l'una mano è bene obligata a essere come l'altre, e l'uno ochio come l'altro per rispetto degli lati e de' riscontri. A però ècosa certa che le membra dell'architettura dipendono dalle membra dell'uomo*". (This statement, incidentally, is among the considerations that call into question the claim of Vasari [1996, II, p. 709] that Michelangelo "never consented to be bound by any law, whether ancient of modern, in matters of architecture". Ramsden [Michelangelo 1963, v.2, p.290*f.*] is mistaken in attributing Michelangelo's argument to Vitruvius' doctrine of symmetry, for the latter, to repeat, addresses the proportions of a structure and is not about bilateral symmetry. Cf. the discussions of this letter in Ackerman 1986, p. 37*ff.*, and Summers 1981, pp. 418 - 446.) Vasari regarded the human body as symmetric and stated that a building's façade must be "*compartita come la faccia dell'uomo*". His anthropomorphism extends to the entire human body, and to its functions as well as its form. "*Bisogna poi*", he writes, "*che rappresenti il corpor dell'uomo nel tutto e nelle parti similmente*" including even "*un centro che porti via tutte insieme le brutezze ed i puzzi*" – see the introductory section *dell'Architettura* cap. vii. Comp. Palladio (1570, p.6): "*gli edificii habbiano da parere uno intiero e benefinito corpo, nel quale l'un membro all' altro convenga*". Wilson (1977, p. 51) quotes the opinion of Sir Roger Pratt (1620 - 1685) that rooms within a house should be arranged symmetrically "as we find it to be in our own bodies". For comparable expressions by Pascal, Hogarth, Diderot *et al.* see Hon and Goldstein (2008, pp. 128*ff*); see also Taylor 2003, p. 25. This view is implicit in Montesquieu's (1825, p. 616) likening of an asymmetric building to a deformed body: "*Un batiment avec ... une aile plus courte qu'une autre est ...*

Another word used from the outset for the mirroring of the left and right halves of a form was "symmetry", which in the end would of course become the standard term for it. "Symmetry" was not a new word, however. As we saw in an earlier Chapter, it had been used by Classical writers for a broad range of altogether different meanings, among them the concept of proportions that are pleasing or harmonious. It would continue to be used in this sense during the Renaissance and for long thereafter, even in modern times.[59] Its adoption in the fifteenth century for the concept of a form whose left and right halves mirror each other was therefore by no means a foregone conclusion. It was moreover both a baffling and an inauspicious choice, for it meant that the lexicon of aesthetic theory now employed the same word for two altogether unrelated phenomena, which inevitably led to a blurring of the distinction between the two. The confusion was particularly acute in Renaissance architectural thought, where issues of symmetry (in the new sense) and of proportion often had to be addressed more or less concurrently.

The earliest use that we possess of "symmetry" for the new concept occurs in the autobiographical *Commentarii* of Pope Pius II, who died in 1464. On the exterior side walls of his new palace in Pienza, Pius wrote, a doorway is set off to one side of the central

---

peu fini qu'un corps avec ... un bras trop court".   Edmund Burke in the eighteenth century, and the nineteenth-century French architectural theorist Durand (2000, p. 113), are rare voices rejecting the human body's alleged symmetry as the model that architecture must emulate.

[59] Cristoforo Landino, a younger contemporary of Pius and Alberti, equated symmetry with *"vera proportione"* (quoted Panofsky 1972, p. 27); cf. *"le loro simetrie"* of Manetti (1970, p. 51, I.316) which is clearly about the proportions of buildings (though the 20th-century English translation renders it nonsensically as "symmetry"); also to be understood as "proportion", Caesariano's edition of Vitruvius (1521, Bk.I, xv) has the Milan cathedral designed *"secundum Germanicam symmetriam"*, echoing Cortesi's references to *"germanica symmetria"* and *"prisca symmetria"* in his *De Cardinalatu* (c. 1510) Bk II, cap. 2; text in Weil - Garris 1980, p. 76 and fn. 28, p. 102. Pomponio Guarico's *De Sculptura* (1504) has a section entitled, *Ubi agitur de symetriis.* I have not seen a copy of this very rare book but suspect that "symmetry" there also refers to harmonious proportions. Baldinucci's *Vocabolario* (1697) defines "Simetria" as "Proporzione". This usage persisted, so that, for instance, Guiseppe Bossi's monograph of 1811 entitled *Delle opinioni di Leonardo da Vinci intorno alla simetria de' Corpi Umani,* concerns Leonardo's ideas about the *proportions* of the human body, not its "symmetry".

axis and it is matched by a fake (but visually identical) doorway at an equal distance on the other side of the center (*figs.* 6.15 -16).[60] This arrangement was required, Pius explained, "*ad servandam simmetrie gratiam*" – for the sake of symmetry - and it is clear that he intended the term to be understood in the modern sense.[61] We do not know how widespread this meaning of *simmetria* was at the time the pope wrote his memoirs, or who first used it in this way (there is no reason to suppose that it was Pius who did so); but we may perhaps infer from the fact that Pius did not explain what he meant by the term that in his day "*symmetria*" had already become a recognized meaning for the mirroring of the two lateral halves of a shape. This of course raises the question of why Alberti preferred to use a different term, or different terms, for the concept: but that is a question, unfortunately, to which we have no answer.

It is would seem that with Pius, more so than with Alberti, we are very near the inception of the concept, in the period when old tastes and precepts had only just started to be discarded and the new notion of symmetry was not yet widely adopted. Thus Pius, for all his modern sensibilities, also loved the paintings of Giotto, the sculptures of Maitani, and the medieval cathedrals of England.[62] The city that he rebuilt and named Pienza after himself also belongs to both eras, for on two sides of the piazza on which the papal palace stands there are buildings, contemporary with the palace, whose idioms are unambiguously medieval; the piazza itself is very far from being symmetric (*fig.* 6.16).[63] Indeed, the papal palace, which Mack declares "is emphatically

---

[60] I am reminded of the sardonic remark of Pascal (*Pensées*, 1:27): "*Ceux qui font les antithèses en forçant les mots sont comme ceux qui font de fausses fenêtres pour la symétrie: Leur règle n'est pas de parler juste, mais de faire des figures justes.*"

[61] Pius II, 1984, v.2, p. 547: "*in orientali latere quod oppidi plateam respicit cum media porta tenere non posset ad servandam simmetrie gratiam due ianue addite sunt quarum altera muro obducta clause uestigium pre se ferret altera ad quotidianum usum mansit aperta. In occidentali latere idem fecere*". Pius does not explain why the doors could not have been centered, but cf. the suggestion of Mack (1987, pp. 50 - 51). Mack states only that "there were two smaller doors symmetrically placed" (p. 51) and evidently does not recognize that Pius' is the first known use of "symmetry" in the modern sense.

[62] Mack 1987, p. 31; C. Smith 1992, p. 48.

[63] The presence of many Medieval stylistic elements in Pienza could lead one to question Mack's characterization of it as "a Renaissance City".

symmetrical" is not that at all: for as Mack himself reports, the bays of its outer walls, for no apparent reason, vary irregularly in width by as much as 1.7 meters. These considerations do not negate the fact that ideas about symmetric design also played a role in the design of Pienza's new structures. They indicate, rather, that the imperative, *symmetrie gratiam*, was applied selectively - and perhaps was understood imperfectly - by Pienza's builders and patron.[64]

Pius' *Commentarii* were not published until 1584, more than a century after his death. The first known appearance in print of "symmetry" with its new meaning is therefore to be credited instead to *Hypnerotomachia Poliphili*, the captivating erotic novel by Francesco Colonna that Aldus published in Venice in 1499.[65] By my count, the word "symmetry" occurs twenty times

---

[64] Mack 1987, pp. 51, 76. It may be useful to point out here that Wittkower's ideas about the dominant role of theories of proportion in Renaissance architecture find no confirmation in Pienza's structures. And this notwithstanding the letter Biondo Flavio wrote to Pius in September 1462 (Mack, 1961 p. 228) praising the proportions and the harmonious relationship of the individual parts with one another that he discerned in Pienza's new papal palace. Pius himself did not refer in his *Commentarii* to the proportions of the buildings in Pienza except to state, in passing and without elaboration, that the plan of the papal palace is a square: *"Palatium quadratum fuit"* (Pius 1962, v. II, p. 546). We do not know whether Pius' remark points back to the *ad quadratum* of Medieval times or forward to the symmetry of the Renaissance. It could be either – though, as it turns out, the plan, with the porticoes on the southern end, is 36 meters wide but 39 meters long and thus far from square (Mack (1987, pp. 44 - 45). Comp. Pius's description (1961, p. 548), of the length and width of the gallery, that he clearly gives in order to suggest its great size, and not to call attention to its proportions (for which purpose he would have had to – but does not – also give the gallery's height). We therefore have no empirical evidence to support Biondo's letter, which is sycophantic in tone and incorrectly refers to the new cathedral as dedicated to "St Matthew". The architectural principles that guided the construction of Pienza's papal buildings are unclear – perhaps reflecting the eclecticism or tentativeness of the period of transition to Renaissance ideals – and I do not believe that Biondo's references to the proportions of the papal palace do anything to clarify them.

[65] The citations here are from the Methuen facsimile London, 1904 of the first edition by Aldus (Colonna, 1499). The acrostic by which Godwin (Colonna 1999, pp. xiii ff.) convincingly identified Francesco Colonna as the author was known at least as early as 1546, when it appeared in the French translation published in Paris. It also appears in the Venice, 1635 edition (Vershbow 2013, p. 42). The attempt of Lefaivre (1997) to amplify

in this book. It does not always have the same meaning, however. In nine instances it appears to be intended in the modern sense of "bilateral symmetry."[66] We read, for example, that the channels cut through a pyramid are "symmetrically... distributed" – *symmetriatamente ... distributi* – in the structure (b, ii'); that the wreaths on one wall of a courtyard are "arranged symmetrically" – *symmetriato congresso* - with those on the opposite wall (f, iv'); that box - trees in a garden are shaped as "symmetric moons" – *symmetriate lune* (u, iii'); and that palm trees are planted symmetrically – *symmetriatamente* - (m, iii') in a row along a hedge.[67]

The "symmetry" in these and a number of other passages almost certainly refers to forms whose two lateral halves mirror each other – the new meaning of the term - and not to their proportions. In other passages, though, the term is clearly used to refer to the proportions of a form. Sometimes it occurs without any implication that the proportions are attractive. Thus, all the parts of the Pantheon are built *cum observabile Symmetria* – with "the same symmetry", that is, in accordance with a single system of proportion (c, v'); and the "entire symmetry" - *tuta la Symmetria* - of a portal is derived from the size of the squares on which the structure's columns rest (c, i'). On other occasions however it is used with an adjective to indicate that the proportions are beautiful: as in Poliphilo's appreciation of the *exquisita Simmetria* of a pyramid (a, viii'), or of the *elegante & symmetriato* design of a building (y, iii'). There are also, however, some passages in which it is not possible to determine which of the meanings of "symmetry" Colonna intended. When he wrote that a palace is "symmetrical in its architecture" – *pallatio ... di symmetriatta architectura*

---

a suggestion of von Schlosser (1929, p. 9) by attributing authorship of the *Hypnerotomachia* to Alberti is, to say the least, not at all persuasive.

[66] Colonna did not use Alberti's *collocatio* for "symmetry" though, given his interest in architecture, he almost certainly would have known *De re aedificatoria*, whose first printed edition had appeared fourteen years before the publication of *Hypnerotomachia*.

[67] In his translation of *Hypnerotomachia Poliphili* (Colonna 1999) Godwin twice paraphrases with "symmetric" passages in which the term is not directly indicated in the original Italian: the "symmetrical" courtyard (p. 96: *vide* f,iiii') is *aequabile* in the text; while the "upright symmetry" of the obelisk (p. 131: *vide* h,vi) is *aequalitate statario*. This is not to criticize the translator, however, for the context seems to indicate that *aequabile* and *aequalitate* could indeed do duty as synonyms for "symmetry" (in the modern sense), though I am not aware of other instances in which they are used in this way.

- (e, viii'), for example, we do not know whether he was referring to the structure's proportions or to the mirroring of its two lateral halves. At no point did Colonna explain what he meant by the term or acknowledge that he used it in two quite different ways. We can take this as an indication that Colonna believed that *"symmetria"*, in both meanings, would be familiar to the readers of his book.[68] In neither sense, therefore, was it a usage that he had just coined.

Colonna's ambiguous use of "symmetry" indicates, as one might have expected, that the word became a source of confusion from the moment it acquired the additional meaning of shapes whose two lateral halves mirror each other. In a fanciful novel such as *Hypnerotomachia Poliphili* this confusion was perhaps not of much consequence.[69] Sebastiano Serlio's treatises on architecture, on the other hand, were among the most widely read of the Renaissance. Their influence extended beyond Italy to France, Spain, England and the Low Countries, and by the end of the sixteenth century several had been translated into the principal west-European languages.[70]

In his *Fourth Book of Architecture* (1537), discussing one of his designs for a city gate, Serlio stipulated that the postern on one side of the main portal must be matched by a false (*"finta"*) postern on the other side of it. This, he explained, was required *"per servar la simmetria"* - "for the sake of symmetry"; and because his drawing of the structure shows that the false postern was identical to the real one both in appearance and in its distance from the center (and because the presence or absence of the

---

[68] Stewering (2000), who allows herself to think that Nature's forms are symmetric, misrepresents Colonna's use of "symmetry", and fails to recognize his frequent use of the term to mean "proportions", or "beautiful proportions".

[69] It should be pointed out, however, that by the standards of the day the *Hypnerotomachia*, with its ambiguous *"symmetria"*, was quite widely diffused. An edition was published in Venice by Aldus' heirs in 1545, three printings of a French translation appeared in 1546, 1554 and 1561, respectively, and an English translation (see fn. 82, below) in 1592. It would not be until the third decade of the 17th century that "symmetry" in the modern sense was used in French, however, and almost a century after that that it occurred in English, and so we have no reason to believe that the transmission to French and English of the new meaning can be attributed to the *Hypnerotomachia*. (I regret not having been able to examine the French editions for their use or uses of "symmetry".)

[70] Serlio 1996, I, xxxi - xxxv; pp. 470 - 471.

additional postern would not have changed the overall structure's *proportions*), we can be confident that in this passage Serlio intended "*simmetria*" to be understood in the modern sense.[71]

The context and phrasing of Serlio's remarks are of course strikingly similar to Pius' description of the real and fake doorways in the walls of the papal palace in Pienza. We have no grounds for supposing, however, that Serlio derived his ideas and terminology from Pius' *Commentarii* (which at the time existed only in a few manuscript copies). The resemblances between the two passages must therefore be explained either as a coincidence or, perhaps more plausibly, as an indication that both men drew from the same (though to us unknown) school of thought.

There is, though, a significant difference between the two accounts. Pius, as we noted, evidently did not believe it was necessary to explain what he meant by "symmetry". Serlio on the other hand, as part of his discussion of the fake postern, evidently *did* feel the need to do so, from which we may perhaps infer that the term was no longer as widely known as it had been in Pius' day, three-quarters of a century earlier (nor, for that matter, three or four decades earlier when Colonna wrote the *Hypnerotomachia*). Serlio therefore gave his readers a definition of the term. "*Simmetria*", he told them, means "*corrispondenzia proportionata*" or, "proportionate correspondence".

This was, to say the least, a highly confusing definition. "*Corrispondenzia proportionata*" is an opaque phrase that is without precedent in the literature of aesthetic theory, and suggests that Serlio may not have been entirely clear in his own mind whether he intended to use "symmetry" for the *correspondence* of a structure's two lateral halves (the modern meaning, and the one applicable to the gate about whose symmetry he was writing in this passage) or for its *proportions* (the Classical meaning and the one, as already noted, that is inapplicable to that gate).[72]

---

[71] Serlio 1996, v. 1, p. 260: "*Ma per servar la simmetria, che vuol dir correspondenza proportionata, è necessario farre un' altra finta. La misura della porta cosi è da fare, che quanto serà la larghezza dell'apertura, sia la metà di esse aggiunta all'altezza*". It was of course a *non sequitur* to declare that the false postern was needed to make the design symmetric; a functioning postern would have had the same effect!

[72] Hon and Goldstein (2008, p. 118), who recognize the problematic nature of the term, suggest that Serlio may have intended to merge the two concepts. "According to Serlio", they write, "symmetry involves both proportion ... and correspondence", but this is contradicted by Serlio's unambiguous use of "symmetry" in the *Seventh Book* to mean "bilateral

Later, to be sure, in his *Seventh Book*, which was published posthumously in 1575, Serlio used "symmetry" in a way that leaves us in no doubt that he understood the word in its modern sense.[73] But this was of little consequence for the history of the word, because the *Seventh Book* was not translated into other European languages until the twentieth century, and its unambiguous use of "symmetry" in the modern sense is not present in Serlio's earlier books. There, indeed, immediately after defining symmetry as "proportionate correspondence", Serlio went on to describe the optimal *proportions* of the gateway he was writing about, thus giving the misleading impression that "symmetry", as he was using the term, was not about bilateral correspondence at all but about proportions. This confusion was heightened a few years later, but still during Serlio's lifetime, when Flemish and French translations of the *Fourth Book* were published in Antwerp by Pieter Coecke van Alst.[74] In these editions Coecke dropped the mention of the false postern in Serlio's Italian text and merely declared that, "for the sake of symmetry", the opening's *proportions* would have to be such - and - such.[75] (In Coecke's versions the structure has but a *single* opening.) For Coecke's readers, therefore, "symmetry" had nothing to do with the modern sense of the term, but simply meant "well - proportioned".

Both Serlio's reference to the *need* for a second – if fake – postern, and Pius' statement that a fake doorway had been inserted "for the sake *of* symmetry", remind us that, from its inception, the *concept* of symmetry was prescriptive or normative (unlike the merely descriptive *idea* of symmetry), requiring that the things we make *should be* shaped symmetrically.

In its prescriptive aspect, (which I call the concept of symmetry), symmetry makes a startling appearance in Serlio's

---

symmetry". As we have already observed, moreover, the addition of a second postern in Serlio's gateway, while it made the design symmetric, did not affect its proportions, which would have been the same whether or not either one or both flanks housed a postern. As we have also seen, both meanings are attached to the term in the *Hypnerotomachia*, but, as far as one can tell, are never used for *the same* object.

[73] Book VII, chap. LXVI.

[74] He was the father-in-law, as it happens, of the elder Brueghel.

[75] "*Toute porte de cité a besoin de poternes: mais pour garder la symétrie, c'est - à - dire, correspondance proportionnée, est besoin d'ensuivre par cette manière, qu'autant que sera la latitude ou largeur de l'ouverture de la porte soit ajutée la moitié d'icelle à sa hauteur*".

Seventh Book. There, in Chapter 62, we learn that, in a certain Italian city whose citizens cared greatly about architecture, a rich but miserly man lived in the house his grandfather had built back in the days when good architecture still lay buried – *ancora sepolta.* This house was flanked on either side by newer buildings whose fine design only made the miser's house seem all the uglier (*piu brutta*) by comparison. So offensive was this deformed (*difforme*) house, indeed, that the very sight of it induced feelings of nausea and queasiness (*nausea & fastidio*) in the prince who ruled the city. What caused the prince to react in this way is not a mystery. The façade of the house was very obviously asymmetric. The twenty principal windows on it comprised at least eight different shapes and sizes and were arranged asymmetrically. Moreover, the entrance to the building was not centered on the façade, something that (Serlio tells us) is very contrary to good architecture ("*cosa che e molto contraria alla buona Architettura*").

The prince tried to persuade the miser to improve the building's appearance but he, caring (in Serlio's telling of it) more for his money than for the decorous appearance of the city, always replied that he would like to do as the prince asked, but that for the time being he was short of funds. Eventually the prince's patience ran out, and he warned the miser that he would confiscate and demolish the house unless it were rebuilt within one year in the style of its neighbors. The miser capitulated. He called in the city's best architect and ordered him, without regard to cost, to alter the house's appearance in a way the prince would find pleasing. The architect tore down the entire façade and built a new one in its place; the asymmetric plan of the interior of the house was modified only to the extent that a room was partitioned to create an entrance hall for the newly - centered doorway to lead into. The drawing (*fig.* 6.1) with which Serlio illustrates this account shows that the main objective of the renovation was to transform the house's façade from an asymmetric to a symmetric layout. This involved rebuilding the windows (whose appearance was modernized during the process), and moving the entrance to the center of the structure.[76] In his description of this

---

[76] The fenestration of the new façade is pleasingly varied, albeit in a format of symmetry and order, both lacking in the earlier structure. Also as part of the renovation, a heavy cornice was placed between the two principal floors. Serlio says that this episode took place during his lifetime, which means that a date even before the end of the 15th century is possible; and that it could have been Serlio himself, born in 1475, who designed the new façade. It should be noted that Serlio does not mention

project, Serlio does not use the term "symmetry" though it is obvious that the main purpose of his work was to create a symmetric façade.

Albeit less colorfully, Serlio described two other projects in which older houses were given symmetric facades. In the first of these, in Chapter 66, he does indeed use the term "symmetry" in the modern sense. The project involves a well-built house that suffered from a number of drawbacks, "the clearest error being that the door of this house is not in the center, as it should be – and also the windows have a certain inequality" (*fig.* 6.17). The owner of his house wanted "not to appear inferior to his neighbors", who "unerringly build with good arrangements". By this, Serlio explained, he meant that "as a bare minimum they adhere to *simmetria*". The façade was rebuilt in symmetric form.[77] In the second project, Chapter 67, Serlio described how the owner of two adjoining asymmetric houses wished to build a single façade onto the two structures, "putting the door in the center, as is necessary".[78] Where there were formerly two interestingly asymmetric facades there is now a single, rather bland, symmetric one instead (*fig.* 6.18).[79] In neither case were the structure's proportions altered.

Serlio's narrative, as we see, is set in a time when enthusiasm for the principles of "good" architecture was not only fervent but was matched by revulsion – strong enough to induce feelings of nausea! – at the taste of an earlier generation. The recent changes had at their core the unequivocal rejection of asymmetric design. Feelings on this point ran so high that the presence of even a single asymmetric building was thought to detract from a city's appearance, and social and political pressure could be

---

the *proportions* of the building – evidently they were not among the things that had made it seem *difforme* – and they were not altered as part of the renovation.

[77] We can infer that the new facade would have won less grudging approval from Serlio if its windows had been equidistant from each other, but they had been installed only recently and the owner, wanting "the least disruption and expense possible", did not wish to re-arrange them. Although not equidistant, their arrangement on either side of the central axis – the newly-built doorway – is now symmetric.

[78] Serlio 2001, Bk. VII, caps. LXII. LXIII

[79] Ironically, Serlio's illustration of the rebuilt façade shows that the window above the main portal is *not* centered over it but is somewhat to its left. This is surely an error by the engraver.

brought to bear to compel its owner to spend substantial amounts of money on the merely cosmetic transformation of its façade from an asymmetric to a symmetric design. As we see from this instance, the intensity of the feelings that could be aroused by asymmetric design is quite startling. It suggests that the desire for symmetry was not based on aesthetic considerations alone, but may have had deeper and more complex and irrational origins – possibly, the ones discussed in the first portion of this Chapter.[80]

Pius II, Colonna and Serlio are evidently the only Renaissance writers who used "symmetry" in the sense of bilateral symmetry. As we have already noted, when other Italians used the word they invariably did so with its Classical meaning of (harmonious) proportion.

And it was more often than not in this sense too that "symmetry" was used in England before the end of the nineteenth century:

- Thus for John Shute, whose architectural treatise was published in 1563, the dimensions of a Tuscan column were to be determined by "the order and rule of Symetria" that, as he explained, require a "parfaicte knowledge" of arithmetic.[81] Clearly, he was referring here to the mathematical calculations needed to determine the optimal proportions of a column.[82]

---

[80] But the rejection of asymmetric design may not have been quite as widespread as Serlio implied. In Bk. III (90v) Serlio wrote of the "disordered and discordant" effect created by the asymmetries of the Baths of Diocletian, which he compared unfavorably to the Baths of Antoninus, with their "correspondence in all parts" and the "harmony" that resulted from it. He went on however (94v) to apologize to "supporters and defenders of ancient things" if they were offended by his criticism of Diocletian's structure, from which we must infer that in Serlio's day there was still some sentiment in favor of asymmetric design, which was associated with the "ancient" way of doing things.

[81] Shute 1563, p. iiii. He added, borrowing from Vitruvius, that the rule of "Symetria" is based upon "the Simetrie of a strong man". At about the same time John Dee, in an obscure passage of his *Preface to … Euclid* (1570, pp. 32 - 33), refers to "*Symmetrie* mentall" as a constituent of Alberti's "Mathematicall perfection", which seems to imply the Classic meaning of the term.

[82] Half a century after Shute, Dallington (1605, p. 13) described the limbs, feet and arms of the giant statue in the Pratolino garden as being "symmetricall to his head". It is an obscure statement, and the fact that it refers to paired members of the body *may* mean that Dallington was using the term in the modern sense (with the head representing the central axis)

- In 1611, when Pike published his English translation of Serlio (it was based on Coecke's Flemish translation) he stated what the dimensions of a structure must be if it is to conform to "semetry, that is due measure" – in other words, if it is to have good proportions.[83]
- For Henry Wotton, too, a decade after Pike, symmetry had to do with good proportions: as he put it, "*Symmetria* is the convenience that runneth between the Parts and the Whole".[84] When he wished to refer to what we now call symmetry Wotton used the term "Uniformitie" instead. This, he claimed, is "the great Paterne of Nature" and is exemplified by our bodies "than which there can be no Structure more uniforme ... each side agreeing with the other, both in the number, in the qualitie and in the measure of the Parts".[85]
- Somewhat later in the seventeenth century we find the writer of an unsigned letter stating that in Roger Pratt's work at Raynham Hall "there was somewhat in it divine in the symmetry of proportions of length, height and breadth which was harmonious to the rational soul".[86] Clearly here too "symmetry" refers to the apt proportions of a design.[87]

---

rather than to suggest that those members were well - proportioned in relation to the head. There is no way of being sure, however. It has been suggested that this Dallington is the "R.D." who translated the first, 1592, English version of *Hypnerotomachia Poliphili*. The translator's difficulty with the term "symmetry" is suggested by his omission of it in a number of passages where it occurs in the original – e.g., *cum observabile Symmetria* (c, v[1]). Elsewhere he adds a gloss to ensure that the reader understands the term in its older meaning. Thus, he renders *"mirada & exquisite symetria"* [a, viii[i]] as "a marvelous and exquise symmetrie *and due proportion*" [my emphasis].

[83] Serlio 1982, Bk IV, cap. 1, fol. 5v.

[84] Wotton 1968, p. 119. Comp. *ibid*, p. 54: "in some windowes and doores the symmetry of two to three, in their breadth and length".

[85] *ibid*, p.21. We saw earlier that Wotton, following Vasari, advanced the case for symmetric design without using any term for it by arguing that buildings must resemble the well - shaped male body with the principal entrance, like the mouth, in the center and windows, "like our eyes, be set in equall number and distance on both sides".

[86] Text in Gunther (1928, p. 133), who dates the letter *circa* 1663.

[87] In his modestly-entitled *True Intellectual System of the Universe* (1678) the Cambridge neo-Platonist Ralph Cudworth listed "symmetry and

- In Evelyn's translation of Freart (1664) we read that the Vitruvian "*Eurythmia,* and *Venusta species Aedificii* ... creates that agreeable harmony between the several dimensions, so as nothing seems disproportionate, too long for this, or too broad for that, but corresponds in a just and regular *Symmetry* and concent of the Parts with the whole".[88] Here "agreeable harmony"and "symmetry", along with the "consent" of the parts with the whole", arise in a structure when "nothing seems disproportionate": thus here too symmetry refers to proportion. [89]

Addison's equation, in 1712, of "symmetry" with "the beauty of an object" also appears to use the term in the older sense.[90] John Gwynn, too, evidently had this use in mind when he wrote of

> a tasteless structure where each part
> is void of order, symmetry or Art.[91]

Although (as we will see presently) Burke used "symmetry" in the newer sense, he reverted to the earlier meaning when he wrote of someone who had "symmetrized every disproportion".[92] The lexicographers stayed with the older meaning,

---

asymmetry" among "the basic essences" (Cudworth III, p. 588). It seems however that he used "symmetry" in the traditional sense: as in his reference (III, p. 598) to "beauty and pulchritude and symmetry". The first unambiguous use of "symmetry" in the modern sense by an English author occurs seven years after Cudworth's work was published, in Temple's *Garden of Epicurus* (1685), which is discussed further on in this Chapter.

[88] Freart 2017, p. 146.

[89] The same point is apparently made by Freart (*op. cit.,* p. 207) in the following passage: "...By means of these measures it may easily be computed what proportions all the parts and members of the Body have one by one to the whole length of the Body; and what agreement and symmetric they have among themselves, as also how they vary or differ one from another; which things we certainly conclude most profitable and fit to be known". The use of "symmetric" as a noun is not recorded in OED.

[90] *Spectator* No. 411, June 21, 1712.

[91] Gwynn, 1742.

[92] Burke 1796, #55. William Mason (1772) also used the term for both meanings. In Book II (50 - 55) the "lambent flow" of Nature's lines

alone. Dr. Johnson, notably, defined symmetry as the "adaptation of parts to each other; proportion; harmony; agreement of one part to another".[93] The *Encyclopedia Britannica*, in its initial appearance in 1771, also stuck to the traditional usage, defining symmetry as "the just proportion of the several parts of any thing, so as to compose a beautiful whole". One of the best-known instances of the earlier usage is in the "fearful symmetry" of Blake's tiger (1794).

In the middle of the nineteenth century Ruskin, in *The Stones of Venice*, used the term in the earlier sense with his reference to the "exquisite symmetry and richness" of the canopy over a tomb.[94] The older meaning is also employed by Charlotte Bronte who, in *Villette*, described her heroine's "pale small features, her fairy symmetry, her varying expression".[95] The older meaning of symmetry, indeed, has survived into our own day and age, as we see from P. D. James' description, in one of her fine detective novels, of the "attractive symmetry in the ... proportion between the strong walls and roof" of a house.[96] "Symmetry" would retain its Classical meaning in England until well into the nineteenth century, and it was not generally understood in the modern sense until the twentieth century.[97] Moreover, the

---

"charms us at once with symmetry and ease"; while in Book IV (70), Norman architecture is unspoiled by the modern predilection for "misplac'd symmetry".

[93] The last of these is too ambiguous to allow us any confidence that what the great Doctor had in mind was intended in the modern sense. It seems more plausible here that the parts "agree" because they are in some sense consistent with each other in their appearance or proportions, rather than because they mirror each other.

[94] Ruskin 1903 - 13, v. XI, p. 84. Ruskin also used "symmetry" in the modern sense.

[95] Bronte 1853, p. 279.

[96] James 2005, Kindle version at 5763.

[97] In the 19th century, among the more curious explanations of "symmetry" in the Classical sense is that of J. B. Ker (1840, pp. 104 - 5) who defined it as "complete order, proportion" and claimed that it implied "a state devised by the Supreme Being". He derived the word from the Dutch "*sij'm met rije*", which he translated as "we come into a duly regulated state of things".

use of "corresponding parts" for bilateral symmetry continued in England until at least the first decade of the nineteenth century.[98]

In 17th-century France, too, the term occurs, confusingly, with both meanings. Its earliest appearance is in 1623, in Pierre Le Muet's influential *Manière de bâtir pour toutes sortes de personnes.* Le Muet seems to have been uncertain about which of the word's two meanings he really had in mind. Under the heading, *"La belle ordonnance consiste en la simmetrie"* – beautiful arrangement consists of symmetry – he first defined symmetry as that which requires "the parts equally distant from the middle to be equal to one another".[99] Yet on the very next line (and still under the same heading) he went on to declare that the parts of a building that are equally distant from the middle must be correctly proportioned – *"proportionnées"* - in relation to each other and to the overall structure.[100] In the English version of *Manière de bâtir,* published in 1670, the ambiguity of the French original was abandoned, as was the modern meaning of the term, in favor of the original definition of symmetry as harmonious proportion. "Fair ordering and comeliness", it reads, "consisteth in the *symmetry or equal proportions"* [my italics] of the parts of a building.[101] Another consideration indeed leads one to question the extent to which Le Muet ever really *did* understand "symmetry" in its modern meaning. A number of engravings that appear in both the French and English editions of *Manière de bâtir* show designs of buildings whose main entrances as well as certain other features are in a manifestly asymmetric relation to the overall façade

---

[98] e.g., Gandy 1805, p. vii: "architectural design in general should be uniform, that is, having corresponding parts on each side of a centre".

[99] Le Muet 1623, p. 4.

[100] *"Selon la largeur, elle consiste à faire que les parties esgalement esloingnees du miliere soient esgalees entre elles. Que les parties soient proportionnées au total & entre elles. Selon la hauteur elle consiste à faire que les parties esquelles mesmes symétrie aura esté observee pour le regard de la largeur, soient aussi de mesme niveau en leur hauteur.Car il peut arriver qu'une partie symétrique en largeur ne le sera point en hauteur. Pour exemple, les demies croisees, lesquelles vous pouvez asseoir en pareille distance du milieu de l'edifice, neantmoins les frontons qui leur feront imposés n'arriveront pas à la hauteur de ceux des croisees entieres; ainsi ce qui fera symétrie en largeur, ne le sera pas en hautcur ; partant tels ouvrages sont à eviter."*

[101] Le Muet 1670, p. 3. "Equal proportion" is not to be understood in the modern sense of "all of the same size" but as a consistent system of proportion throughout a structure.

(*fig.* 6.19). "Symmetry" in these buildings, because it cannot mean *bilateral* symmetry, is therefore likely to refer to their *proportions*.[102]

Another influential architect and theorist, Claude Perrault (1613 - 1688), attempted to resolve the confusion between symmetry as "proportion" and symmetry as "correspondence" by declaring that there are "two kinds of proportion". One of these, that he says is "very difficult to discern", consists of "the proportional relation of the parts". The other, that is "very apparent", is "called symmetry", and is "a balanced and fitting correspondence of parts that maintain the same arrangement and position".[103] It is not clear why Perrault regarded symmetry as a "kind of proportion", too – what is gained by doing so? – but the confusion engendered, ultimately, by Serlio's *corrispondenzia proportionata* may be the explanation.

In the early eighteenth century the same confused double meaning of "symmetry" appeared in Sebastien Le Clerc's *Traité d'architecture* (1714), an English translation of which, by Ephraim Chambers, the encyclopedist, was published in 1732.[104] Le Clerc required that there be pilasters on both sides of a window "to make a symmetry", which implies that he understood the word in the modern sense; yet he also used the word "symmetry" to refer to the proportions of certain arrangements of columns.

The earliest use by an English writer of "symmetry" with the meaning of "bilateral symmetry" appears to have been in Temple's "Garden of Epicurus" (1685). In this essay Temple introduced the concept of *sharawaggi*, which he claimed was a Chinese method for designing buildings and landscapes that intentionally dispensed with "any order or disposition of parts that shall be commonly or easily observ'd".[105] Temple contrasted this

---

[102] Fréart's use of the term is also ambiguous, but seems to have tended toward the earlier meaning: Fréart 2017, pp. 28, 36, 974, 88.

[103] Perrault 1993, pp. 53, 50.

[104] I have seen only the English translation.

[105] Temple 1731, v. I, p. 186. Honour (1962, p. 144) suggests that Temple may have learned about the design of Chinese gardens from Far-East travelers whom he met during his embassy in The Hague. Lang and Pevsner (1949) on the other hand declare that according to Sinologists "*sharawaggi*" is not a Chinese word. Similarly *OED* on "Sharawaggi". Honour, (*op. cit.*, p. 145), though without citing authorities, finds Chinese and Japanese origins for it; Clark 1944. Lewis (1960, p. 102,n.) makes the entertaining suggestion that the word was probably invented as a hoax

type of design (which of course was necessarily asymmetric) with English designs whose "beauty... is placed chiefly in some certain proportions, symmetries or uniformities". That these three latter terms are not merely synonyms, and that Temple here means "symmetry" in the sense we understand it today, can be inferred from the context, which contrasts the English manner (by which trees are "ranged so, as to answer one another") with the irregular, *sharawaggi*, way of doing things.[106]

The next occurrence of "symmetry" in the modern sense is in Evans' poem, *Vertumnus* (1713), which we encountered on an earlier page. The Oxford Botanic Gardens, Evans declared, were laid out with "perfect symmetry", and there can be no doubt that the reference is to the severely symmetric layout of the beds within a square compound that still characterizes the Gardens.

It was not until the middle of the eighteenth century, however, that the new meaning of "symmetry" started to be used with some frequency in English literary circles. We find Walpole, for example, telling his friend Horace Mann in 1750 that he was "almost as fond of the ... want of symmetry in buildings as in grounds of gardens".[107] Hogarth used it with the modern meaning in his influential *Analysis of Beauty* (1753). Symmetric shapes gratify us, he wrote, "not from seeing the exact resemblance which one side bears the other, but from the knowledge that they do so on account of fitness, with design, and for use". Beyond such practical considerations, however, symmetry can "but little

---

by Temple. Murray, 1998, gives the concept a Japanese rather than Chinese origin, and claims (not quite convincingly) to have "resolved" the whole problem. For the "insistent asymmetry" of Japanese architecture (for which however there seems to be no term), see Ramberg (1960) and fn. 42, p. 359, *below*; also Morse 1972, p. 135, *ff.*

[106] According to Temple the Chinese "scorned" roads laid out in straight lines and flanked by regularly-spaced trees. It is curious to note that, almost 100 years after Temple, William Chambers (1772, p. 52) reported that roads in China are laid out with "exact order and symmetry". See Lovejoy 1955a, pp. 122 - 135, for a good account of the reservations Chambers had about his contemporaries' understanding of Chinese design.

[107] Quoted Tunnard 1978, p. 83. Some years later Walpole (1891, v. IX, p. 70) wrote a catty description of the Duke of Modena having "a mound of vermilion on the left side of his forehead to symmetrise with a wen on the right". (A wen is a benign cyst of the skin.) In his book on gardening Walpole (1995, pp. 25 - 6; 29) deplored the "fantastic admirers of symmetry" who "corrected" the shapes of trees, and he dismissed "symmetrical" gardens as "unnatural".

serve the purpose of pleasing the eye … and soon grows tiresome".[108]

Burke's role in the adoption of this new use was especially important. In his "Essay on the Sublime and Beautiful",
which first appeared in 1756, Burke denounced the fashion of
"disciplining" Nature and "teaching her to know her business"
by reducing her forms to "squares, triangles and other mathematical figures with exactness and symmetry" – the latter term
clearly referring to symmetry in the modern sense. [109]

---

[108] Hogarth, 2015, p. 33. Comp. *ibid*, pp. 33-38, disparaging equation of
"symmetry" with "uniformity".

[109] "Essay on the Sublime and Beautiful", Part III, Section 4: "And certainly nothing could be more unaccountably whimsical, than for an architect to model his performance by the human figure, since no two
things can have less resemblance or analogy, than a man and a house, or
temple: do we need to observe, that their purposes are entirely different?
What I am apt to suspect is this: that these analogues were devised to
give credit to the work of art, by showing a conformity between them
and the noblest works in nature; not that the latter served at all to supply
hints for the perfection of the former. And I am the more fully convinced,
that the patrons of proportion have transferred their artificial ideas to nature, and not borrowed from thence the proportions they use in works of
art; because in any discussion of this subject they always quit as soon as
possible the open field of natural beauties, the animal and vegetable king-
doms, and fortify themselves within the artificial lines and angles of architecture. For there is in mankind an unfortunate propensity to make
themselves, their views, and their works, the measure of excellence in
everything whatsoever. Therefore, having observed that their dwellings
were most commodious and firm when they were thrown into regular
figures, with parts answerable to each other; they transferred these ideas
to their gardens; they turned their trees into pillars, pyramids and obe-
lisks; they formed their hedges into so many green walls, and fashioned
their walks into squares, triangles and other mathematical figures, with
exactness and symmetry; and they thought, if they were not imitating,
they were at least improving nature, and teaching her to know her business. But nature has at last escaped from their discipline and their fetters;
and our gardens, if nothing else, declare we begin to feel that mathematical ideas are not the true measure of beauty." It should be noted too that
Burke disdained simplicity. In his view, anything that can be grasped in
its totality is not grand: a certain obscurity is a necessary condition of the
sublime. "Any thing which we can grasp fully and wholly perceive sinks
to the level of the commonplace … a clear idea is therefore another name
for a little idea'". This is a view that was anticipated by Pico della Mirandola in the 15th century; Wind (1986, p. 16) excoriated it as a "pernicious
axiom"

Joseph Heely (1775, pp. 15, 9), lamenting how Nature is "mutilated and torn by the pencil of art" in the gardens of Hampton Court and Kensington Palace, and the "tiresome sameness" that is the result, declared, "Beauty in gardening is not to be considered by a perfect symmetry, as in a palace; it is composed, and ever delights in the wildness, of fancy and a sympathizing irregularity". The antonyms of symmetry here are wildness and irregularity, so that Heely's "symmetry" is clearly to be understood in the modern sense.

In France, the modern meaning of symmetry appears to have been securely established by the third quarter of the eighteenth century, when Montesquieu discussed the concept at some length in the section "Des Plaisirs de la Symétrie" of his *Essai sur le Gout*.[110] The English translation of this essay (published in 1777) reflects the continuing confusion in England between the two meanings, sometimes rendering "symétrie" as "symmetry" and at others times, nonsensically, as "proportion".

It is perhaps indicative of the difference between the two countries with respect to this word that, although the French crystallographer Haüy reported in 1800 that snow crystals quite often have *"un charactière particulier de symétrie"*, the English explorer Scoresby was still describing them as "regular" in 1820: and it was not until 1880 that the term "hexagonal symmetry" was first used for them, in a paraphrase by Huxley of Scoresby's description.[111]

In Germany from a relatively early date "proportion" was used for what, in Italy and elsewhere, was being called "symmetry". Thus, Albrecht Dürer's book, published posthumously in 1532, carries the title, *Vier Bücher von menschlicher Proportion,* while the Italian edition, published in 1591, is entitled *Della Simmetria dei corpi humani.* It is clear however, as we saw from Hollando's dialogs with Michelangelo (p.283, fn. 11, *supra*) that the two terms – symmetry and proportion – do not always mean quite the same thing. A hint of what this distinction *may* mean can perhaps be found in the pattern book by Wendel Dietterlin which appeared in Nuremberg in 1598: in German under the title *Architectura von Austheilung Symetria und Proportion der fünf Seulen,* and in Latin as *Architectura: de constitutione, symmetria, ac*

---

[110] Montesquieu 1825, pp. 616 - 7. We have already cited Pascal's use of "*symétrie* " almost a century earlier (fn. 60, *supra*).

[111] Huxley 1887, pp. 60, 62.

proportione quinque columnarum.[112] Although Dietterlin's use of the terms is far from clear, it seems that for him the one may refer to what is a necessary condition for the other. Thus, (in the section on Doric columns) he offers an historical account of the discovery of symmetry. There was a time, he declared, when no one knew about correct symmetry (*"damahlen aber noch von keiner rechten Simetri* [sic] *wusten..."*). Eventually, however, people learned to arrange and proportion the column according to the length of a man's foot (*"entlichen die Seuln in eines Mannsfussleng ... abtheilen und proportioniern lassen"*. Thus, symmetry was discovered[113]. May we perhaps infer from this that correct proportions (and arrangements) produce symmetry – a harmonious result?[114]

The German poet Goethe (1749-1832) attached great importance to symmetry.[115] "Nothing we look at can be attractive without symmetry", he declared; adding that symmetry is an indispensable component of "the most perfect standard of art". Thus, although he admired the paintings of Rogier van der Weyden, he thought that the lack of symmetry in their composition meant that "they fail to satisfy the strict demands of art". Goethe believed that Byzantine art was "always" symmetric, and that it was from Byzantium that the Latin West acquired the concept of symmetry.

---

[112] Dietterlin, 1968

[113] That Dietterlin did not understand symmetry in the modern sense is also evident from the more than 200 designs in his book, that are invariably – and often very boldly – asymmetric. Krufft (1994, pp. 169 *ff*) makes the interesting suggestion that in the late sixteenth century Dietterlin and other German architects ignored the neo-Classical motifs of the Italian Renaissance because of their association with the Roman church. Possibly this helps account for the persistence of asymmetric design in the somewhat misnomered "Renaissance" architecture of Germany.

[114] A much later German, Camillo Sitte (1843-1903), recognized the problem posed by these terms and suggested a possible solution to it: "Proportion and symmetry were essentially one and the same for the ancients, with only this difference: under proportion in architecture was understood merely a certain generally pleasing quality of relationship based on feeling (for example, that of the height of a column to its thickness), whereas symmetry involved the same relationship but expressed precisely through numbers".

[115] This discussion of Goethe's views on symmetry is excerpted from Chapter 3 of Selzer, 2021. Goethe's principal statement on symmetry is in the section on Heidelberg in his *Aus seiner Reise am Rhein, Main und Neckar* (1816).

It is not at all clear, however, what Goethe meant by "symmetry". According to the Grimm's authoritative dictionary of the German language, symmetry as the mirroring of a form's two lateral halves had been a recognized meaning in German since the early 18th century. In his principal discussion of symmetry, however, Goethe did not use this definition of the term, but introduced one of his own devising. For a composition to be symmetric, he wrote, its various parts must relate to one another in a specific way that he called "interchangeable" (*sich wechselweise aufeinander beziehen*). This is achieved, he went on, when a composition has "a middle, an above and a below, a here and a there" (*eine Mitte..., ein Oben und Unten, ein Hüben und Drüben*). When this requirement is met, symmetry comes into being (*entsteht*). Goethe then went on to classify different grades of symmetry. In the lowest grade, symmetry is "wholly comprehensible and immediately recognizable". What distinguishes the higher grades is the progressively greater number and complexity of their design elements, which make their symmetry increasingly difficult to detect. The "early symmetry" gives way to a kind of "spiritual symmetry" – "a visible mystery", as Goethe also calls it – and finally to symmetry that is "concealed". The very invisibility of a form's symmetry attests to it being the highest form of symmetry! It is impossible to take these lucubrations seriously. Goethe's criterion of "a middle, an above and a below, a here and a there", for one thing, could be applied to *any* type of form (in the unlikely event that it were ever thought worth applying), and is not a characteristic, specifically, of a symmetric form. It will come as no surprise to learn that there is no record of Goethe ever using his definition of symmetry in describing a work of art.[116]

Nor is there any evidence that Goethe's views influenced later 19th-century German theoreticians of art and architecture who were interested in symmetry. One of these was Camillo Sitte (1843-1903), who recognized, as we have already seen in Chapter Three, that the modern concept of symmetry was unknown to the Ancients, and that when they used the word, they meant something else by it. Sitte attributed the advent of the modern concept to the use of "real" architectural drawings in the lodges of the Gothic master architects. These, he suggested, led builders to become "more and more concerned with symmetrical axes in the modern sense [and] ... increasingly conscious of the theoretical

---

[116] For another advocate of invisible symmetry, see Chapter 2, fn. 79.

notion of the right and left. For this new concept the old word was chosen". He acknowledged, however, that it was only since the Renaissance that the concept of symmetry has "conquered the world".[117] Sitte's explanation is a provocative one, but it seems inconsistent with the clear fact, which we documented in an earlier chapter, that there is no evidence of intentional symmetric design in medieval ("Gothic") structures.

He is surely on a firmer footing when he dates the concept, or more specifically its "conquest" of the world, to the Renaissance. As a town planner, Sitte took a very dim view of what he called the "obnoxious" preference for "the artificial regularity, the pointless symmetry and uniformity" that characterized town planning in his day, which he excoriated as the product of "weak, meager and unfortunate taste".[118] His analyses of the deeply satisfying effects of irregular, asymmetric medieval town squares, with roads leading to but not passing through and continuing from them, are masterful.

Gottfried Semper (1803-1879) and Heinrich Wölfflin (1864-1945), by contrast, regarded symmetry as one of several essential components of beauty.[119] For Semper, who is perhaps still the preeminent authority on the history of design aesthetics, symmetry, along with "proportionality" and "direction", is one of the "three necessary conditions of formal beauty".[120] He distinguished "strong symmetry", with its "complete identity of elements right and left", from "*Ebenmass*", which his translator renders as "weak symmetry", where there is only a "balance of masses".[121] Semper claimed that the "symmetrical principle" is present, albeit "in latent form" in "snowflakes, flowers and so on" as well as in "the lesser parts of trees".[122] (Latent symmetry does not seem to be a viable concept; in any case, Semper does not explain what it is.) His insistence on the desirability of symmetric design was nonetheless not inflexible. The fact that weapons are "almost without exception independent of symmetry", he wrote,

---

[117] Sitte 2006, p. 190.

[118] *idem.*

[119] The introduction to *Style* (Semper, 2004) by Mallgrave is an excellent starting-point for anyone wishing to understand Semper's complex and subtle ideas.

[120] *ibid,* p. 83. I have not seen the original, German, text.

[121] *ibid,* p. 85.

[122] *ibid,* pp. 85, 88.

"enlivens their forms and makes them so attractive to the vigorous warrior"![123] As far as I can tell, he did not call for this "enlivening" effect to be present in buildings or other artifacts.

Wölfflin presented his notion of symmetry in a precocious
and not always intelligible doctoral dissertation that he wrote before he turned 22.[124] Where Semper identified three necessary
conditions for formal beauty, Wölfflin followed the construct developed by the philosopher of art, Friedrich Theodor Vischer
(1807-1887), who posited two external and four internal aspects
of "Form". The latter were (1) regularity; (2) symmetry; (3) proportion; and (4) harmony.[125]

These, in Wölfflin's terms, "are nothing other than the
conditions of organic existence".[126] Wölfflin's ideas about architecture are, indeed, rooted in the view of buildings as analogs of
organic forms, above all, the form of the human body. The "un-
challengeable" demand for symmetry, he declared, is derived
from the arrangement of our bodies. "Because we are built symmetrically, we believe that we are entitled to demand this form
for architectural bodies, too. And this not because we regard our
species as the most beautiful but because it alone seems right to
us". That, he explained, is why we feel physical discomfort when
we encounter an asymmetric object; "it is for us as if the symmetry of our own body has been disturbed, or as if a limb were
mangled." By way of demonstrating this point, Wölfflin gives the
example of a cup that has only one handle. "Even without thinking about it", he claims, "we make the side with the handle be the

---

[123] *ibid*, p. 865.

[124] Wölfflin 2017. The best critical discussion of this work is Hart, 1981.
See also Mallgrave 1994, pp. 39 – 56.

[125] Wölfflin 2017, pp. 23-24, citing Vischer, *Kritische Gänge. Neue Folge*
(1860), Pt. V. Vischer had defined symmetry as "a juxtaposition of identical parts around a separate and dissimilar middle point", but Wölfflin
(p. 26) rejected this on the grounds that we do not regard forms that have
juxtaposed halves but lack a defined center as asymmetric. Later however (p. 39) he seems to contradict himself with the claim, "what is important is that [the] center stands out as dominant and thereby establishes the parts on either side of it as dependent" – a position into which
he forced himself by taking as his starting point (the fallacy) that the human body with its dependent arms and legs, is symmetric: cf., "the prominent center, not resembling the parts, represents the inner coherence that
is analogous to the construction of our own organization and that of every
animal" (p. 39, emphasis supplied).

[126] *idem*.

rear of the cup, so that the symmetry is preserved. But if there are two handles, the relationship changes again and we regard it as analogous to our arms".[127] This is, surely, a bizarre and unconvincing claim.

There is, clearly, much that is nonsense in this discussion. Where asymmetry is discrete, he goes on, it appears merely as "a displacement of balance". However, where it is pronounced, asymmetric design "obliges us to take note of each part individually and to regard the whole more as a fortuitous collection than as an organic amalgam".[128] In other words (by the logic of Wölfflin's argument) trees, with their branches and foliage, which some will consider to exemplify "organic amalgams", are rather in Wölfflin's view, mere fortuitous collections of forms!

Wölfflin declared that "we demand absolute symmetry in monumental buildings: a grave and measured bearing". As an architectural historian, however, he conceded that in the Middle Ages and Renaissance, Germans "thought otherwise". "They reckoned each part should function for itself in its own location, and seem not to have paid attention to the overall structure, which because of its disunity usually makes a lively rather than a worthy or serious impression on us. We tolerate such freedom only in private or rural buildings":

> An idiosyncratic need draws our times toward asymmetry in interior and decorative arts. The restfulness and simplicity of stable equilibrium have become boring; we emphatically seek movement and agitation, in short, the condition of disequilibrium; people no longer want pleasure, as Jacob Burckhardt once said, "but relaxation or distraction, and so the most formless or the most colorful are welcomed"... The modern taste for tall mountains, for the most powerful masses without rule or law, can also be traced back in part to a similar requirement.[129]

There seems to be no end to the challenge of understanding symmetry, and its place in aesthetics. Finnish-American architect Eliel Saarinen, (1873-1950), whom we met in Chapter One, had an unusual approach to the subject. He believed that the Parthenon and "almost every" other Greek temple was symmetric,

---

[127] *ibid,* p. 27.

[128] *ibid.,* p. 40.

[129] *ibid,* p. 40.

and that the same was true of the human body and the bodies of other animals: and indeed of "most of the flowers".[130]   He doubted however that Vitruvius understood *simmetria* in the modern sense, though he recognized that during the Renaissance and thereafter Vitruvius' authority was used to justify that meaning: "'symmetry' was put on a pedestal and ... became the guiding star of all building design".[131] It is clear that Saarinen did not approve of symmetric design *per se*. Frequently, he argued, symmetry "is artificial ... is apt to sterilize the living quality of form". It is unclear how this perfectly sensible observation can be reconciled with what we have just seen is Saarinen's belief that everything from the human body, most Greek temples and most flowers are symmetric...

In Saarinen's view, the "most essential property" which makes a building beautiful is not symmetry but balance. "If a building is lacking in balance it is lacking in that most essential property which might make it beautiful. The Parthenon has both symmetry and balance. That exquisite Erechtheion has a perfect balance, whereas symmetry would have deprived it of much of its intimate charm. And surely, symmetry would have meant a complete destruction of beauty in the case of Mont Saint Michel with its superbly balanced building masses."[132] "Beauty of balance", he stressed, "is the fundamental thought in all design, for the principle of balance has a universal significance ... in nature as well as in human art". Sometimes, "when the 'balance' line of design happens to coincide with the middle-line of symmetry", symmetry is "logical and thus beautiful". In all other cases, however, "symmetry is artificial".[133] In sum, then: "'balance' is the primary idea, 'symmetry' being its by-product – yet *only in those circumstances in which symmetry is essential for balance*. Naturally then, when we speak about beauty in this connection, we must speak about 'beauty of balance' rather than about 'beauty of symmetry'".[134]

The muddle over the meaning of the word "symmetry" has continued into modern times. In a chapter entitled "The Aesthetics of Proportion", Umberto Eco wrote that "mainly due to

---

[130] Saarinen 1985, pp. 282 - 285.

[131] *ibid.*, p. 283.

[132] *Ibid*, p. 286.

[133] *Ibid*, p. 284.

[134] *Ibid*, p. 285.

the influence of Vitruvius ... the concept of symmetry was very common" in medieval plastic art: but on the following page – this is a chapter about *proportion!*– he referred to double - headed eagles and two - tailed mermaids as examples of medieval symmetric design.[135] We may be confident that whatever inspired the creators of those designs it was not "symmetry" in the Classical sense used by Vitruvius when he set forth his ideas about proportions.[136] Nor is Eco the only modern scholar to mistake Vitruvius' concept of symmetry for bilateral symmetry. In their fine edition of Serlio's *Books*, Hart and Hicks cite Vitruvius, I, ii, 3 - 4 on symmetry to explain Serlio's "proportional correspondence".[137] Vitruvius of course used the term to indicate proportions, not the mirroring of a form's two lateral halves; and as we have seen it is clear from the context that it was in the latter sense that Serlio used "symmetry" when he referred to the necessity of balancing the postern on one side with a postern on the other. In any case, bilateral symmetry - we must not tire of emphasizing this! - is not a Vitruvian concept, and therefore Serlio could not have acquired it from Vitruvius. Another relatively recent instance of the confusion between the two concepts of symmetry is Wittkower's characterization of bilateral symmetry as "a primary aspect of proportion".[138] His definition of bilateral symmetry as "the balance of parts between themselves and the whole", is *sui generis* and not one that I think anyone else would recognize, except perhaps as a distant reflection of the ideas of Perrault that we discussed above. If anything, of course, Wittkower's formulation could be understood as a definition of harmonious or pleasing proportions, rather than of bilateral symmetry.

Aside from the confusion engendered by the two meanings of the term, "symmetry" in the sense of the mirroring of the right and left halves of a form remained an elusive concept well into the 19th century. A drawing that the architect Downing presented to readers of his *Cottage Residences* clearly shows an asymmetric structure (*fig.* 6.20).   Downing however, although he acknowledged that his design was not symmetric "in shape", nevertheless insisted that it was symmetric "in bulk and in the

---

[135] Eco 1986, pp. 39 - 40.

[136] Eco's supposition that Vitruvius enjoyed great influence in the Middle Ages is perhaps a rather idiosyncratic one.

[137] Serlio 1996 vol.I, p. 449 fn.72.

[138] Wittkower 1978, p. 123.

mass of composition". Fortunately, "symmetrically irregular", the term he devised for such structures, never caught on. In any case, as we see from the illustration, the distribution of the building's bulk is *not* equal, the wing on the right being both taller and broader than the one on the left. Downing predicted that this design would bring its owners "more intense and enduring pleasure" than a "regular" house.[139]

It took the better part of five hundred years for the meaning that Pope Pius II attached to the word "symmetry" to become generally accepted. Ironically, by the time this happened scholars had come to realize that there are in fact numerous distinct types of symmetry, to which they have given such names as "translational symmetry", "rotational symmetry", "helical symmetry", etc.[140] Thus "symmetry" in the sense introduced in the fifteenth century is now only one of several types of symmetry, and to be literally correct one should refer to it specifically as "*bilateral* symmetry". Most people nowadays, however, mean "bilateral symmetry" when they say "symmetry", and I have followed popular usage here.

Before concluding this Chapter, we should briefly consider the word "asymmetry", which, as we saw earlier, literally refers to a shape that is "without symmetry". An insidious aspect of these terms is their implication, not only that asymmetry lacks that which symmetry possesses (rather than, perhaps, the other way around) but also that symmetry is somehow the appropriate standard, the norm, while asymmetry is an aberrant – somehow defective or erroneous - version of it: a deviation from the norm.[141] (Perhaps it is this that explains why we sometimes hear

---

[139] Downing 1844, p. 3. Comp. Etlin 1994, pp. 81-2.

[140] Weyl 1952, pp. 41*ff*. We discussed modern science's very different meaning of symmetry as invariance (in Nature's operations) in Chapter One.

[141] Arnheim (1977, p. 36) has expressed as clearly as words allow the opinion (which he does not even *attempt* to substantiate) that symmetry is the norm from which asymmetry deviates. "Matter", he wrote, "is grouped symmetrically around the vertical axis unless intervening forces modify this simple equilibrium... We may say that what requires explanation about any particular shape is not its symmetry but its asymmetry". Similar preposterousness comes from Ekrami et al, (2018), who allege that, "Perfect bilateral symmetry [in humans] is the optimal outcome of the development of bilateral traits in the absence of developmen-

that a shape is "almost symmetric" but never that it is "almost asymmetric".) In view of what we have seen in previous Chapters to be the overwhelming predominance of asymmetric forms both in Nature and in human artifacts, it is surely the presence of symmetry and not its absence that ought to seem remarkable and set us in search of an explanation. In the real world it is asymmetric shapes, not symmetric ones, that constitute the norm!

(This is perhaps as good a point as any to return to the question of why the tendency to see asymmetric shapes as symmetric is not matched by the opposite tendency, which would be to see symmetric shapes as asymmetric – an error that is just about unheard of. Perhaps the best answer we can give is that those who are invested in the importance and value of symmetry inevitably prefer to see asymmetry as seldom as possible. Their commitment to the fallacy that symmetry is (just about) universal encourages them to see asymmetric forms as symmetric; but it also discourages them from seeing symmetric forms as asymmetric. Their pertinacity prevents them from recognizing that they are fighting a losing battle!)

The negative connotation of the word asymmetry is apparent in the earliest known use of the term in English, when "symmetry" was still used to refer to good proportions.[142] It was with this sense that John Evelyn used it when he associated "asymmetrie" with the "want of decorum and proportion" ex-

---

tal perturbations." It is also still quite common for symmetry to be regarded as normative in an aesthetic sense, as in *Roget's Thesaurus'* equation of symmetry with "shapeliness, finish, beauty"; or the statement in the *Encyclopedia Britannica* on - line edition that "symmetry in nature underlies one of the most fundamental concepts of beauty". Weyl (1952, p. 16) associated symmetry with normative values in another way, suggesting that symbols of justice and everlasting truth are "naturally" presented in symmetric, frontal view. This, he says, is the reason why "public buildings and houses of worship, whether they are Greek temples or Christian basilicas, are bilaterally symmetric". But he is factually incorrect. Some of the most notable "public buildings", such as the Curia Julia or Senate building, in Rome, the Palazzo della Signoria and the Bargello in Florence, the Palazzo Publico in Siena, or the Doge's Palace in Venice, have asymmetric fronts. So, among many others, have the Parthenon, the Pantheon, the basilica of St Mark in Venice and the cathedral of Notre Dame in Paris.

[142] In fact, asymmetry was used negatively, as an antonym of "harmony" – here, for sickness, weakness and ugliness - as early as the 6th century by John Philoponus of Alexandria in his *Commentary on Aristotle's de Anima 145.1-4*, Suda Adler number: alpha 3977.

emplified by Gothic buildings, which he thought of as "fantastical and licentious" and lacking "any just proportion".[143] A few years later Boyle, in his essay "Discourse of things above reason"(1681) contrasted "harmonious" truths with others that are "not symmetrical", and went on to refer to the latter as "asymmetrical or unsociable" – obscure terms in this context, to be sure, but clearly not intended as laudatory.[144] Asymmetry continued to be used for that which is unharmoniously proportioned and irrational into the nineteenth century.

It was not until late in the nineteenth century that "asymmetry" started to be used without reference to proportion but as the antonym of "symmetry" in the sense of "bilateral symmetry". But just as the positive connotation of the older meaning of symmetry was absorbed into the new meaning of the term, so too did the negative connotation of asymmetry as the antithesis of symmetry-as-proportion become absorbed in some measure into the new meaning of asymmetry. We have seen that Montesquieu likened an asymmetric building to a deformed body. For Wölfflin the mere sight of an asymmetric form was "for us as if the symmetry of our own body has been disturbed, or as if a limb were mangled."[145] Echoing Wölfflin on this point, too, Wittkower declared that an asymmetric human body (as if there are any that are symmetric!) "evokes reactions such as pity, irritation, or repulsion".[146] That, of course, is nonsense!

Appendix: Alberti's *Collocatio*

Leon Battista Alberti's *collocatio* is noteworthy as the earliest extant formulation of both the idea and the concept of symmetry, and also because Alberti regarded it as one of the three

---

[143] Freart 2017, p. 8 (i.e. Evelyn's introduction to Freart). Friedman (1998, 153*ff*), incorrectly assumes that Evelyn used "symmetry" and by implication, its opposite, in the modern sense.

[144] "Discourse of Things Above Reason" in Boyle 1979, p. 235.

[145] Wölfflin 2017, p. 26.

[146] Wittkower 1978, p. 123.

essential components of good design.[147] With rare exceptions,[148] historians however have either (i) ignored *collocatio* altogether, or (ii) minimized its scope, or (iii) characterized it in terms that, of doubtful meaning in themselves, bear absolutely no resemblance to anything Alberti ever wrote. [149]

---

[147] Alberti 1966a, Bk. IX, cap. 5 – *"praecipua esse tria haec in quibus omnis quam quaerimus ratio consumetur numerus, et quam nos finitionem nuncupabimus et collocatio"*. The requirements of *collocatio* must be attended to at the very outset of work on a new project: *"quare in primis observabimus ... ita ut muto dextra sinistris... aequatissime conveniant"* (*ibid.*, Bk. IX, cap. 7.) Alberti concluded his discussion by declaring that he had now shown what beauty is and what it consists of: *"itaque et quidnam sit pulchritudo et quibus constet partibus..."*

[148] Oechslin (1985) is one of these exceptions. He declared, without reservation, that "Alberti set forth the rules of mirror-reflection [i.e. bilateral symmetry]."

[149] (i) e.g.Wittkower 1949, etc; Borsi 1986; Burckhardt, 1985. Heydenreich (1996, p. 44), refers to "the fundamental laws of architecture contained in ... *collocatio*": but then says nothing further about it. It is curious to note Wittkower's skittishness about discussing, not just Alberti's *collocatio*, but symmetry in general: "Renaissance architects", he wrote, (*ibid*, p. 70), "always regarded symmetry as a theoretical requirement, and rigidly symmetrical plans are already found in Filarete, Francesco di Giorgio and Giuliano da Sangallo". It is curious that Wittkower did not mention Alberti, or *collocatio*, in this context, nor explain what the theory in that "theoretical requirement" may have been. He added, though without explanation, that "in practice this theory was rarely applied", a statement that will probably startle most people who are familiar with Renaissance architecture.

(ii) Kruft (1994, pp. 46 - 7) mistakenly limits *collocatio* to architectural design. Poeschke (2008, p. 186) described *collocatio* as *"die Symmetrie eines Gebäudes als spiegelbildliche Anordnung seiner Teile"* adding that it *"erhaelt sonnst quasi den Rang eines Naturgesetzes"*, an accurate definition as far as it goes though also mistakenly limiting *collocatio* to architecture. Gadol too (1969, pp. 109 - 110), with her definition of *collocatio* as "architectural symmetry in the modern sense", limits *collocatio* to architecture. Perhaps echoing Wittkower, and also without explanation, she adds that Alberti "did not intend his remarks [on *collocatio*] to be carried out": a curious claim, curiously worded.

(iii) Hersey (1977) characterizes *collocatio* as "a question of choice and distribution". Rykwert's two shots at defining the term both fall short. In Alberti 1965 (p. 252) he translated *collocatio* as "relation", while in Alberti 1988 (p. 422) he rendered it, no less vacuously, as "decisions that determine the arrangements of a building". Tavernor (1998, p. 46), solemnly intones that *collocatio* is "what would be called planning

Some of the blame for this surely rests with Alberti himself. His presentation of the concept is often turgid and inept, and is scattered haphazardly (and without always being identified by name) in different sections of *De re aedificatoria*. He began his discussion of it with the oracular pronouncement that *collocatio* pertains to "*situm et sedem*" ("siting and seating"), whatever that may mean; and then added that it is easier to see when it is done poorly ("*ubi male habita est*") than to indicate rules for doing it well ("*quam intelligatur per se qui decenter ponenda sit*"). It is doubtful, he went on to say, whether rules can help a person meet the requirements of *collocatio* if he lacks innate good judgment ("*ad iudicium insitum natura animis hominum*").[150]

These remarks might lead one to suppose that *collocatio* is a subtle concept: indeed, one that can be grasped only by people with superior minds. Yet it is nothing of the sort. In Alberti's own telling of it, in fact, *collocatio* is a very simple notion that can be understood at once by just about anyone. It consists of two elements. One is descriptive: it is the idea of bilateral symmetry – the precise mirroring of the appearance of the left and right halves of a form. The other is prescriptive, or normative: it is the axiom that if we want the things we make to be beautiful then we must make them symmetric.

Alberti had little to say about the first – the descriptive – element of his construct other than to emphasize that with *collocatio* the mirroring of the left and right lateral halves of a shape is exact, so that every feature of one of the halves is accurately mirrored by the other.

He discussed the prescriptive element at greater length. The ancient Greeks, he claimed, learned what beauty is by studying Nature's forms; and because those forms are invariably symmetric ("*dextra sinistris ... convenirent*"[151]), Greek architects always took great pains to ensure that, in whatever they made,

---

today", but with the proviso that its results be "such that man, society and cosmos are in complete harmony". This stipulation marks another appearance of Tavernor's – seemingly incurable - addiction to extreme nonsense; for, if it were adopted, it would cause the permanent cessation of all construction work everywhere! See also Pelt and Westfall 1993, p. 275; Mateer, 2000, p. 48; Pennick 2012, p. 67; and Luecke 1994, p. 82; Summerson 1963, p. 36.

[150] Bk. IX, cap. 7.

[151] Alberti appears never to have stated explicitly that by *convenire* he meant "mirroring", and not "duplicating" (he may not have been aware of the difference between the two) but his designs, such as the facades of

even the smallest details (*"in minutissimis"*) on one side were matched precisely (*"exactissime"*) by those on the other. The imperative of imitating Nature's symmetry, Alberti declared, was also obeyed by artists, who understood that in their statues, pictures and ornaments all details must appear as twins (*"gemella videantur"*).[152]

We are invited to believe, therefore, that *collocatio* is the restatement of a long-forgotten principle of Classical art. Alberti claimed that he discovered it by carefully examining "every building of the ancients, wherever it might be, that has attracted praise".[153] This is an implausible claim on the face of it, and is not made less so by the fact that Alberti – unlike Serlio and other architectural writers in the next century - did not publish any details of his research, or even name the buildings he examined; no drawings or notes he may have made in connection with those studies are known.[154] There is indeed no evidence that Alberti ever inspected an ancient Greek building, not even the temples at Paestum which are only 65 miles or so south of Naples. [155] The only independent knowledge we have of Alberti's familiarity with Classical remains is a remark by Pius II, who referred to him as "a scholar and a very clever archaeologist", and quoted a report from Alberti about his unsuccessful attempt to raise the Emperor Caligula's boats from the bed of Lake Nemi.[156]

---

the churches in Rimini and Mantua, make no other understanding of the term possible.

[152] This confirms my point that *collocatio* is not limited, despite what Krufft and others have stated, to architecture alone.

[153] Alberti, 1966, VI:1, 3, 6. Notwithstanding its title, Alberti's *Descriptio Urbis Romae* is a guide to the mapping or surveying of the city, and does not describe in detail any of the structures within it.

[154] The only ancient structure he described by name in *De re aedificatoria* (IV: 3) were the ruined walls of Antium Anzio. His comments about it are limited to the banal statements that the walls followed the winding line of the coast and that they must have been very long. Grafton (2000, pp.233, 254-6) says Alberti "presumably" and "very likely" made notes and drawings of ancient buildings, but these are empty claims, and he himself admits (p.255) that "the documents do not state as much".

[155] Kunst (n.d.) properly rejects the notion that the ruins of Paestum were only discovered in the middle of the 18th century.

[156] Pius II, 1962, p. 316. See also Biondo 2005, pp. 188 - 192.

But even if we suppose for the sake of argument that Alberti *had* examined all the buildings that he claimed to have examined, his view that symmetry is a principle of Classical art and architecture would not be tenable. For despite what Alberti - and so many after him - supposed, symmetry (as we saw in Chapter Three) was known in the ancient world, if it was known at all, only rather briefly and as an esoteric doctrine shared by a small number of people: even Euclid appears to have been ignorant of it. The extreme rarity (if not complete absence) of symmetric design in the surviving physical relics of the ancient world bears this out. It is simply not credible, therefore, that any of the ancient buildings Alberti examined – assuming that he actually *did* examine any – were symmetrically designed. Alberti also claimed to have seen perfectly "twinned" statues. This too is a baffling report, for Alberti did not identify any of these statues, and none like them appear in the standard work on Classical sculptures known in the Renaissance.[157]

Alberti's claim that the Ancients regarded Nature's forms as symmetric can also be dismissed. There is no literary evidence to support this claim, and natural forms are not shown as symmetric in surviving works of art from the Classical era. In reality, of course, Nature's claims are *not* symmetric. Indeed,

---

[157] Bober and Rubenstein (2010). (The Warburg Institute's *Census of Antique Works of Art and Architecture known in the Renaissance* is now available in a searchable database at www.census.de.) The reliefs on Trajan's Column and the Arches of Titus and Constantine, would have been known to Alberti, but they too do not uphold his claims about the symmetry of Classical sculpture. Alberti's younger contemporary, Mantegna (1431 - 1506), whose "definite endeavour" it was, according to Saxl (1979, p. 69; similarly Clark 1981, p. 122) to represent ancient history "with archaeological faithfulness", clearly did not think that Roman architecture was symmetric. In his *Saint James addressing the Demons* (Tietze - Conrat, 1955, pl. 9) the saint is standing under an arch whose interior is decorated with irregularly-shaped and spaced coffers. The triumphal arch in *Saint James before Herod Agrippa* (*ibid,* pl. 11) is almost astonishingly asymmetric (the right-hand side of the architecture has *not* been cut off. Note the much narrower – yet more broadly-fluted – column on the right). In *The Martyrdom of Saint Christopher* (*ibid,* pl. 14) the large brick house has markedly asymmetric fenestration and its eaves are of different lengths (the one on the right is shorter). Note too that the funerary monument in pl. 14 is set lower than the one in pl. 15 in which we see the other half of the house's façade. Obviously, then, Alberti and Mantegna had very different views of Classical architecture, at least in regard to its asymmetry.

it seems as if Alberti himself inadvertently acknowledged as much. In a curious passage, he remarked that the resemblance between the right and left halves of some statues he had seen was greater than one could expect to find in Nature's own creations, in which "we hardly ever see one feature so exactly like the other" (*"in cuis operibus ne nesum quidem naso similem intueamur.*) The requirement of *collocatio*, it will be remembered, is that the two lateral halves of a form must mirror each other precisely – and it is from this standard, evidently, that according to Alberti himself Nature falls short. If Nature's forms are not symmetric, it must follow that the Ancients , diligently imitating Nature, would have created forms that are also not symmetric.[158] To be sure, this is not an implication that Alberti acknowledged.

Separately, Alberti also argued that any combination of two disparate shapes is inherently offensive. We would be repelled, he wrote, by the sight of a dog that had one ear like that of a donkey, or of a man who had one hand or foot that was much larger than the other: or again, by the sight of a horse that had one gray and one black eye. Whether asymmetric arrangements are usefully characterized by such images may be doubted. Moreover, as he presented it, Alberti's point is to deplore the asymmetry of the dog's two ears, not the inappropriateness of one of them. Does this mean that we would not be repelled by the sight of a dog whose two ears were – identically – shaped like those of a donkey?

In explaining why we should make our artifacts symmetric, then, Alberti can do no better than to say that that is how Nature makes things, which is untrue; and how things were made in ancient Greece and Rome, which also is untrue; and that asymmetric forms cannot be beautiful, which is not only a matter of uninformed opinion, but one that is contradicted by the undoubted pleasure we get from looking at many asymmetric forms, whether natural or man-made, whose loveliness, indeed, is very often enhanced by their asymmetry.

Alberti's rationale for the symmetry norm thus rests on the ineptest of arguments. Yet it is also the case that the advocacy of symmetric design has not climbed to a higher intellectual level

---

[158] Wotton (1958, p. 94), ever the lover of paradox, cited Quintilian to justify his remark that sometimes the imitation of nature can lead to results that are "too naturall".

since his day – I am thinking, for example, of Montesquieu's argument, which we noted earlier, that we enjoy symmetric designs because it is easier for us to comprehend them.

Although Alberti probably completed his architectural treatise some years before Pius built his palace in Pienza, we have no reason to believe that it was from Alberti that the pope acquired his knowledge of the concept of symmetry. Indeed, we may infer from the fact that Pius used *simmetria* and not *collocatio* for the concept, (as well as from the likelihood that the reason Pius did not explain the term *simmetria* was because he believed his readers would be familiar with it) that another version, or perhaps even several other versions, of the concept of symmetry, in addition to Alberti's, may have been current in the middle of the fifteenth century. It is also probable that Francesco Colonna, author of *Hypnerotomachia Poliphili*, derived *his* knowledge of the concept from another source than either Alberti or the pope. What that source may have been it is impossible for us to determine. We are unable to trace the origins of the concept any further back than the mid- to late- fifteenth century writings of these three men.

And just as we cannot attribute the origin of the concept of symmetry to Alberti, so too are we unable to attribute its subsequent history to his influence. There is little or no evidence of the influence of *collocatio* either during Alberti's lifetime (he died in 1472) or later. Not a single writer before the twentieth century is known to have referred to Alberti's views on bilateral symmetry; and it is significant that no one else ever adopted his term for it. Perhaps nothing illustrates more vividly Alberti's lack of influence over the subsequent history of the concept of symmetry than the fact that although Wotton owned an exceptional copy of *De re aedificatoria* (it had belonged to Alberti, and had his annotations) it was not Alberti but Vasari whom he cited as the source of his ideas on the subject.[159] If anyone played an influential role in propagating the idea of symmetry, that person was probably Serlio, whose books, popular in France and the Low Countries, may well have been the medium by which the concept of symmetry first reached beyond the Alps.

The historical significance, if any, of Alberti's *collocatio* does not lie in its influence, accordingly, but in the fact that it is

---

[159] Wotton 1968, p. 117. Wotton, who became provost of Eton, gave his copy of *De re aedificatoria* to the Eton College library, on whose shelves it is still to be found.

the earliest surviving statement of the idea of symmetry, as well as the earliest surviving attempt to formulate a rationale for the concept of symmetry.

It may be useful briefly to consider Alberti's influence and importance more generally. Writers in the first century or so of the Italian Renaissance are restrained in their references to him. Manetti mentions him merely as someone who "gave precepts" about the Classical manner of building.[160] Biondo characterized Alberti as "*geometra nostro tempore egregius*" and as the author of "*elegantissimos … libros*" on architecture, i.e., the *De re aedificatoria* - far from fulsome words, but rather more so than those of Vasari who, in his *Life* of Alberti, criticized much of his architectural work, and commented that writers on architecture receive more attention than architects, even though the latter may know more about the subject than the former![161] Palladio listed Alberti among "other excellent writers who came after Vitruvius" but cited him only on a few relatively minor matters, such as the design of roads.[162] Only Serlio pulls out the stops. Alberti, he writes, was "expert in both theory and practice" and he was "greatly praised as an architect". (His admiration for Alberti did not lead him, however, to adopt *collocatio* as his term for symmetry. As we saw earlier, he chose instead to refer to it as *simmetria*.) The Tudor polymath John Dee praised Alberti's "Arte" for its "Mathematicall perfection" without however stating of what its perfection consisted.[163] Alberti's fame reached its peak in the 17th century. Freart referred to him as "the one who has particular esteem above the rest", though he referred to Palladio, Scamozzi, Serlio and Vignola far more frequently and on several occasions criticized Alberti's judgement.[164] Evelyn's enthusiasm for Alberti was unqualified. No lover of architecture, he wrote, "does not greedily embrace all that bears the name of Leon Battista Alberti".[165]

For the next two centuries we hear almost nothing about Alberti. More recently, however, the view has gained ground of Alberti's towering influence and achievement – an axiom that

---

[160] Manetti 1970, pp. 55, 1. 384.

[161] Biondo 2005, v. I, p. 191.

[162] Palladio 1570, p. 6.

[163] Dee 1570, pp. 32-33.

[164] Freart 2017, p. 55.

[165] *ibid*, p. 181. But the work by Alberti that he notices in the Fréart volume is not *de re aedificatoria* but, *de statua*.

scholarly slavishness does not challenge - and he has been the subject of a seemingly endless outpouring of favorable but not altogether plausible treatises. Indications of the reluctance of modern scholars to acknowledge Alberti's relatively minor stature and influence are not hard to come by. Saalman, for example writes of the "profound" influence of Alberti's work on Manetti's *Life of* Brunelleschi, but soon acknowledges that Manetti is "in as profound a contrast with Albertian principles and practice of architecture as was Brunelleschi himself".[166] Similarly, Panofsky in his great essay on the concept of the Renaissance, declares, but without giving evidence for it, that Leonardo da Vinci and Dürer were among "Alberti's followers" and then seemingly contradicts himself by acknowledging that Alberti's "method of determining and recording human proportions was, and even remained for some time, unique" (we are not told when or under what circumstances the period of its uniqueness ended)[167]. Rykwert and his colleagues, in their edition of *De re aedificatoria*, declare that "Alberti's contemporaries accepted *De re aedificatoria* as a model of learned Latin writing immediately", but offer no evidence to support their claim, whether we understand it to refer to Alberti's command of Latin prose or to the substance of his book.

Equally, Weil-Garris and Damico, writing of the influence of Alberti as well as of Vitruvius on Cortesi, cope with the fact that the evidence for that influence is not really there by allowing themselves to say: "In the traditional way Cortesi uses sources without identifying them. It is instructive that the names of Vitruvius and Alberti are never mentioned"![168] Boase, also

---

[166] Manetti 1970, pp. 28 - 29.

[167] Panofsky 1972, p. 27. Krautheimer (1956, p. 251 - 253) states rather dubiously that it is an "inevitable" conclusion that the perspective construction which Alberti presented in *Della Pittura* was applied "verbatim" by Ghiberti in the *Isaac* and *Joseph* panels of the *Porta del Paradiso*. He dates those panels to the very year in which *Della Pittura* was published, and then notes that in the later panels Ghiberti "slipped out of Alberti's perspective system as suddenly as he had slipped into it". Pope-Hennessy (1980, pp. 39-70), in his essay on the occasion of Ghiberti's sixth centenary, also cites the influence of *Della Pittura,* though he dates the panels that supposedly reflect that influence to years that precede the book's publication in 1435.

[168] Weil - Garris and Damico, 1980. I am tempted to ask whether it is "instructive" that the name of Donald Duck, too, is not mentioned!

without documentation, claims that "Vasari seems to have absorbed Alberti's theories" but acknowledges that there is "little direct trace" of them in Vasari's works.[169] Clark claims that Uccello, Piero della Francesca and Leonardo da Vinci were greatly influenced by Alberti's *della Pittura*, but acknowledges that the evidence for this too is slim.[170] Most recently Eriksen asserts Alberti's "pervasive" influence on Wotton while acknowledging that the latter refers to him "only occasionally"[171]. Dissenting from the academic consensus is Krufft, who remarks that Alberti "never attained the widespread impact that Vitruvius, Serlio and Vignola enjoyed for centuries".[172] Consistent with my opinion that Alberti's influence is a shallow modern contrivance, it is perhaps worth noting that the statue to him in the church of Santa Croce was erected no more recently than at the beginning of the 20th century, and with funds contributed by a member of the Alberti family. The scholarly "Societe International Leon Battista Alberti" seems now to be defunct; its journal *Albertiana* has not appeared since about 2014.

---

[169] Boase 1979, p. 53.

[170] Clark 1983, pp. 100- 105.

[171] Eriksen 2010, p. 16.

[172] Krufft 1994, p. 49. The distinguished German art historian Julius von Schlosser (1866 - 1938), in his *Ein Kuenstlerproblem der Renaissance – L. B. Alberti* (1929) dismissed Alberti's architectural work as "pedantic" and even as "a masquerade". Alberti was the forefather of the Renaissance, Schlosser writes, and the first theoretician of the visual arts; otherwise he belongs in the shadowland of 'antiquarian' research" (*"ohne diese seine Wirksamkeit wandelte er heute wohl gänzlich im Schatten der 'antiquarischen' Forschung"*. Something in us, he adds, opposes placing him among the great figures of his time, even in his own specialty [architecture], where he was far less original than some suppose. See also Grafton, 2002.

CHAPTER SEVEN
THE NATURAL GARDEN
IN ENGLAND

7.1 High Wycomb park (designed by Repton)

"The shift to Nature in the form of landscape gardens sprang
from an eagerness to break free from symmetry."
- Max Friedländer (1969, p. 29)

"Nor is there any Thing more shocking *than a* stiff regular Garden*;*
where after we have seen one-quarter thereof, the very same
is repeated in all the remaining Parts, so that we are tired instead
of being further entertain'd with something new as expected."
- Batty Langley (1728),p.iv.

In England at the beginning of the eighteenth century the formal garden of the Renaissance began to give way to what became known as the "natural" garden, sometimes also as the "English" garden.[1] The change was by any standard revolutionary. For two centuries gardens had been characterized by stark geometric shapes, artificial contrivances and symmetric arrangements (*figs.* 6.5 – 6.7). Now informal, free-flowing, design became the norm, along with an emphasis on variety of texture, color and form; intimate vistas; the discretely contrived interplay of light and shade; and unabashed visual complexity (*figs.* 7.1, 7.2).[2]

---

[1]"Natural garden" begs the question, to be sure, of how "natural" anything that is designed and cultivated by humans can be. In the Aristotelian view (*Physics,* 192:68; Hardie and Gaye trans.), which enjoyed much currency during the Renaissance and for long thereafter, "art partly completes what nature cannot bring to a finish, and partly imitates her". Spenser (*Faerie Queene* Bk. 4, Cant. 10) expressed the thought thus: "All that Nature did omit, Art (playing Nature's second part) did supply it". Comp. Shakespeare's paradoxical, "This is an art which does mend nature, change it rather, but the art itself is nature" ("Winter's Tale", Act 4, Sc. 4). In other words: Nature is fulfilled, not falsified, when she is improved by Art. Sir Joshua Reynolds (1997, Discourse XIII) however insisted that gardening "is a deviation from Nature".

[2] Comp. Pevsner 1964, p. 174: "The English garden … is asymmetrical, informal, varied, and made of such parts as the serpentine lake, the winding drive and winding path, the trees grouped in clumps and smooth lawn (mown or cropped by sheep) reaching right up to the

The natural garden, for all that it was meticulously designed, gave the impression - *non murato ma veramente nato*[3] - that it stemmed from Nature's plan and not, like the Renaissance garden, from the will of Man. Here Man was Nature's patient and enthusiastic collaborator - perhaps merely her helper - and certainly not her conqueror.

The spirit of Renaissance horticulture, that had regarded Nature as a threat and aimed at her subjugation, was nowhere evident.

We suggested in Chapter Six that the Renaissance gardener's belligerence toward Nature was a response to the deadly plagues that ravaged Europe for more than three centuries. Only an end to that terrible threat, we may speculate, could allow a more benign perception of Nature to take hold, and to be manifested in horticulture. And indeed, giving substance to the thesis of these pages, *the emergence of the natural garden, and the radically changed view of Nature that made it possible, coincided with the end of the era of the plagues.*[4] It was, as the great art historian Max Friedländer noted (in the motto of this Chapter), a shift back to Nature.

The change can be observed in the works of two seventeenth-century contemporaries. We have already referred to Charles Cotton (p. 288f.) Cotton's dyspeptic remark in *Wonders of the Peake* (i.e., the Peak District in northern England, through which the Derwent River flows) that the beauty of Chatsworth's gardens was achieved, not *by* Nature but *despite* her, places him and those gardens in a familiar Renaissance context.

What helped form Cotton's perception is suggested, I believe, by the chilling metaphors, drawn from the lexicon of disease, that he used to describe the valley of the Derwent and,

---

French windows of the house", ratherlike the features that are clearly seen in the frontispiece to this chapter, *fig.* 7.1.

[3] "Not built but, in truth, born". This felicitous phrase is Vasari's. He coined it, however, for the Palazzo Farnese in Rome (*fig.* 6.4), a structure in whose decidedly *murato* appearance it is impossible to discern even a hint of the *nato*.

[4] White (2014, p. 311) is one of the few to have recognized the connection between the two. "[When] the threat of the plague was finally lifting" he writes, "only then were people ready to burst through the walls around their gardens and invite wild nature in. The landscape gardens of Capability Brown ... were *horti inconclusi,* embracing and celebrating the reality of nature, no longer a defense against Nature's real world".

by extension, Nature herself.[5] They bring to mind the fact that Cotton wrote the poem a scant fifteen years after the Great Plague of 1665-1666. He would have known that in the Peak District's "plague village" of Eyam, which quarantined itself for 14 months during the Great Plague, only 83 of 350 inhabitants survived the outbreak (in which about one-fifth of London's population also perished).

William Temple was born in 1628, two years before Cotton; his essay, "Upon the Gardens of Epicurus" appeared in 1685, four years after the publication of *Wonders of the Peake.* Nevertheless, the difference between the two works is fundamental: it is the difference between two eras' perception of Nature. For Cotton, Beauty and Nature were antithetical; for Temple, Beauty was the happy outcome of Man's cooperation with Nature. (His essay, indeed, would play a role in the transition to the new style of gardening.) Cotton's poem seems haunted by the catastrophe of 1665-1666, but nothing Temple wrote so much as hints at it. Neither Cotton nor Temple could know, of course, that with the Great Plague the era of plagues in England had at last come to a close.[6] But that knowledge, completely absent from Cotton's verse, is implicit in the fearless embrace of Nature that we find in Temple's essay.

The end of the era of the plagues encouraged people to acquire a new view of Nature and, with it, freed them to

---

[5] For Cotton the Derwent was a "blue scrofulous scum"; and the land through which it flowed was "so deform'd" that it resembled nothing so much as "impostumated boyles", "warts and wens" - and even "Nature's pudenda"! Not everyone saw the Peak District in this way, however. In 1636, only 8 years after the plague killed his patron the Earl of Cavendish, Thomas Hobbes published a lengthy poem *De Mirabilibus Pecci* – i.e., "the wonders of the Peak [District]"- which extolled the pleasure to be had from exploring natural landscapes: "*Quam dulce est, inter circumque nitentia stagna/Insternete vias, aestivâ semper, arena/Discipulum memet naturae tradere rerum*" (rendered somewhat boldly in a translation of 1678: "How sweet it is upon the Sandy shore/of Crystall Pooles, great Nature to explore".) Cotton's scrofulous Derwent is the very one that, a little more than a century later, Wordsworth would celebrate as "the fairest of all rivers".

[6] Defoe reverted to the earlier epoch with his *Journal of the Plague Year* (a fictional account of the Great Plague), published in 1722. Mullett (1936) suggests that Defoe may have written this work in the expectation that the outbreak of the plague that was then devastating Marseilles would soon spread to England. It turned out however that the Marseilles outbreak was the last to afflict western Europe.

recognize the visual, intellectual and emotional barrenness of the Renaissance garden. What brought them to the vital, free-flowing forms of the natural garden, in particular, was a re-awakened enthusiasm for the aesthetic principle of variety, a principle that had been consciously negated during the Renaissance by the prevailing dogmatic insistence on simplicity and symmetry in design.[7] The principle of variety, indeed, would become one of the reigning ideas of eighteenth-century horticulture. "All the beauties of gardening", Spence recorded Alexander Pope as saying, "might be comprehended in one word, variety".[8]

Its rediscovery began as early as the end of the sixteenth century, when variety, in the form of disorder and disharmony, came to be extolled as an instrument for providing the contrasts that heighten a person's pleasure in beautiful things.

We find, for example, "E. K.," in his dedicatory epistle to Spenser's *Shepeardes Calendar* (published in 1579), remarking that we are often "singularly delighted with the shewe of naturall rudeness, and take great pleasure in ... disorderly order". For "E. K." the show of disorder was not valued in itself, however, but because it acted as a foil to, and thereby enhanced our pleasure in, what he called "the daintie lineaments of beautye" that were the true *raison d'être* of a garden. As he explained: "So oftentimes a dischorde in Musick maketh a comely concordance: so great delight tooke the worthy Poete Alceus to behold a blemish in the joynt of a wel shaped body". Or as Spenser himself noted in *The Faerie Queene* (1590-1596), "dischord oft in Musick makes the sweeter lay". [9]

Sir Henry Wotton, in *The Elements of Architecture* (1624), records how the principle of variety inspired the owner of Ware Park to heighten the contrast between flowers in his gardens:

> [He] did so precisely examine the tinctures, and seasons of his flowres, that in their setting, the inwardest of those which were to come up at the

---

[7] Ogden (1949) states that "aside from the principle of decorum there was probably no aesthetic principle ... regarded as more important" during the sixteenth and seventeenth centuries. Humphreys however (1937, pp. 39-69) finds the precursors of the natural garden in a revolt against "the sheer boredom" of the formal garden, along with a number of other factors, of which delight in Nature's "infinite variety" is only one.

[8] Spence 1964, p. 48.

[9] *The Faerie Queene* III. ii. 15.

same time, should be always a little darker than the outmost, and to serve them for a kinde of gentle shadow, like a piece not of Nature, but of Arte. [10]

At about the same time there also came into fashion the idea, which can be traced back to Pliny,[11] of including an area of wilderness in an otherwise formal garden. The main part of the garden, Bacon wrote in 1625, "is best to be square ... not too busy or full of work" – in other words, it was to be laid out in the formal fashion – but fully one-third of the garden's overall acreage should be given over to what Bacon called a "heath". This was to be "framed, as much as may be, to a natural wildness", and with flowers strewn all over it, "here and there, not in any order". It appears, though, that Bacon's own application of this concept was quite tentative, for according to Dutton his wilderness was "a phantasy of meretricious artificiality" that would have evoked "the withering scorn" of proponents of the natural garden.[12]

More than a century later the idea of antithetical arrangements within a single confine remained current, as we can see from the epistolary novel, *Felicia to Charlotte* by Mary Collyer, that was published in 1749. Writing in breathless style to Charlotte, Felicia reports that one side of her garden in the country exemplifies "the beauty of art" - "The hedges, that are on each side of the principal walks, are formed of ever-greens, resembling walls, adorned, at proper distances, with pilasters which, with eternal verdure, branch into all the decorations of architecture *etc. etc.*" On the other side of the garden by contrast there stands "the triumph of nature"- "The uncultivated wildness, which pleases without method and without design, charming most where the easy confusion and agreable [*sic*] disorder render art superfluous and labour vain."[13] One gets the

---

[10] Wotton 1968, pp. 110-111. For Wotton, here, it is in art rather than in Nature, that variety is to be found.

[11] Pliny (1952, vol. 1, p. 390): "*et in opere urbanissimo subita velut illati ruris imitatio*".

[12] Dutton 1950, p. 50. For a detailed description of one "wilderness", that bears out Dutton's remark, cf. Malins 1966, p. 13. For a charming account of the curious and complex history of another wilderness - the mound of New College, Oxford's gardens - see Robin Lane Fox, "Mound of the Muses", *Financial Times* (London) 18 January, 2013.

[13] Collyer 1749, pp. 38-9. The term "agreeable disorder" came originally from *The Villas of the Ancients* (Castell 1728, p. 27), as the culmination of a sequence that led from "E. K. 's" "disorderly order" in 1579 to Wotton's

distinct impression that for Miss Collyer the difference between
the two sections was not a device to call attention – by contrast -
to the fineness of the cultivated garden but was welcomed for its
own sake.

In fact, from the outset we find instances in which
variety was celebrated for its own sake, its complexity valuable
inherently, rather than for the contrasts it provided.

The earliest example of the appreciation of variety that
Ogden cites is one such instance. It occurs as early as 1541 in the
statement by Coverdale that God made the grounds of the
Garden of Eden "not like on every side, but in many places set up
pleasantly".

The suggestion here is evidently that it was the
variations in the garden that made it pleasant. Ogden also cites a
French work, the English translation of which was published in
1594, that celebrated the fact that the world is filled with "things
of dislike and contrarie qualitie". According to the author of this
work, "nature is so desirous of contraries, making of them all
decency, and beautie; not of things which are of like nature". In
somewhat the same vein, Ogden notes the remark by Henry
Peacham in 1606 that he found a certain landscape painting "most
pleasing, *because* [my emphasis] it feedeth the eye with varietie".[14]
The supreme expression of pleasure in the diversity of Nature's
forms, colors and textures – and not altogether coincidentally, the
earliest recorded attack on the formal garden – is the blind
Milton's vision (1667) of the first garden, with its boundless
glories "powr'd forth profuse":

> … from that Saphire Fount the crisped Brooks,
> Rowling on Orient Pearl and sands of Gold,
> With mazie error under pendant shades
> Ran Nectar, visiting each plant, and fed
> Flours worthy of Paradise, which not nice Art
> In Beds and curious Knots, but Nature boon
> Powr'd forth profuse on Hill and Dale and Plain
> Both where the morning Sun first warmly smote

"delightful confusion" in 1624, and to Pope's "harmoniously confused"
(1713: Windsor Forest). The less felicitous "easy confusion" appears to
have been of Collyer's own devising.

[14] There seems to have been only a very moderate degree of "varietie" in
this garden, which evidently was composed of a number of zones, each
with a distinct *formal* design. It was in other words a collection of small
formal gardens joined to each other in a rectangular grid.

The op'n field, and where the unpierc't shade
Imbround the noontime Bowers:
Thus was this place,
A happy rural seat of various view;
Groves whose rich Trees wept odorous Gumms and
    Balme,
Others whose fruit burnisht with Gold'n Rinde Hung
    amiable... [15]

Note how "Nature boon" in her profusion ("powr'd
forth profuse") is unhindered here by the "nice Art in Beds and
curious Knots" that characterized the Renaissance garden. Note
too Milton's description of the "various view" – not to be
captured with the first glimpse – that presented itself to the eye
from this happy rural seat: and how the brooks (of nectar!)
flowing through the grounds do not run in straight channels but
in "mazie error", i. e. , they wander about in maze-like fashion.[16]
Virtually the entire program of the 18th century's natural garden
is anticipated in Milton's description of the first Garden.

How radical it was for its time can be seen from the fact
that two of Milton's contemporaries thought that the trees in the
Garden of Eden had been planted either in straight rows (Evelyn)
or in quincunx (Sir Thomas Browne).[17]

We have already mentioned Sir William Temple's essay,
"Upon the Gardens of Epicurus". Lovejoy regarded it as "the
probable beginning of the new ideas about [gardening] which
were destined to have consequences of such unforeseen range".[18]
It should be noted, however, that Temple was not at all one of
those who,

---

[15] *Paradise Lost*, Bk. IV, 237-249 (Milton 1958, p. 79). Milton's vision of the
Garden's free-flowing and lushly varied forms was anticipated in the 17th
century by some Continental artists, among them the Brueghels, Rubens
and Poussin.

[16] Sherbo 1972.

[17] Prest 1981, p. 90.

[18] Lovejoy 1955a. As we have seen, however, the origins of the natural
garden predate Temple's essay. *Paradise Lost* was published 18 years
before the appearance of Temple's essay.

Tir'd of the scene parterres and fountains yield,
He finds at last he better likes a field.[19]

Indeed, his essay contains a lengthy passage celebrating the formal gardens of his time, one of which he described in detail and deemed "the perfectest figure of a garden I ever saw."[20] Among the points Temple argued in favor of the formal garden was that "in regular Figures 'tis hard to make any real or remarkable Faults".[21] This is surely an extraordinary criterion, and helps one to sympathize with Walpole's acerbic remark about the "want of ideas, of imagination, of taste" shown by Temple "when he dictated on a subject that is capable of all the graces that a knowledge of beautiful nature can bestow".[22]

Nevertheless, Temple *did* acknowledge that under certain circumstances gardens entirely laid out on irregular lines can be beautiful, and it is his brief statement to this effect that gives his essay its particular interest for us here.

"Forms wholly irregular", he wrote there, if done well, could combine "many disagreeing Parts into some Figure which shall yet, upon the whole, be very agreeable". Indeed, he conceded that such arrangements might "for ought I know have

---

[19] Pope, "Moral Essays" Epist. IV, 87-88.

[20] Temple 1731, v. 1, pp. 170-190.

[21] *Ibid*, p. 186.

[22] Walpole 1995, p. 33. An anonymous poem of 1767, "The Rise and Progress of the Present Taste in Planting Parks..." comments even more acerbically on how Temple's gardens "tortur'd Nature sore in every part":

Temple the easy, learned and polite,
Who thought as freely as his pen cou'd write;
No garden plans from graceful nature drew,
His trees by pairs in nuptial order grew;
Or plac'd like sentinels at each corner stand,
To guard Pomona's gifts from Rapine's hand;
Pleas'd still with fountains, and with gay alcoves,
With statu'd Venus, and her train of loves;
Fix'd round parterres in regular design,
And gravel walks as level as a line.
From want of taste for undulating hills,
Bustles of oaks, fine vales, and murmuring rills;
Extensive lawns, and close embracing shades,
Long lakes, bright spiry rocks, and opening glades;
He tortur'd nature sore in every part...

more Beauty" than the conventional gardens of his day. Temple claimed to have seen gardens designed along these lines (unfortunately, he does not tell us where they were) and then added that he had heard from people who had been in China about the way gardens, houses and decorative objects there were designed.

In England, he wrote, "the Beauty of Buildings and Plantings is placed chiefly in some certain Proportions, Symmetries, or Uniformities; our Walks and our Trees ranged so, as to answer one another, and at exact Distances". By contrast people in China "...scorn this Way of Planting, and say a Boy that can [count to] an Hundred may plant Walks of Trees in their strait Lines, and over against one another, and to what Length and Extent he pleases... But their greatest Reach of Imagination is Employed in contriving Figures, where the Beauty shall be great, and strike the Eye, but *without any Order or Disposition of Parts, that shall be commonly or easily observ'd* {my emphasis]. And though we have hardly any Notion of this Sort of Beauty, yet they have a particular Word to express it; and where they find it hit their eye at first Sight, they say the *Sharawadgi*[23] is fine or admirable, or any such Expression of Esteem.[24]

Temple advised his readers not to try the Chinese method in their own gardens. "I should hardly advise any of these Attempts in the Figure of Gardens among us; they are Adventures of too hard Achievement for any common Hands; and though there may be more Honour if they succeed well, yet there is more Dishonour if they fail, and 'tis Twenty to One they will".

Thus, although Temple *can* be credited with introducing the concept of *Sharawaggi* to England he was not an advocate of its adoption for English landscapes or gardens.[25]

---

[23] On Sharawaggi or sometimes Sharawadgi, see pp. 317, fn. 102, *above*.

[24] Temple 1713, p. 186. Temple added that the fabrics of "Indian Gowns or the Paintings upon their best Skreens or Purcellans" have a "Beauty ... of this kind without Order". Temple's criterion of beauty hitting the "eye at first sight" shows how deeply that requirement – which as we saw originated in the early Renaissance as a corollary to the requirement for symmetry - was rooted in the aesthetics of his time. The application of this requirement to *Sharawaggi* is implausible; the two are fundamentally in contradiction to each other.

[25] Walpole was in the end not convinced by Temple's description of Chinese design. In the Chinese garden, he wrote, (1995, pp. 38-9), "nature, it seems [is] as much avoided, as in the squares and oblongs and

The real herald of the new movement in English gardening was not Temple but the third Earl of Shaftesbury (1671-1713), who in 1710 announced his dislike of the formal garden and his yearning for Nature's untrammeled forms. "I shall no longer resist the Passion growing in me for Things of a *natural* kind" he wrote, "where neither *Art,* nor the *Conceit* or *Caprice* of Man has spoil'd their *genuine Order,* by breaking in upon that *primitive State.* Even the rude *Rocks,* the mossy *Caverns,* the irregular unwrought *Grotto's,* and broken *Falls* of Waters, with all the horrid Graces of the *Wilderness* itself, as representing Nature more, will be the more engaging, and appear with a Magnificence beyond the formal Mockery of princely Gardens."[26]

These sentiments were echoed, if in somewhat more measured terms, by Joseph Addison (1672-1719). Addison's statement on the natural garden comes as something of a surprise because his ideas about architectural beauty have about them, on the contrary, the distinct tone of the Renaissance's aesthetic minimalism. In *The Spectator* No. 415 of June 26, 1712, he expressed his appreciation of the interior of the Pantheon and his disapproval of the interiors of Gothic cathedrals. His reasons were essentially those of Filarete 250 years earlier. No figures have a greater "Air" to them, he declared, than the concave and the convex, for in them "we generally see more of the Body". The Gothic interior, by contrast, is one of "Confusion", for the sight of it is "split into several Angles [and therefore] does not take in one uniform Idea, but several Ideas..." Sentiments like these, when applied to horticulture, had led to the rigidity and artificiality of the formal garden. [27]

---

strait lines of our ancestors". Chinese landscapes he added, are a "gaudy scene ... the work of caprice and whim". Where he got this information from is not known. Chambers (1772, p. 157) went further and rejected the opinion that Chinese design is always irregular. The Chinese, he wrote, are "no enemies to straight lines because they are generally speaking productive of grandeur...nor have they any aversion to regular geometrical figures".

[26] "The Moralist" pt. III, sect. ii (Shaftesbury 1749 v. 3, p. 235). How far Shaftesbury's enthusiasm would have carried him is an open question. As Sir Joshua Reynolds pointed out (Discourse XIII) only a habitat that man has not shaped has a "genuine order" untouched by human art or caprice: and therefore it is not a garden. Paradoxically, Shaftesbury also favored *simplicity* of design, for which see Havens 1953.

[27] Addison had earlier (*Spectator* No. 62, May 11, 1711) expressed his scorn for the "Goths in poetry", as he called them "who, like those in

Just the previous day however, in the *Spectator* No. 414 of June 25, 1712, Addison had written in an entirely different vein about the design, not of buildings but of gardens. He contrasted the natural, free, forms of the Chinese – no doubt he had learned about them from Temple's essay – to those of British gardeners:

> Our British Gardeners, on the contrary, instead of humouring Nature, love to deviate from it as much as possible. Our Trees rise in Cones, Globes, and Pyramids. We see the Marks of the Scissars upon every Plant and Bush. I do not know whether I am singular in my Opinion, but, for my own part, I would rather look upon a Tree in all its Luxuriancy and Diffusion of Boughs and Branches, than when it is thus cut and trimmed into a Mathematical Figure: and cannot but fancy that an Orchard in Flower looks infinitely more delightful, than all the little Labrynths of the most finished Parterres.

Addison had begun his essay by discussing the relationship between Art and Nature. Standing on its head the Renaissance view that the task of art is to improve nature by simplifying her forms, he declared that Art is "very defective" in comparison with Nature. Art may attain the beautiful, the strange, the polite or the delicate, but it has nothing of the "Vastness and Immensity which afford so great an Entertainment to the Mind of the Beholder". Its designs moreover are incapable of reproducing the "August and Magnificent" qualities of Nature herself:

> There is something more bold and masterly in the rough careless Strokes of Nature, than in the nice Touches and embellishments of Art.

Architecture, not being able to come up to the beautiful Simplicity of the old Greeks and Romans, have endeavoured to supply its Place with all the Extravagances of an irregular Fancy". However, ten years earlier, in his *Remarks on...Italy* ("From Rome to Naples"), Addison promiscuously enjoyed both simplicity and variety. Of the Rotunda, or Pantheon, he wrote, "I must confess the eye is better fill'd at first entering the Rotund, and takes in the whole Beauty and Magnificence of the Temple at one view", while in the interior of St Peter's he enthused over its "greater Variety of Noble Prospects".

> The Beauties of the most stately Garden or
> Palace lie in the narrow Compass, the
> Imagination runs them over and requires
> something else to gratifie her; but, in the wild
> Fields of Nature, the Sight wanders up and
> down without Confinement, and is fed with
> an infinite variety of Images, without any
> certain Stint or Number.

Recalling the "artificial Rudeness" of some French and
Italian gardens he had seen, Addison declared it to be "much
more charming than the Neatness and Elegancy" of the formal
garden.[28] He went on to make the radical – and in the event highly
influential - suggestion that an entire estate might be "thrown
into a kind of Garden" in such a way as to increase both the
pleasure and the profit that its owner derives from it. "A marsh
overgrown with willows", he wrote, "or a mountain shaded with
oaks":

> are not only more beautiful, but more beneficial,
> than when they lie bare and unadorned. Fields of
> Corn make a pleasant Prospect, and if the walks
> were a little taken care of that lie between them, if
> the natural Embroidery of the Meadows were helpt
> and improved by some small Additions of Art, and
> the several Rows of Hedges set off by Trees and
> Flowers, that the Soil was capable of receiving, a
> Man might make a pretty Landskip of his own
> Possessions.

In these few sentences Addison set down the program
that, in the ensuing decades, would transform the appearance of
the English landscape and its gardens.

Some of the most prominent writers in England added
to its intellectual foundations. Pope's role is well known, though
his enthusiasm for Nature's freely flowing forms and visual
complexity was probably less wholehearted than is often
supposed. As a very young man Pope found himself disliking
both the ordered artificiality of the formal garden *and* the

---

[28] We do not know which these gardens were. Addison's statement,
composed in 1712, is valuable for the light it casts on the prevailing view
– e. g. Hays (Benes and Harris 2001) - that "irregular" gardens did not
appear in France until the end of the eighteenth century.

untamed aspect of Nature, whose creations he described as "chaos-like together crush'd and bruis'd". He placed himself between these two extremes in a man-made world that is –

> ... harmoniously confus'd:
> Where order in variety we see,
> And where though all things differ, all agree. [29]

- a paradoxical formulation, this, that perhaps does not invite close analysis. Pope's reference on another occasion to "the amiable simplicity of unadorned nature", moreover, casts doubt on whether the middle ground he sought really exists, for nature is neither simple nor unadorned, and often (as Pope's contemporaries Addison and Burke, among others, recognized) is anything but "amiable".[30]

In the fourth of his "Moral Essays" (dedicated, one notes, to Lord Burlington, whose avid Palladianism was far from compatible with the ideals of the natural garden[31]), the mature Pope offered a manifesto of the revolution against the formal garden. He invited his reader to spend a day with him at the villa of "Timon", which has been identified as Sir Robert Walpole's Houghton Hall:

> At Timon's villa let us pass a day;
> Where all cry out, "what sums are thrown away;"
> So proud, so grand; of that stupendous air,
> Soft and agreeable come never there,
> Greatness with Timon dwells in such a draught
> As brings all Brobdignag before your thought.
> To compass this, his building is a town,
> His pond an ocean, his parterre a down:
> Who but must laugh, the master when he sees,
> A puny insect shivering at a breeze!
> Lo, what huge heaps of littleness around !

[29] "Windsor Forest", Pope 1831, v. 1, p. 51.

[30] Pope's "amiable simplicity" is perhaps to be understood as an echo of Renaissance ideas. Some 18th century aesthetic thinking resurrected those ideas with regard to every medium of design, including Nature's designs: see Havens, 1953.

[31] Dr. Johnson (1810, v. 3, p. 155) remarked that Pope "can derive little honour" from his penchant for associating his work with the names of aristocrats such as Burlington. See also the discussion of Burlington's Chiswick House and its garden on p. 360, below.

The whole a labour'd quarry above ground.
Two Cupids squirt before: a lake behind
Improves the keenness of the northern wind.
His gardens next your admiration call;
On every side you look, behold the wall!
No pleasing intricacies intervene,
No artful wildness to perplex the scene;
Grove nods at grove, each alley has a brother,
And half the platform just reflects the other.
The suffering eye inverted Nature sees;
Trees cut to statues, statues thick as trees;
With here a fountain never to be play'd,
And there a summerhouse that knows no shade;
Here Amphitrite sails through myrtle bowers,
There gladiator fight or die in flowers;
Unwater'd, see the drooping seahorse mourn,
And swallows roost in Nilus' dusty urn.

All this empty excess, Pope predicted, will be unavailing, and the
land will one day return to pursue its authentic purposes:

Another age shall see the golden ear
Imbrown the slope, and nod on the parterre,
Deep harvests bury all his pride has plann'd,
And laughing Ceres reassume the land.[32]

Pope to be sure did not reject the notion of refashioning Nature's
works to create gardens. But there were limits to what ought to
be done:

Something there is more needful than expense,
and something previous e'en to taste–'tis sense.
Good sense, which only is the gift of Heaven

To someone designing a garden he offered this advice:

To build, to plant, whatever you intend,
To rear the column, or the architecture to bend,
To swell the terrace, or to sink the grot,
In all, let Nature never be forgot;
But treat the goddess like a modest fair,
Nor overdress, nor leave her wholly bare;
Let not each beauty every where be spied,

---

[32] The "golden ear … imbrowning" is very poor!

Where half the skill is decently to hide.
He gains all points who pleasingly confounds,
Surprises, varies, and conceals the bounds.
Consults the genius of the place in all
Tells the waters or to rise or fall,
Or helps th' ambitious hill the heaven to scale,
Or scoops in circling theatres the vale:
Joins willing woods, and varies shades from
    shades;
now breaks,or now directs, th'intending lines.[33]

... Good advice indeed. But Pope never decided how completely he wished to disown the principles of the formal garden or to affirm those of the natural garden. His own garden, of which a plan survives (*fig.7.3*), reflects only dimly the principles that are manifested in the gracious tranquility of the natural garden. Although irregular and asymmetric in its layout, Pope's garden had too many straight rows of paths and trees to be more than an unhappy compromise between the conflicting values of the formal and the natural garden.[34]

No one had a more telling critique of the formal garden or was a more effective spokesman for what he called "the modern taste in gardening" than Horace Walpole (1717-1797). Witty, eccentric, energetic, Walpole's accomplishments include the first Gothic novel (*The Castle of Otranto*, 1764), as well as the design of his house, Strawberry Hill, which helped to stimulate the Gothic Revival; the books that he published at his Strawberry Hill Press are still valued by collectors for their fine design. Walpole's abhorrence of the formal garden has a curious aspect to it, for he was the son of Sir Robert Walpole, whose gardens at

---

[33] In a slight variation – different perhaps because not constrained by the requirements of meter – Pope told Spence (1964, p. 159) that "All the rules of gardening are reducible to three heads: - the contrasts, the management of surprises, and the concealment of bounds". Variety, he added in reply to Spence's question, "is included mostly in the contrasts". He then quoted the two lines of verse above and said they were inspired by Horace's *omne tulit punctum qui miscuit utile dulci* (*Ars Poetica* v. 343), which however refers specifically to mixing the useful with the pleasant.

[34] Thacker (1979, p.182), calls Pope's garden at Twickenham "tame and restrained ... relatively formal". Pope was not the only one to hesitate. Streatfield (1981, p. 36) makes the point that even Kent used symmetric layouts, such as in the groupings of trees in Holkham and Euston and in his unexecuted hillside scheme for Chatsworth.

Houghton Hall may have been the ones pilloried by Pope. In *The History of the Modern Taste in Gardening* (1780) the younger Walpole did not refer to the Houghton by name but one wonders whether he did not have it, too, in mind:

> When the custom of making square gardens inclosed with walls was ... established, to the exclusion of nature and prospect, pomp and solitude combined to call for something that might enrich and enliven the insipid and unanimated partition. Fountains, first invented for use ... received embellishments from costly marbles, and at last to contradict utility, tossed their waste of waters into air in spouting columns. Art, in the hands of rude man, had at first been made a succedaneum to nature; in the hands of ostentatious wealth, it become the means of opposing nature; and the more it traversed the march of the latter, the more nobility thought its power was demonstrated. Canals measured by the line were introduced in lieu of meandering streams, and terrasses were hoisted aloft in opposition to the facile slopes that imperceptibly unite the valley and the hill... Statues furnished the lifeless spots with mimic representations of the excluded sons of men. Thus difficulty and expence were the constituent parts of those sumptuous and selfish solitudes; and every improvement that was made, was but a step farther from nature... To crown these impotent displays of false taste, the shears were applied to the lovely wildness of form with which nature has distinguished each various species of tree and shrub. The venerable oak, the romantic beech, the useful elm, even the aspiring circuit of the lime, the regular round of the chestnut, and the almost molded orange-tree, were corrected by such fantastic admirers of symmetry. The compass and square were of more use in plantations than the nurseryman. The measured walk, the quincunx, and the etoile imposed their unsatisfying sameness on every royal and noble garden ... and symmetry, even where the space was too large to

permit it being remarked at one view, was ... essential. [35]

The "tiresome and returning uniformity" and "symmetrical and unnatural design" that Walpole castigated had characterized the gardens of the nobility and gentry everywhere in England. [36] It was literary men like Shaftesbury, Addison, Pope, Walpole and Burke who provided the ideas that led to the rejection of the formal garden and pointed the way back to designs incorporating natural forms. But it was a series of hugely talented and entrepreneurial garden designers who drafted plans for putting them into effect. The singular achievement of Charles Bridgeman (1680-1738) and William Kent (1685-1748) was to reverse the previous order of things. Instead of extending the formal garden into the countryside, they brought the countryside into what now became the natural garden. It was by these men, Walpole wrote, that the garden was "set free from its prim regularity, that it might consort with the wilder country without".

Bridgeman, Walpole wrote, "banish'd verdant sculpture, and did not even revert to the square precision of the previous age. He ... disdained to make every division tally to its opposite, and although he still adhered much to strait walks with high clipped hedges, they were the only great lines; the rest he diversified by wilderness". Walpole's highest admiration was for Kent who, as he declared, "leaped the fence and saw that all nature was a garden":

> He felt the delicious contrast of hill and valley changing imperceptibly into each other, tasted the beauty of the gentle swell, or concave scoop, and remarked how loose groves crowned an easy eminence with happy ornament, and while they called in the distant view between their graceful stems, removed and extended the perspective by delusive comparison. Thus the pencil of his imagination bestowed all the arts of landscape on the scenes he handled. The great principles on which he worked were perspective, and light and shade. Groupes of trees broke too uniform or too extensive a lawn; evergreens and woods were opposed to the

[35] Walpole 1995, pp. 25-27..

[36] *ibid*, p. 29.

glare of the champain, and where the view was
less fortunate, or so much exposed as to be
beheld at once, he blotted out some parts by
thick shades, to divide it into variety, or to make
the richest scene more enchanting by reserving
it to a farther advance of the spectator's step...

But of all the beauties he added to the face
of this beautiful country, none surpassed his
management of water. Adieu to canals, circular
basons, and cascades tumbling down marble
steps, that last absurd magnificence of Italian
and French villas. The forced elevation of
cataracts was no more. The gentle stream was
taught to serpentize seemingly at its pleasure...
The living landscape was chastened or
polished, not transformed. Freedom was given
to the forms of trees...[37]

The revolution in taste that celebrated Nature - embracing
her freely-flowing forms, her pervasive asymmetry, her
seemingly boundless visual variety of colors, shapes and textures
- changed the appearance of English gardens and parks. Beyond
that, though, its scope proved to be surprisingly limited. For,
while it eradicated *the hostility* toward Nature that prevailed
earlier during the era of plagues, it did not entirely eradicate *the
fear* of Nature that had been the cause of that hostility.

Rather, and most paradoxically, it joined this fear to the
new enthusiasm for Nature, creating an ingenious blend that
combined an acknowledgement of the objective fact of Nature's
menacing potential with pleasure in the loveliness of Nature's
forms.   Thus, while Nature remained "dark, uncertain,
confused",[38] her menacing aspect was now experienced as
something – pleasurable! It became, as Addison said, an

---

[37] *ibid,* pp. 43-45.

[38] Burke 1939, Part 2, Sect. ii. During the Renaissance people sometimes
referred approvingly to the sense of foreboding that the gloom of a
church's interior could instill in a person. Alberti, for example, wrote of
the *"horror qui ex umbra excitatur"*, (quoted, with other examples,
Germann 1973, pp. 56 – 57). It is possible that these sentiments anticipate
the eighteenth century's enthusiasm for Nature's terrors. In our banal age
we can hear a debased variant of this paradoxical pleasure in the shrieks
emanating from people as they thrill to the terror of plunging earthward
on a roller-coaster.

"agreeable kind of horror"[39]. It was Burke who, most influentially, called it the Sublime, and described it as "delightful horror". Its ruling principle, he wrote, arises "in all cases" from the "ideas of pain and danger" that Nature's creations can arouse in us. These, he insisted, give us more intense pleasure than we can get from looking at Nature's conventionally beautiful forms.[40]

Lovejoy claimed that in the eighteenth century "the primacy of irregularity was no longer limited to the theory of landscape-design" but was extended to architecture, too.[41] But this greatly overstates the matter. By and large, *the advent of the natural garden was not matched by a similar revolution in the design of English buildings*. The nobility and gentry, which must give Nature free rein in their gardens, continued as before – or perhaps with all those Palladian designs, went on even more resolutely than before – to build their houses in rigidly formal, symmetric style.

Joseph Heely, for one, while greatly admiring gardens marked by "wildness of fancy and a sympathizing irregularity" was adamant (though without explaining himself) that the houses standing in them should have "perfect symmetry".[42]

---

[39] Addison, *Remarks on Italy* ("Geneva and the Lake").

[40] Burke 1939, Part 1, Sect. viii. Burke's contemporary, Hogarth (2015, p.57) wrote of "a pleasing kind of horror". An indication of how far Burke's formulation spread is in a comment (*Diary of Studies*) made in 1820 by the 15-year-old Benjamin Disraeli on reading Virgil. The *Georgics*, he complained, is "a garden beautifully cultivated and elegantly laid out, but no beautiful distance pleases & no lofty precipice terrifies the eye all is alike same & consequentially tiresome". A century later Rilke would write (*Duino Elegies*), "For beauty is nothing but the beginning of terror which we are barely able to endure, and it amazes us so because it disdains to destroy us". (*"Denn das Schöne ist nichts als des Schrecklichen Anfang, den wir noch grade ertragen, und wir bewundern es so, weil es gelassen verschmäht,uns zu zerstören"*, etc.)

[41] Lovejoy 1955b.

[42] Heely 1775, p.9. Asymmetric design does not always preclude symmetric design. Wilson (1977, p. 17) makes the curious point that in the Elizabethan era it became increasingly common for a preference for symmetric house fronts to be accompanied by a preference for asymmetric interiors. Very different are the Japanese, who seek to imbue both their architecture *and* their gardens with the profound values they see in Nature, and are led by this to eschew symmetric forms. "The hidden richness ... which the Japanese appreciate in nature, they emulate

Thus in 1726, when the innovative principles of natural gardening were already widely accepted, Lord Burlington began building Chiswick House (*fig. 7.4*), which inaugurated the so-called Palladian revival in England, renewing the fashion for symmetric architecture. Yet the *gardens* of Chiswick House were largely designed by William Kent, the selfsame gardener whom Horace Walpole greatly admired as one of the foremost practitioners of the natural garden movement! This is not to say that there were not voices advocating asymmetric architecture. In one of his lectures Joshua Reynolds commented on the design of additions to existing buildings:

> ... As such buildings depart from regularity, they now and then acquire something of scenery by this accident, which I think might not unsuccessfully be adopted by an architect, in an original plan, if it does not too much interfere with convenience. Variety and intricacy is a beauty and excellence in every other of the arts which address the imagination; and why not in architecture.[43]

A few asymmetric houses were indeed built. Vanbrugh, despite the relentless symmetry of his most notable buildings, gave asymmetric form to the house he designed for himself, the grandiloquently-named Vanbrugh Castle.[44] But these are

---

in architecture", writes Ramberg (1960). "The Japanese have been concerned to make the architectural intrusion into the natural scene an easy and gracious one, and this concern has done much to preclude symmetrical organization of buildings. Symmetrical arrangement bears very much the stamp of a human and intellectual order. It involves the interruption of the natural pattern and implies man's intent to improve nature in terms other than its own. It insists on man's divine dispensation, his special place between nature and the gods. Its endorsement seems testimony to the propriety of man's rude refashioning of the world in the image of that passion for regularity he calls reason. The Japanese have been much more disposed to defer to nature than to project its improvement in a systematic intellectual manner. They have been more interested in discovering the order in the face of nature than in accomplishing an ideal in an imperfect world".

[43] Reynolds 1997, p. 243.

[44] Somewhat later, in the Regency era John Nash designed a number of asymmetric structures, most notably Cronkhill in Shropshire (Pevsner 2010, pp. 141-2).

exceptions. Almost invariably, the buildings set in the graciously free-flowing, asymmetric landscapes and gardens of the eighteenth century (figs.7.1,7.2) were symmetric.[45] This inconsistency is quite paradoxical. Part of an understanding of it may lie in the fundamental difference between the antecedents of Renaissance horticulture and the antecedents of Renaissance architecture.[46] The Renaissance architect had access to an almost inexhaustible supply of Classical artifacts and ruins that could provide him with authentic models on which to base, or on which he could *claim* to base, his designs.

The designer of a Renaissance garden or landscape, on the other hand, had no such guides available to him. At best, he could hope to infer the principles of Classical garden design from a few ambiguous literary passages, such as Pliny's descriptions of his villas and the grounds in which they stood, or from the

---

[45] In his "Essay on the Beautiful", the Scots philosopher and mathematician Dugald Stewart (1753-1828) implicitly rationalized the use of both asymmetric and symmetric design. Asymmetric designs offend us, Stewart declared, when they lack "Sufficient Reason", namely purpose or utility. "The beauty of winding approach to a house, when the easy deviations from the straight line are all accounted for by the shape of the ground, or by the position of trees, is universally acknowledged; but what is more ridiculous than a road meandering through a plain, perfectly level and open"? In architecture, too, Stewart argued, we are "offended" by asymmetric design if we cannot conceive "how the choice of the architect could be thus determined, where all circumstances appear to be so exactly alike. This disagreeable effect is, in a great measure, removed, the moment any purpose or utility is discovered; or even when the contiguity of other houses, or some peculiarity in the shape of ground, allows us to imagine, that some reasonable motive may have existed in the artist's mind, though we may be unable to trace it" (Stewart 1855, p. 210; see also Etlin 1994, pp. 77-80.) It would seem that for Stewart the "Sufficient Reason" requirement existed only for asymmetric designs, and not also for symmetric ones. In his view, evidently, symmetry is the default, the norm, and requires no justification.

[46] A simpler but perhaps not very helpful explanation of this paradox was offered by Viollet-le-Duc (1987, vol. 2, p. 268): "Many have ventured to disregard the laws of symmetry in building houses in the country", he noted, "but it seems there are more difficulties in the case of mansions in the city". He suggested that this might be understood "a matter of fashion. People think that their houses in the city should be irreproachable in point of symmetry, in the same way as they make it a matter of conscience not to appear in the streets without the usual hat on their head".

skeletal remains of one or two ruined gardens such as those of Hadrian's villa in Tivoli.

The Renaissance garden, in consequence, could never rival the Renaissance building in its Classical authenticity.

It suffered from a further disadvantage. Compounding its questionable Classical paternity was the fact that it was only by the suspension of disbelief, and by accepting fanciful claims regarding the ideal forms of Nature's creations, that a person could overlook the Renaissance garden's contrived and artificial appearance, and believe that there was anything natural about it. It was easy to accept that a Renaissance building looked "classical", but the notion that the Renaissance garden mirrored Nature's forms was a very tenuous one, indeed.

The Renaissance garden, then, was neither manifestly Classical nor manifestly natural. It had served the important function of providing Man with a symbolic victory over Nature during those long, long years when Nature was identified with the unspeakable terrors of the plague. But when the time came that that function was no longer needed, the aesthetic itself – inauthentic and ungratifying as it was – could be discarded with few if any misgivings.

That time came, of course, with the end of the era of plagues, which removed the psychological (and perhaps the principal) need for this aesthetic. The formal garden was now rapidly dismantled, as though it had never really been more than the stage set that John Evelyn thought it resembled.[47]

Appendix: Symmetry and Political Power

In conclusion, now, and only briefly, we must at least touch on symmetry's role as a tool for asserting, not just man's power over

---

[47] I do not mean to suggest, of course, that the inner need for order and predictability – indeed, for the control of Nature - arose solely in response to the plague, or disappeared once the era of plagues had passed. In some measure it is inherent in the human condition. For example, Baetjer and Links (1989, p. 20) suggest that the "essence" of the creed of Canaletto (d. 1768), was that of "man imposing his artistic will on fluid, formless Nature... The city ... symbolizes the order that man alone can bring to the environment, making sense of previous chaos. To build ... is to defeat Nature, and Canaletto ... apparently [had] no doubts or regrets about the defeat".

Nature, but also man's power over Men.   Nowhere is this more apparent – this "setting forth of state", as Ruskin called it – than in the overpowering scale and symmetry of the palace and gardens built for Louis XIV, "the Sun King", at Versailles. In England, the end of the era of the plagues coincided with the rise of the Whig aristocracy, and it is during this time that we find the final and boldest manifestation of Renaissance horticultural principles, with immense estates laid out in geometrical shapes, including arrow-straight roads flanked by symmetric rows of trees that stretched out in every direction from the main house (itself invariably symmetric). The rhetoric of domination, not just of nature but of man, inherent in this idiom is finely described by John Prest: "We feel ourselves to be in the presence of autocracy, for if anything is out of place in an avenue or formal garden, the autocrat can spot it at once, and so too can the functionary whom he employs, and of the restless search for domination which Hobbes attributed to mankind".[48]

The introduction of symmetry occurred, as we have seen, during the plagues that devastated Europe. This was also the time when, not altogether coincidentally, Europe witnessed the emergence of the powerful and intrusive centralizing State, and its auxiliaries and sometimes rivals, the centralizing Church, and the Corporation.[49] Symmetric architecture was employed by State, Church and Corporation as a means of demonstrating the power of the ruler over the ruled. What structure demonstrates symmetry's potential in this regard more forcefully, more sinisterly, than Bernini's appalling piazza and colonnade in front of the Vatican basilica?

The principal instrument of control used by these entities is bureaucracy, and it will perhaps occasion no surprise to learn that the first buildings erected expressly to house government bureaucracies were relentlessly symmetric in their appearance – I refer to the Medicis' Uffizi ("offices") and the procuratorial buildings in Venice's Piazza San Marco. Each of these structures adjoins earlier ones – the Palazzo della Signoria and the Basilica San Marco – whose unabashed asymmetry evokes an altogether different era.

Symmetry is the aesthetic of power. Its service to the State and its auxiliaries is described in a historical perspective by

---

[48] Prest 1981, pp. 94-5.

[49] Strayer 1970, chap. 3.

Wölfflin.[50] In earlier times, he wrote, builders thought that each component of a structure "should function for itself in its own location, and seem not to have paid attention to the overall structure". This disunity, he suggested, made for "a lively" impression.

In this description we see buildings that reflect the structure and values of a decentralized society, one in which each unit functions in a largely autonomous fashion ("for itself in its own location") and largely oblivious to the wider world and its institutions ("the overall structure"). The society, too, is a "lively" one: somewhat unregimented and somewhat disorderly.

Architecture like this does not reflect the structures and values of modern society. The autonomy of the earlier period has been largely subverted by instruments of centralized and omnipresent and ever-more intrusive power. The "freedom" of the earlier dispensation, Wölfflin declared, is "tolerated" nowadays "only in private or rural buildings". It is decidedly *not* acceptable in public buildings, which he calls "monumental". These must convey an impression that, far from being "lively", is "worthy or serious"; and is evidently intended by Wölfflin to be seen by the reader in contrast to the "freedom" of earlier times.

And how is this "worthy or serious" impression to be achieved? Wölfflin did not hesitate. For monumental buildings, he declared, "we demand absolute symmetry, a grave and measured bearing" (...*verlangen wir ... unbedingte Symmetrie: würdige gemessene Haltung*)! Here then we have an affirmation of the view of symmetry as the aesthetic of power, an aesthetic that asserts the primacy of the whole and the subordination of its parts – the primacy of the State (or its auxiliaries) and the subordination to it of its subjects. And that this view is not merely an academic's construct but reflects realities of the modern state, there are the building regulations issued by the kingdom of Bavaria in 1864 which, Sitte reports, demanded "as their main aesthetic requirement that in facades everything should be avoided 'that might offend *symmetry* or *morality*', it being apparently left open to interpretation which of the two would be considered the greater offense".[51] The State's values – its "morality" – are to be enforced by the power of the State, and are, in this extraordinary ordinance, explicitly associated with symmetry!

---

[50] Wölfflin 2017, p. 40.

[51] Sitte, p. 190.

We noted earlier the intimate connection between the principles of symmetry and simplicity. This connection has gained significance in modern times, when simplicity overtook symmetry itself as an aesthetic imperative in general and as an integral component of the aesthetic of power.

Simplicity is understood as a correlate of unity and order, and it is its effectiveness in combating *dis*order that enables it to serve as an adjunct to symmetry in the aesthetic of power.[52]

Prominent among twentieth-century architects who espoused this principle was Le Corbusier (1887-1965), whom we find echoing characteristic Renaissance misconceptions about Nature's forms. These, he claimed, are "cleanly and clearly formed; they are organized without ambiguity".[53] And as with Nature, so too with architecture. Seeing the Parthenon for the first time, Corbusier exulted in "the precise relationships of its pure forms", which – incredibly! – gave him "the impression of naked polished steel".[54]

In Corbusier's view it was imperative that a structure be homogenous: a unity. This is to be achieved by ensuring that the plan is reflected throughout the building, including its elevations. Failing that, "we have the sensation ... of shapelessness, of poverty, of disorder, of willfulness ... *Where order reigns, well-being begins*".[55] Elsewhere, he explained the need for order on the grounds that "otherwise there would only be chance, irregularity and capriciousness". Corbusier's ideals included, "absence of verbosity, good arrangement, a single idea"; flooring that "stretches everywhere it can, uniformly and without

---

[52] An augury of the modern, authoritarian, assault on intricate, asymmetric, design is Lewis F. Day's warning about the "lawless" spirit of some artists. Day conceded the naturalism of Japanese design ("there is always something to find out; which is just what there would not be in a simple and orderly geometric pattern of the European type"). He acknowledged, too, that Japanese design is "a relief from the monotony of absolutely formal disposition". He went on then to warn, however, that the artist's "impatience of order" is not sufficient justification for "license...We have to be on guard against a certain spirit of anarchy, which appears to have taken possession of so many artists. There is a class ... which will repudiate, not only the laws of art, *but the need for all law whatsoever*" (my italics). Lewis F. Day, *Ornamental Design*, 1890, quoted Gombrich 1979, p.58.

[53] Corbusier 1986, p. 212.

[54] *ibid.*, p. 217.

[55] *ibid*, pp. 48, 54. (My emphasis.)

irregularity"; trees "planted in ordered patterns"; and "ordered forests of pillars".[56] He denounced as "transgressions" buildings that "one can see only in fragments and as one moves about".

The complexities, contradictions, ambiguities, and inconsistencies that are inherent in human experience and are indispensable to the human spirit receive short shrift in Corbusier's scheme. The homogeneity and simplicity that he advocated in their place – the unambiguous subjection of the parts to the whole - are of course fundamental requirements of any authoritarian system: *Ordnung* – order - *über Alles!*

Nevertheless, Corbusier designed great buildings that ignored these imperatives – among them the chapel at Ronchamp, the Carpenter Center at Harvard, the Mill Owners' Association building in Ahmedabad, India.

No such ambivalence, no such occasional dabbling with asymmetry and complexity, no such triumph of sensibility (or just, decency) over dogma, is to be found, however, in the work of Corbusier's still more influential contemporary, Mies van der Rohe (1886-1969) who, despite his Dutch name, was in fact a German.

The task of architecture, Mies wrote, is to "create order out of the desperate confusion of our time".[57] Closely related to this imperative is the dictum for which Mies is best known – "less is more". [58] The connection between these two statements, that is to say, between order and simplicity, is of course a fundamental one. Complex structures may be orderly but their order is not readily perceived. What Mies called for, and produced, was orderly and instantly obvious simplicity: a carefully-fashioned unity that, in the case of Mies, at least, was the architectural echo of "*ein Volk, ein Reich, ein Führer*".

It might be tempting to attribute such expressions, and his architecture, to Mies' infamous collaborations with the Nazis. The fact however is that his beliefs and skills were welcomed, even enthusiastically celebrated, in the United States – one might call him the architectural cousin of Werner von Braun! – and that

---

[56] *ibid,* pp. 158, 186, 59-60.

[57] Quoted Venturi 1977, p. 41.

[58] Most architectural writers seem unaware that Mies lifted the phrase (without acknowledgment, of course) from Robert Browning's poem of 1855, "Andrea del Sarto".

his influence and celebrity here were greater than they had ever been in his native Germany, which he left in 1937.[59]

The remarks on the preceding pages constitute a very brief presentation of some ideas regarding the aesthetic of power. Cursory though these remarks are, however, I trust that they have sufficed to persuade the reader of the supporting role that the principle of symmetry, and its correlate, the principle of simplicity, play in shaping the authoritarian environments that seem to encroach upon us, more and more, with each passing day.

---

[59] Notwithstanding his infamous association with the Nazis, Mies was in the last years of his life awarded the highest medals of the British and American associations of architects; and Lyndon Johnson awarded him the Presidential Medal of Freedom.

CHAPTER EIGHT
WITTKOWER AND THE
SANTA MARIA NOVELLA FAÇADE

8.1 In this image Wittkower's drawing of Santa Maria Novella is laid over
a photograph of the facade, with the scale of the drawing adjusted so that
the widths of the upper storey in both are identical. The combined image
shows that the upper level does not fit into a square based on the
structure's width, as Wittkower claimed; moreover, the 15 "squares" (as
Wittkower mistakenly called them) in the attic, as well as the main portal,
are not centered on the structure. Also contrary to Wittkower's claims, the
two volutes on the upper level differ in size and the superimposed frame
enclosing them is not a square but a rectangle (1.15 : 2.22). On the lower
level, the actual sides of the façade extend well beyond Wittkower's
drawing, thus establishing that the overall façade does not fit into a square.
The image here, accordingly, refutes Wittkower's hypothesis about the
design of the façade, and its place in Renaissance architectural history.

"... ut par sit non scripsisse hunc nobis,
qui ita scripserit ut non intelligamus."

-- Leon Battista Alberti 1966a, VI, i

According to Rudolf Wittkower, the concept of symmetry was ignored by most Renaissance architects, among them Leon Battista Alberti. The few who took any notice of it at all treated it as a merely "theoretical" idea, and "rarely applied" it in their designs.[1]

(That symmetric design is only found "rarely" in Renaissance architecture will come as a surprise to many people. Wittkower's insistence on this point was not only factually incorrect, of course, but blinded him to the significance of Alberti's *collocatio* as the earliest known attempt to provide a theoretical basis for the notion that good design, whether of buildings or of other artifacts, must always be bilaterally symmetric.)

Consistent with his opinion, Wittkower did not refer to issues of symmetry in his celebrated analysis of the façade – attributed to Alberti[2] – of the church of Santa Maria Novella in

---

[1] Wittkower 1949, p. 70.   Gadol, (1969, pp. 109-110), a follower of Wittkower, departs from him by defining *collocatio* as "architectural symmetry in the modern sense", but she then returns to the fold by claiming that Alberti "did not intend his remarks [on *collocatio*] to be carried out". How she can have determined Alberti's unstated intention regarding this matter is baffling; and as I point out at the end of this Chapter, the three church facades that are known to have been Alberti's work are all symmetric.

[2] Although the attribution of the Santa Maria Novella façade to Alberti is generally accepted, the evidence for it is far from conclusive. It rests primarily on a letter to the humanist scholar Landino, a contemporary of Alberti, (quoted by Mancini 1882, p. 461), which establishes a connection between Alberti and the Santa Maria Novella façade but leaves the nature of that connection unclear. In his first, 1550, edition of the *Lives*, Vasari stated only that Alberti designed "the door on the façade of Santa Maria Novella"; in the 1568 edition he added that Rucellai, patron of the project, received from Alberti, his friend, "not only advice but the actual model" for the new façade. That statement is an ambiguous one and its accuracy a century or more after the event is open to question. No model of the new façade is known to exist and no contemporary source mentions it.

Florence. That façade, Wittkower wrote, was "the most important" of the Renaissance, and it "set the example" that would be followed by architects for centuries to come. What made the façade so significant, Wittkower said, and established Alberti as a pre-eminent Renaissance theorist, was that its design was based on Classical theories of proportion that Alberti himself had revived.

Wittkower's thesis was that the Santa Maria Novella façade is "exactly circumscribed by [an imaginary] square"; that "the place and size of every single part and detail" of the entire design is "fixed and defined" by a system of proportion based on the progressive halving of that square; that this system of proportion is "the Leitmotif of the whole façade"; that Alberti's *concinnitas* is its intellectual foundation; that the "chief characteristic" of *concinnitas* is *eurhythmia*, which Wittkower described as "the Classical idea of maintaining a uniform system of proportion throughout all parts of a building"; that this idea had been an "axiom of all Classical architecture"; and that Alberti's "strict" application of this system of proportion established the Santa Maria Novella façade as "the first great Renaissance example of *eurhythmia*".[3]

(Nor was the façade altogether "new". What we see now was a modification of parts of the medieval façade.) The suggestion of Millon (1994, p. 24, quoting Alberti 1966a, v. II, pp. 860 – 862) that for Alberti models were "not a vehicle to present an idea to a client but a means to study and realize an idea" possibly casts some further doubt on Vasari's report. I would mention here, too, that the common attribution of the Palazzo Rucellai to Alberti is not confirmed by Filarete (Bk. VIII), and is called into question by Mack (1974).

[3] Wittkower *op. cit.*, pp. 45-7; 33; p. 33, fn. 5; and Wittkower, 1940-1941, p. 1, fn. 4 and p. 8, fn. 2. Wittkower never indicated how he determined what the dimensions of the Santa Maria Novella façade are, and it is remarkable that he was never challenged to do so. Wittkower did not acknowledge his debt to Heinrich Wölfflin (1966, p. 42), who claimed, though without providing evidence, that Alberti's "facades" (i.e., in the plural) are in a 1:2 proportion. Wölfflin in turn acknowledged that his analysis was based in part on the statement of Thiersch (2017, p. 6) that "Harmony arises from the repetition of the [proportions of the] main figure of the work in its subdivisions" – "*das Harmonische entsteht durch Wiederholung des Hauptfigur des Werkes in seinen Unterabteilungen*" - a concept that bears more than just a passing resemblance, even if unacknowledged, to the one Wittkower presented as *eurhythmia*.

It deserves to be pointed out that the line drawings of numerous structures that Thiersch used to demonstrate his thesis are of

These propositions became - and remain - the standard account, not only of the façade of Santa Maria Novella itself, but of some of the basic principles that are said to underlie Renaissance architecture; and they established Wittkower as one of the most influential art historians of modern times.[4]

Yet Wittkower's thesis, for all its elegance and notwithstanding the acclaim it has enjoyed, was deeply flawed at the outset and remained deeply flawed even after the four revisions of it that he published over the course of thirty years.[5]

---

doubtful accuracy. For its part, Wölfflin's line drawing of the Santa Maria Novella façade (Thiersch 2017, p. 105, *fig.* 5) is not accurate, as I have shown by superimposing it on a photograph of the facade (see *fig.*8.2). To compound matters, the proportions of the drawings presented by Wolfflin and Wittkower are different from each other. Benelli (2015) is a valiant but ultimately unpersuasive effort to find an empirical basis for Wittkower's claims (in part by using Wittkower's own sketches). He acknowledges however that "reliable measurements [were] not always available to Wittkower" and rather damningly adds that "Wittkower intentionally did not acknowledge evidence that Palladio's villas were often built upon preexisting medieval constructions" (and therefore did not conform to the Classical proportional formulae that Wittkower claimed guided Palladio). The Renaissance-era work on the Santa Maria Novella façade was also built on the church's pre-existing medieval structure.

[4] For general surveys of the influence of Wittkower's work see Millon (1972), and Payne (1994). Among those endorsing Wittkower's analysis of the Santa Maria Novella facade are Gadol (1969, pp. 112-114); Borsi (1989, pp. 61*ff*); Evans (1995, p. 248); and Tavernor (1998, pp. 99-106). More recently Hatfield (2004), has referred to "...the remarkable system of proportions that is best described in a memorable discussion by Rudolf Wittkower". For dissenting views cf. Lorch (1999, pp. 45-46) and Ostwald (2000). In the preface to the third edition Wittkower himself attested to the wide influence of his book, though one might question his statement there that Scholfield (1958) was among those who "took [his] cue" from it: and perhaps even more so his modified statement, in the fourth edition (Appendix III), that Scholfield's book was "partly derived from" *Architectural Principles in the Age of Humanism.* Scholfield himself (*op. cit.*, pp. 35, 39, and comp. pp. 51-2) argued that Wittkower's stress on the importance of neo-Platonism in Renaissance theory of architecture "has little to recommend it" and that Wittkower's musical theory "does not...explain all the facts of proportion as it was practiced in the Renaissance". Scholfield, (*ibid.*, p. 55 fn. 3), did however accept Wittkower's analysis of the Santa Maria Novella façade.

[5] The first appearance of the study was as part of Wittkower 1940-1941; this essay is referred to here as "the initial version". Successive revised versions appeared in the first edition of *Architectural Principles in the Age*

We can start by noting some problems with his attempts to link Alberti's ideas to Classical architectural theory.

Wittkower's claim that *eurhythmia* was an "axiom of all Classical architecture" ignores the fact that our knowledge of ancient Greek and Roman architectural theories is too limited to sustain plausible generalizations about *any* of their axioms. In particular, Wittkower's claim ignores the uncertainty of scholars today about the meaning of *eurhythmia*. Schofield, indeed, declares that *eurhythmia* as it has come down to us has "no recognized meaning at all".[6] Uncertainty about what *eurhythmia* is applies *a fortiori* to Vitruvius, Wittkower's sole source for the meaning of the term.[7] Although one would not know this from

---

of Humanism (Wittkower 1949, pp. 36-41); second edition (Wittkower 1952, pp. 36-41); third edition (Wittkower 1962, pp. 36-41); and fourth edition (Wittkower 1971 pp. 41-47). My remarks here are based primarily on, and cite the pagination of, the fourth and final edition, which was published shortly before Wittkower's death in 1971, and which differs from the third edition only in minor stylistic alterations and in having a new introduction. I identify differences between the fourth and earlier versions when they illustrate significant changes in Wittkower's analysis. Wittkower noted in the original version that it was written in wartime London without access to a copy of the Latin text of Alberti's *De re aedificatoria,* and that he relied instead on what he rather curiously called "the still unsurpassed" (but in fact at that time the only) English translation of 1755 by Leoni, and on the Italian translation of 1750 by Bartoli. A short train ride however would have brought Wittkower to the library at Eton, where Wotton's extraordinary copy of *De re aedificatoria,* annotated by the author himself – see M. R. James, *A descriptive catalogue of the manuscripts in the library of Eton College,* 1895 – remained on the shelves until at least August, 1941, when Eton's rare books were sent for safety to the vaults of the Bodleian in Oxford (email communication from an Eton College librarian, Dec. 20, 2011). The book-length editions of Wittkower's study were published between 1949 and 1971, and there would have been plenty of opportunity for him to get hold of the full Latin text of *De re aedificatoria* in those years. The fine edition by Orlando and Portoghesi, moreover, containing the Latin text and Italian translation, was published in 1966, or well before the last edition of *Architectural Principles* in 1971.

[6] Scholfield 1958, p. 18; he adds that, although "not obviously nonsense", translations of the term "unfortunately... convey very little sense"- a point illustrated by such renderings of *eurhythmia* as a "nameless grace", Foat (1915); as "abstract beauty, but not necessarily visual beauty, a sense of fine crafting", Wilson Jones (2003, p. 43); and as "shapeliness ... simple, inherent proportions of each element", Taylor (2003, p. 25).

[7] Vitruvius I,ii,3.

reading Wittkower, it is widely agreed that Vitruvius used *eurhythmia* "so sketchily that his entire concept of it is not clear".[8] What *does* seem clear however is that whatever *eurhythmia* may mean, it assuredly is not the same thing as "proportion", a concept that is served in Latin by the word *proportio*.[9] And for Vitruvius at least, *proportio* is definitely not the Latin equivalent of the Greek *eurhythmia* – as Wittkower would have us believe – but of *analogia*.[10]

---

[8] Scranton 1974.

[9] It is far from clear what role ideas about proportion played in Classical architecture. Coulton (1982, p. 66) calls into question whether *any* modular system was used by Greek architects, at least before the Hellenistic period. Addison (1705, at 2158-9) questions the adherence of the ancient Romans to rules of proportion: some say, he wrote, that "the Ancients, knowing Architecture was chiefly design'd to please the Eye, only took care to avoid such Disproportions as were gross enough to be observ'd by the Sight, without minding whether or no they approach'd to a Mathematical Exactness: Others ... say the Ancients always consider'd the Situation of a Building, whether it were high or low, in an open Square or in a narrow Street, and more or less deviated from their Rules of Art, to comply with the several Distances and Elevations from which their Works were to be regarded." Addison, in the same discussion, quotes Desgodetz, with regard to old Roman pillars, "that the Ancients have not kept to the nicety of Proportion, and the Rules of Art, so much as the Moderns in this Particular" and he cited opinions that blame the allegedly defective proportions of those pillars on the workmen of Egypt and other nations who sent their pillars, already shaped, to Rome. The tables compiled by Perrault from Desgodetz' measurements of Roman monuments "seemed to demonstrate irrefutably that the ancients did not adhere to mathematical rules after all" (Wilson Jones 2003, p. 6).

[10] "... *a proportione quae graece* analogia *dicitur*" - Vitruvius III.i.1 – settles the matter conclusively. Wittkower's erroneous equation of *eurhythmia* with "proportion" is taken directly (though without acknowledgment) from Granger's flawed translation in the Loeb Classical Library edition of Vitruvius. Granger, in turn, evidently derived it from Lewis and Short (1879) who, venturing into Greek etymology, declared that *eurhythmia* means "beautiful arrangement, proportion, harmony of the parts" - for which their sole citation is the Latin text of Vitruvius I,ii,3, the passage in which Granger translated *eurhythmia* as proportion! Wittkower's use of Granger's translation, it must be said, raises a question about his command of Classical Latin. I would add that one may perhaps infer from Vitruvius III.i.1 ("*ex qua ratio efficitur symmetriarum*") and I.iii.3 ("*et ad summam omnia respondent suae symmetriae*") that the objective of

Wittkower's attempt to link Alberti's ideas to the concept of *eurhythmia* was of course also not helped by the fact that Alberti himself never used the term. At first, Wittkower addressed this awkwardness by declaring (though without explanation) that "Alberti's definitions coincide to a larger degree with those of Vitruvius than is generally admitted".[11] Sensibly, Wittkower deleted this claim from the later versions of his work.

He did however retain his initial assertion that Vitruvius' *eurhythmia* "is covered by" Alberti's *concinnitas*. "Covered" is an ambiguous term, to be sure. In the earlier version, as we have already seen, Wittkower had defined *eurhythmia* as "the chief characteristic" of *concinnitas*. His failure now to identify anything else that *concinnitas* also "covers" obliges the reader to suppose that Wittkower came to regard *eurhythmia* and *concinnitas* as synonyms.

That however was clearly not Alberti's position. For as Wittkower acknowledged in the initial version, Alberti's *concinnitas* comprises the three distinct standards of *numerus, finitio* and *collocatio,* and is achieved only when all three are met.[12] Of these three however only *finitio* refers to a structure's proportions ("*aut maiorem minoremque redegeris*").[13] Clearly, if *concinnitas* encompasses *numerus, finitio* and *collocatio* it could not be a synonym for the alleged system of proportion that Wittkower wanted his readers to recognize as *eurhythmia*![14]

In the later versions Wittkower dealt with this difficulty *tout court* by deleting his earlier acknowledgement of the threefold nature of *concinnitas*. In its place he now substituted the

---

proportio is not to achieve *eurhythmia* but *symmetria* (another confusing term in Vitruvius' use of it).

[11] First version, p. 8, fn. 2. Wittkower did not identify those who were wise enough not to "generally admit" this point.

[12] First version p. 1, fn. 4; p. 8, fns. 1 and 2.

[13] *De re aedificatoria,* Bk. IX, cap. 5.

[14] Alberti (1956, p. 90f., quoted Panofsky 1972, p. 26) offered a somewhat different concept of *concinnitas* in his *Della Pittura,* which was written about 1435, perhaps two decades before *De re aedificatoria.* He depicted *concinnitas* in the earlier work as all parts of a work "agreeing with each other", which they will do if "in quantity, in function, in kind, in color, and in all other respects they harmonize (*corresponderanno*) into one beauty". Without any apparent thought of *numerus, finitio* or *collocatio* this version of *concinnitas* is quite different from the one that appears in *De re aedificatoria,* and Wittkower may mistakenly have relied on it instead.

statement that *concinnitas* is "a correlation of qualitatively different parts – Alberti's *finitio*".[15] Why Alberti would have employed two different terms for what (according to Wittkower) was the same thing remains an open question. It should be said, moreover, that Wittkower's "correlation of qualitatively different parts" is an opaque phrase, and one without precedent.

More to the point though is that with this sleight of hand Wittkower redefined *concinnitas,* disencumbering it of *numerus* and *collocatio* and making of it merely a synonym for *finitio* (and thus by implication of *eurhythmia* as well)! This definition, whatever its merits may be, is not one that was provided by Alberti. Alberti nowhere declared that *concinnitas* refers to architectural proportion, or that it is attained (let alone that it can only be attained) by the application of a single system of proportions throughout a structure.[16]

Wittkower's attempt to link Alberti's concepts to Classical architectural theory has a further flaw. Alberti, as anyone familiar with his work will know, doubted the value of literary sources for the study of Classical architecture. The best way to deduce the theories of the Ancients, he wrote, was to

---

[15] Fourth ed., p. 42.

[16] We may mention here that Wittkower's discussion of Alberti's precepts for calculating architectural proportions is also flawed. Alberti stated that these proportions may be derived not only from harmonic chords but from arithmetic and geometric means, even though the ratios of the latter, such as 4:6:9 or 9:12:16 represent dissonances. In the initial version of his study Wittkower declared that Alberti required proportions to be based only on the ratios of harmonic chords: "Proportions recommended by Alberti are the simple relations ... which are the elements of musical harmony". Wittkower retained this misleading remark in all the postwar versions, but added to them the statement that for Alberti "The ratios of the musical intervals are only the raw materials for the combination of spatial ratios". He also acknowledged that Alberti was "well aware ... that not every proportion using the mean method of calculation results in a musical consonance". These remarks of course do not really correct Wittkower's depiction of Alberti's methods for deriving proportions. Here as elsewhere we see Wittkower's disconcerting propensity to *appear* to correct an earlier statement – sometimes in a way that only adds another layer of error – but without acknowledging the effect of the correction on his overall thesis. It should also be pointed out that, contrary to Wittkower's position, musical chords were used in architecture long before the Renaissance, Cluny being perhaps the best-known example (cf. Prak, 1966).

examine the ruins of their buildings.[17] And he dismissed Vitruvius in particular as a source for understanding Classical architecture on the grounds that Vitruvius was often simply unintelligible:

> [Vitruvius'] speech was such that the Latins might think that he wanted to appear a Greek, while the Greeks would think that he babbled Latin. However, his very text is evidence that he wrote neither Latin nor Greek, so that as far as we are concerned, he might just as well not have written at all, rather than to write something that we cannot understand.[18]

We have seen enough now to conclude that in attempting to derive Alberti's architectural theories from Classical sources Wittkower undertook a task that was bound to fail, even with the questionable procedures in which he sometimes allowed himself to indulge. Unfortunately, Wittkower' tendency to circumvent the evidence would also manifest itself in the empirical portions of his analysis.

It is well known that the friars of Santa Maria Novella had required their architect – presumed by many to have been Alberti - to incorporate extensive portions of the existing medieval façade into his new design.[19] In the initial version Wittkower declared that there is "unambiguous" evidence that Alberti "believed himself to be faithfully continuing the existing

---

[17] *De re aedificatoria* Bk.VI, i - "*Restabant vetera rerum exempla templis theatrisque mandata ex quibus tanguam ex optimis professoribus multa discerentur*". As we saw in the appendix to Chapter Six, however, there is virtually no evidence that Alberti actually *did* survey any ancient structures.

[18] *Ibid.* Fréart (Fréart, Evelyn and Alberti, 2017, p.35) too was frustrated by Vitruvius' obscurity, commenting on a passage, "Another peradventure more subtile and penetrant than I am, might find out the mystery of these words, which I confess I comprehend not". Krautheimer (1969: "Alberti and Vitruvius") declared that to Alberti, Vitruvius was "only a starting point". He also declared, however that "Where Alberti really parts ways with Vitruvius is in his definition of the architect and of architecture", a statement that leaves one wondering what, in fact, that "starting point" could have consisted of, once "architect" and "architecture" have been removed from it!

[19] Borsi 1986, p. 64; Kiesow 1962.

portion of the façade."[20] For this claim to be consistent with Wittkower's thesis, the retained portions of the earlier façade would have had to conform to a single system of proportion, based on progressively-halved squares, which Alberti then extended to the parts that he added to the structure. Although this possibility cannot be ruled out *a priori*, it requires one to believe that the medieval builders of the earlier façade had themselves worked in obedience to the requirements of what Wittkower represented as *eurhythmia*. But to believe that, of course, would be to deprive Alberti's façade of its distinction as the "Renaissance landmark of Classical *eurhythmia*" that Wittkower had bestowed upon it! [21]

It transpires, though, that even by Wittkower's own reckoning not all parts of the Santa Maria Novella façade were derived from the system of proportion based on the progressive halving (2:1) of the façade's overall square, that he alleged was used throughout the structure. In the same paragraph in which he characterized that system as "the Leitmotif of the whole façade", Wittkower acknowledged that the ratio of height to width of the entrance bay is 3:2, and that the "square" incrustations of the attic are one-third of the attic's height. These ratios, of course, contradict Wittkower's fundamental thesis. Moreover, in the initial version Wittkower had also claimed, consistent with his "Leitmotif", that the volutes on the upper

---

[20] The evidence is *not* unambiguous. Wittkower's argument rests in part in part on mistaken assumptions about which were the earlier portions of the façade, a matter on which he changed his original opinion. It also rests on Alberti's perhaps too-frequently quoted letter to de' Pasti about embellishing what has been built rather than spoiling what remains to be built. Ambiguous and possibly platitudinous as this statement is, Alberti did not write it about Santa Maria Novella but about his work on the church of San Francesco in Rimini. In the 3rd and 4th eds. Wittkower dropped his claim that Alberti wished his façade to be "a faithful continuation in idea and form" of the older parts of the structure and substituted for it the bland truism of "Alberti's patent wish to harmonize [Wittkower presumably used this term in a non-technical sense] his own work with the parts already *in situ*".

[21] The same issue arises with regard to Wittkower's work on Palladio: according to Benelli (as noted in fn. 3, *above*), "Wittkower intentionally did not acknowledge evidence that Palladio's villas were often built upon preexisting medieval constructions". The latter show no evidence of harmonic proportions: but if they did show such evidence it would only establish that those proportions could not have been an innovation of the Renaissance!

storey could each be fitted into an imaginary square one-half the size of the principal interior squares. In the postwar editions he moved the discussion of those squares from the body of his text to a footnote, where he now declared – *more suo* without addressing its impact on his proportional thesis - that those squares "are related to the height of the attic as 5:3, or to the height of the upper tier as 5:6". This too of course contradicts his thesis regarding the use of the 2:1 proportion throughout the structure.

Another modification introduced with the 1949 edition was that the square incrustations, as Wittkower called them, of the attic were "related to the diameter of the [central portal's] columns as 2:1".[22] The full significance of this new datum first became apparent only in the third edition, in which Wittkower explained that "it is precisely the derivation of the system from the diameter of the column (Vitruvius' module) that differentiates Alberti's approach from that of the Middle Ages".[23] This modification shifted the entire derivation of the façade from the square in which Wittkower had first claimed that the façade was enclosed to the diameter of the columns flanking the main doorway. But since (according to Wittkower) the attic squares were not identifiably in a 2:1 relation to anything but the diameter of those columns (he said that they were in a 1:3 relation to the attic *height*) what was now left of Wittkower's entire thesis that "the place and size of every single part and detail" of the entire façade was "fixed and defined" by the progressive halving of the overall façade square – or indeed by any other consistent system?

Wittkower's *volte face* is buried in a mere footnote, his preferred *locus* for announcing fundamental alterations of his thesis. In the text above that footnote however he retained the principal statement of the thesis – the one in which he declared that every detail of the design derives from the imaginary square in which the entire façade is contained. It, along with the footnote that introduced the entirely different derivation from the columnar diameter, thus appear on the same page, seemingly as

---

[22] 1st ed., p. 49; and in all later editions. This ratio, it should be clear, is not connected with the halving of sizes that Wittkower referred to elsewhere, for it is one-half of something which is one-third of something else, i.e., the height of the attic.

[23] 3rd. ed, p. 47, fn. 1; also in 4th ed. The reference is to Vitruvius I.2.4, where the diameter *at the base* of the column is given as one of three possible modules for "sacred buildings" (*in aedibus sacris*).

part of the same thesis: whereas in fact the one is a repudiation of the other.[24]

Wittkower's revised thesis however is questionable for another reason, too. Examples of the columnar diameter being used as a module are to be found in pre-Renaissance buildings where not even a reckless scholar is likely to discern the influence of Vitruvius. Santa Maria Novella is often compared to the beautiful church of San Miniato al Monte (circa 1100) that looks down on Florence from across the Arno. The diameter of the un-tapered columns on San Miniato's façade is, by my calculations (based on measurements of my own photographs of the façade) employed as a module on many parts of the structure. It is multiplied by eight to create the diameter of the arches of the blind arcade that runs across the lower storey of the façade; the white marble panels below it are three-tenths the diameter of those arches, which is the same as the width of the recessed panels on the door of the central portal; the two sets of hatch-marks which so brilliantly link the two storeys, are also on this module.[25]

---

[24] Wittkower's revised version of the analysis, based on the columnar module, is perhaps to be understood in the context of the criticism by Ackerman (1951) who, in a review of the first edition, rejected Wittkower's claim that Alberti's design for the Santa Maria Novella façade had been derived from the Classical model because, as Ackerman pointed out, it "lacks the module which connects plan to elevation". The design, according to Ackerman, was not Classical at all but "the rationalized offspring of the Gothic elevation *ad quadratum* which establishes interrelated modular squares within an embracing square". In the revised version Wittkower evidently sought to supply the missing module. However, the one that he now purported to have found fell short of Ackerman's demand for a module that links elevation and plan. (To this day, indeed, no relation between the two has been discovered on the Santa Maria Novella facade.) Ackermann erred in accepting Wittkower's claim that the façade's design is based on the successive halving of squares, a position which, as we have seen, Wittkower himself would implicitly repudiate.

[25] Thiersch (2017, pp. 37-41) found that proportions are repeated in Romanesque and Gothic churches. Panofsky (1957) implicitly challenged Wittkower's view that the application of a single system of proportion throughout a structure is a Renaissance innovation. The entire system of a High Gothic cathedral, according to him, can be derived from as seemingly inconsequential a detail as "the cross section of one pier". Von Simson (1962, p. xvii), while commending Wittkower's work as "brilliant", demolishes his central thesis by pointing out that "a continued tradition" regarding proportion links medieval and

The objections raised up to this point would be less
consequential if the actual dimensions of the Santa Maria Novella
façade were consistent with Wittkower's analysis. Unfortunately,

---

Renaissance architecture. Neagley (1992) makes the case that the design
of the late-Gothic church of St. Maclou in Rouen is based on a 1:2 module.
Yet as Ackerman (1991, p. 225) notes, in the late 14th century design of
Milan's cathedral two different geometrical systems were employed, so
that "the chief purpose of the triangle – to provide a unified correlation
of the parts and the whole – is ignored". If we are to credit Heydenreich
(1996, pp. 16-7, 20), at the threshold of the Renaissance – that is, before
Alberti - the design for San Lorenzo by Alberti's rather older
contemporary Brunelleschi (1377 - 1446) evolved from the square of the
crossing and is "governed by a single scale of proportions"; and his Santo
Spirito is also based on the crossing square, from which the structure's
parts have evolved "in strictly observed proportions" that are "still more
exact in their proportional integration" than in S. Lorenzo. Heydenreich
unfortunately did not disclose the procedures used to arrive at the
measurements on which these statements are based, but the record of his
behavior during the Nazi years does not dispose one to giving him any
benefit of the doubt. Battisti (1981, fig 197, p. 188) implicitly called into
question the role of proportion in Brunelleschi's work by showing that
the arches on the right-hand side of the nave of San Lorenzo vary
inconsistently in height; as do all the nave arches of San Spirito (*ibid, fig.*
216, p. 213; p. 197), where, moreover, the diagonals of the vaulting in a
corner bay are unequal (*ibid,* pl. 214). Battisti (*ibid,* p. 114) also discovered
that the sides of the drum on which the great dome of Santa Maria del
Fiore rests differ in length from one another. These differences have now
been measured with pinpoint accuracy by Dalla Nagra (2004, text vol. p.
36) whose photogrammetric survey shows that the drum is irregularly
asymmetric, with a difference of 57 cm. (22-1/2") between the shortest
and longest side. Brunelleschi, who is known to have supervised work
on the dome very closely, would have been aware of this, of course; and
so we may suppose that he accepted the fact that his greatest architectural
feat would not have a base with Heydenreich's "strictly observed
proportions" to rest on. With regard to the prominence with which issues
of proportion were considered by Renaissance builders I would add that
Vasari – in condemning or praising the design of a building, and in his
lengthy excoriation of Gothic architecture - refers only occasionally and
in passing to issues of proportion. In his comments one does not even
dimly hear references to the criteria stated by Wittkower. And see the
excellent discussion by Anthony Blunt (1940, p. 91). Blunt's important
analysis, which invites reconsideration of the role of rationalistic schemes
of proportion during the Renaissance, is seldom referred to by other
scholars, for reasons which possibly have more to do with Wittkower's
commanding influence than with Blunt's perfidious character and
conduct: (among those he does not scruple to acknowledge in the preface
to *Artistic Theory* is one "Mr. Guy Burgess").

Wittkower never disclosed what, in standard measuring units, the façade's dimensions are or how he determined what they are – a curious shortcoming, surely.   Uncertainty about those dimensions was to some extent put to rest when the façade was measured by professional surveyors working under the aegis of the Institute for the Restoration of Monuments of the Faculty of Architecture of the University of Florence; the results were published in 1970.[26]

The survey's measurements, obtained with the use of photogrammetric techniques, are certainly very accurate. They are incomplete however, for the surveyors measured only the left-hand side of the façade. They did so on the grounds that what they mistakenly believed was "the perfect [bilateral] symmetry of the façade" - *la perfetta simmetria della facciata stessa* - made it unnecessary to incur the expense of measuring the other half!

The division of the two halves was set at a line drawn through the middle of the main door into the church. My own photographic analysis however establishes that this door is not accurately centered on the façade and so, because the surveyors left us with no indication of the width of the "half" that they did not measure, we do not know what the width of the entire façade is. We know the façade's height from the survey but that of course is not enough to tell us whether the façade *is* or is not a square.[27] In *fig.* 8.1 I have proved that in fact the façade is *not* square and that Wittkower's famous diagram of it is inaccurate.

Nevertheless, other measurements obtained by the survey do provide a basis for empirically evaluating Wittkower's thesis. They fail to support Wittkower's claim that the height of the entrance bay is one-and-a-half times its width.[28] They also

---

[26] Bardeschi (1970: plates 1-4 and the section on criteria and methods used in the study on p. 23 of the text volume). By a curious omission Bardeschi does not relate these findings – or even refer - to Wittkower's proportional scheme.

[27] The survey shows that the façade is 35.225 meters tall; the "half" that was measured is 17.825 wide. Assuming that this really was one-half of the façade's width would mean that the width of the structure is 35.650 meters, in which case the façade would be 0.425 meters (16-3/4") too wide to form a square. Correctly centering the measurement on the door (and assuming that the door was centered on the façade) would extend the structure's width by another few centimeters and further increase the discrepancy between height and width.

[28] 4th ed., p. 46. Comp. Bardeschi, pl. 2, showing that the entrance bay is 8.212m. wide and 11.418 m. tall measured from the outer edge of the two

show that the two storeys of the structure are not the same height, as Wittkower claimed they were. The lower storey is 18.000 m. tall, while the upper storey is 17.225 m. (a difference of over 30 inches.) If, as there is no reason to doubt, Alberti was free to determine the height of the upper storey, the fact that he did not make it the same as the height of the lower storey suggests that, whatever his considerations may have been, they are unlikely to have been the ones that Wittkower wanted us to believe guided him.

The survey's measurements also establish that on the left half of the attic the figures which Wittkower identified as "squares" are in fact oblong rectangles whose widths vary irregularly (the largest and smallest differ by about 6%), as do the spaces between them (by up to 62%).[29] Although the survey did not measure the *heights* of these rectangles they seem to be much more consistent with each other than their breadths, *but in no instance is the height of the rectangles one-third the height of the attic*, as Wittkower had claimed. Nor is the width of any of these rectangles twice that of the diameter of the columns, as Wittkower had stated in the later versions. Their average width is 5.79% greater than the columnar diameter. There is therefore no fixed relation between these "squares" and the base diameter

---

columns at their base, and from the base of the columns to the top of the cornice. The actual height of the bay is therefore 90 cm. (35-1/2"), or 7.4%, less than the 12.318 m. (i.e., 3/2 of 8.212) called for by Wittkower. It should be noted, moreover, that the 2:3 ratio given by Wittkower for the proportions of a doorway is not found in *De re aedificatoria*. There, in I. 12, Alberti states that the proportions should be either 1:2 or 1:1.414 (i.e., $x\sqrt{2}$, where x is a square the width of the entrance bay), to accord with which the Santa Maria Novella portal should have a height of either 16.424 m. or 11.612 m. (The actual height is a not-trifling 8 inches lower than the latter option.) That its dimensions do not conform to Alberti's precepts prompts one to question Vasari's report (p. 367, fn. 2, *supra*) that the portal was designed by Alberti.

[29] Bardeschi, *op. cit.*, pl. 2. From the left to the middle (the right-hand side of the attic was not measured) the widths of the rectangles, in meters, are as follows: 1.582; 1.564; 1.507; 1.506; 1.503; 1.520; 1.509; 1.492; and of the spaces between them: 0.944; 0.634; 0.624; 0.622; 0.598; 0.585; 0.585; 0.584. The variations in these measures do not bear out Wittkower's notion (4th ed, p. 158) that "commensurability of measure [is] the nodal point of Renaissance aesthetics".

of the columns on the lower story (or between the dimensions of these "squares" and the height of the attic).[30]

But even if Wittkower's description of the attic's "squares" and their relation to the column diameter and to the attic's height had been accurate, it would not have sufficed to validate his thesis. What would have been needed is evidence of a consistent ratio linking the columnar diameter to *all* the other details of the façade – "the place and size of every single part and detail" of the structure, as Wittkower had claimed. Wittkower fails to offer this evidence. Indeed, most of the details of the facade's design are ignored by him, leaving one to wonder what it was about the misnomered attic "squares" that led him to single *them* out. It should be noted that no later scholar has attempted to relate the columnar diameter to the details of the rest of the structure. My own efforts to do so – I have spent hours on that task – have yielded no such correlation. The empirical evidence for Wittkower's (revised) thesis therefore rests solely on factually incorrect and unexplained assertions about the dimensions of the attic "squares" in relation to the base of the column and the height of the attic.[31]

In 1971, a year after the Florence University survey was published, Wittkower issued the final edition of his *Architectural Principles in the Age of Humanism*. He did not modify his analysis of the façade to reflect the survey's findings, however. Indeed, he did not mention the survey.

Two seemingly empirical studies of the Santa Maria Novella façade have appeared since the publication of the Florence University survey. Franco Borsi, himself a professor of architectural history at the University of Florence, published his book-length study of Alberti in 1973, or two years after the publication of his own department's surveyed measurements of

---

[30] Bardeschi, pl. 2, shows that the diameters of the two columns on the left (those on the right were not measured) vary slightly, within a reasonable margin of error.

[31] Ostwald (2000) speculated that Wittkower's analysis "is traced on inaccurate drawings of the façade", as was the earlier analysis by Wölfflin. The "gross liberties" which Ostwald found in each man's analyses were not however detected by Evans (1995, p. 248) who, on the contrary, described them both as "equally convincing", a surprising statement in view of the fact that the two analyses differ substantially from each other. Ostwald's charge that the two analyses each transposed elements that are well behind the façade onto their elevations is left unexplained, but seems unfounded.

the façade of Santa Maria Novella.[32] In his book Borsi quoted, sympathetically and at length, Wittkower's analysis of the façade's proportions, and accompanied his text with an image in which his own line drawing is superimposed on the surveyed drawing of the façade made by his colleagues at the university. Curiously, Borsi did not refer to this image in his text, but it would seem to have been intended as confirmation of Wittkower's thesis. The result however is a fiasco. The upper-storey square in Borsi's drawing (which in fact turns out not to be a square) seems to be repeated on the lower storey (in fact, it is a shorter rectangle), where however it reaches down only to a point that is well above the threshold of the church; the "squares" enclosing the upper-storey volutes prove to be rectangles, not squares, each of whose exterior dimensions are different; while a seeming semi-circle drawn from the right flank to demonstrate the relation between the upper storey and the volutes proves to be an arc of about 160 degrees.[33] Borsi's drawing too has a "no-man's land" between the "squares" of the two storeys, similar to that of Wittkower, and he too neither explained nor referred to it in his text; he also failed to mention the revised version of Wittkower's analysis relating the structure's proportions to the columnar diameter. One may well wonder at the fact that Borsi did not relate the measurements, taken barely two years

---

[32] Borsi, 1986, p. 68. For the discussion of Santa Maria Novella see *ibid.,* pp. 61-75.

[33] Naredi-Rainer (1997, p. 179, n. 9 and accompanying text), noted that the "square" in Borsi's line drawing of the façade of San Francesco, Rimini, is in fact an oblong. He charged Borsi with "manipulating" the evidence to conform to "the proportions he wanted" to find - *"Wunschproportionen"*. Naredi-Rainer was just as unenthusiastic about Wittkower's work on the Santa Maria Novella façade. He rejected Wittkower's thesis regarding the progressively halved squares and noted that "Wittkower's proportions were inexactly drawn". I have not been able to obtain a copy of Naredi-Rainer's work, and know it only from references to it by Lorch (1999, pp. 45 - 46). Lorch congratulates herself on being the first to notice that Wittkower's system of squares cannot be related to the squares of the great volutes on the upper storey, yet seems unaware that in fact the volutes can only be circumscribed by an oblong and that the dimensions of the right-hand volute are different from those on the left. Wittkower does not report, however, that this volute was clad in marble only in the 1920's. It covered a somewhat smaller trapezoidal buttress that had been in place since at least the 15th century. The volute as we see it therefore tells us nothing about the intentions of the people who reworked the façade in the 15th century.

previously by his own department's surveyors, to Wittkower's analysis of the façade proportions, let alone acknowledge that the former refuted the latter.

A British architectural historian, Robert Tavernor, in a study of Alberti's work, included a line drawing of the Santa Maria Novella façade "overlaid with geometry and proportions derived from an encompassing square".[34] This drawing was based, Tavernor wrote, on a photogrammetric image of the façade that had been made in connection with an international exhibition about Alberti sponsored by the Olivetti Corporation and directed by Tavernor and Joseph Rykwert.[35] Photogrammetry uses sophisticated software to measure an object or parts of it with great precision – it is a very reliable procedure. Tavernor's image however is inaccurate. For example, although the "squares" in the attic and the spaces between them were shown by the University of Florence survey to be irregularly variable in width and in spacing, in Tavernor's image every rectangle has become a square of the same size, except for the outermost ones on the left and right, which however have the same dimensions as each other; the spaces between all of them are identical. Other irregularities in the structure have also been corrected in this image – for example all the horizontal lines of the frames enclosing the white panels on the lower storey are

---

[34] Tavernor 1998, pp. 99-106. An indication of Tavernor's *modus operandi* is his acknowledgment (1998, p. 103) that although "the façade lacks a precise symmetry... there can be little doubt that Alberti intended the composition of number and geometry to be regarded as perfect". How Tavernor came upon his knowledge of Alberti's unrecorded intentions remains a mystery, as does the question of why Tavernor bothered to measure the structure in the first place if he already knew how Alberti intended it to look! Although Tavernor included Naredi-Rainer (1977) in his bibliography he did not mention the latter's reservations about Wittkower's analysis or his view that the measurements on which Wittkower based his analysis of the façade of Santa Maria Novella are incorrect.

[35] Tavernor 1998, p. xi. Precise measurements of any part of photogrammetric images can be made by using shareware programs such as Autodesk Design Review. Tavernor's image had been posted online (www.bath/ac/uk/ace/uploads/alberti/smn-e-l.dwf) but was removed, without explanation, shortly after I sent a message to Tavernor asking about the discrepancies between his image and the results of the Florence University survey. Tavernor did not respond to my query. I have taken the liberty of posting Tavernor's image at www.keep-ahead.org/ SMNisnotsymmetric.dwf

shown by Tavernor as being on the same plane whereas in reality some of the ones on the left slant upward toward the center. Also, the black-and-white stripes on the pilasters, which do not all match each other in reality, do so in Tavernor's image; and the great volutes on the upper storey are the same size as each other.

These anomalies cannot be reconciled with Tavernor's claim to be presenting a *bona fide* photogrammetric image. His image, rather, is a concoction: a fraud: yet another of the reprehensible but all-too-frequent instances documented on these pages of what one might call the academic equivalent of journalism's "fake news". Like Borsi, Tavernor did not address Wittkower's revised version, which has the façade's proportions derive from the diameter of the column. Nor did he attempt to account for - indeed, he did not mention - the discrepancy between the measurements of his photogrammetric image and those made by the surveyors of the University of Florence.

We are obliged to conclude, then, that the evidence - for all that it is incomplete - clearly establishes that no rational system of proportion was used in the design of the façade of Santa Maria Novella.[36] Alberti did not use his standard of *finitio* on the façade, and nothing is gained by trying - as some do - to explain this fact away.[37] It must also be said that Alberti's criterion of *collocatio* too did not help shape the design of the Santa Maria Novella façade. The façade is *not* bilaterally symmetric, though its strong and (more or less) central axis somewhat obscures that fact; and in view of the Dominicans' requirement that the architect retain many portions of the medieval façade, which was asymmetric, it would have been impossible to make it symmetric.[38] In this respect, at least, Wittkower was somewhat

---

[36] Saalman 1959 (p. 94 and fns. 11, 13), referring specifically to Alberti's contemporaries Brunelleschi and Filarete, suggests that the elevation may be inherent, as it were, in the plan, but this suggestive idea clearly does not apply to Santa Maria Novella, the relation of whose plan to elevation is altogether irrational.

[37] For example: "*a dimostrare che le misure necessariamente offerte dall' Alberti per realizionare le varie pari della composizione all'insieme di facciata non sottostanno rigidamente alle prescrizione del de re aedificatoria, ma si adattano con sufficiente elasticita a rendere communque un'immagine estremamente bilanciata e matura degli elementi architettonici utilizzati,constituendo un insieme basta sulla forma quadrata e sulle sue aggregazioni*" - Nocentinni, 1992, p. 43.

[38] Ironically, it is implicit in Wittkower's thesis of progressively-halved squares that the façade would be (but of course in reality is not) bilaterally symmetric: and indeed, Wittkower's line drawings show a

structure whose two halves mirror each other in every detail that he recorded on them. A simple test of whether or not the façade is symmetric – one readily available to Wittkower – is to trace the outline and other details of a photograph of the façade on tracing paper and then to reverse the sheet and determine whether the drawing still conforms to the details of the photograph. Today, this can be done even more accurately (as I have done: see *fig.* 8.1) with any simple computer graphics program. The result of such procedures shows that the façade is not symmetric. Using this procedure, we also find that a square derived from the width of the upper storey extends from the peak of the pediment only to a point well above the attic, where it intersects the bottom of the circular window; that a somewhat wider square which contains the lateral tips of the pediment extends half way into the attic; and that a square derived from half the width of the lower storey is altogether too large for the upper. The irregularity of the attic "squares" is immediately apparent from this procedure. So are the differences between the two great volutes on the upper storey. These, it transpires, are of different widths, reflecting the fact that the upper storey is not centered on the lower; the volutes also differ from each other in height. Imaginary frames encompassing the volutes would therefore be of different sizes, and both would be oblong, not square. The procedure also shows that asymmetric arrangements abound on the façade. Specifically, we see that neither the main door nor the portal which contains it – both generally thought to be by Alberti - are centered on the façade or for that matter on each other; that the three smaller arches of the arcade on the right not only differ in width from each other but are significantly narrower than those on the left, which also differ in width from each other; that the arch over the right-hand portal, though wider than the others on its side, is not as wide as the arch over the left-hand portal; that the vertical white panels are all of different widths and that the horizontal lines of the upper row of them on the right-hand side of the façade are higher than those on the left, but that the bases of the four blind arches on the left-hand side are higher than those on the right; and that the black-and-white bands on the pointed arches of the *avelli* on the right side do not meet the horizontal bands on the walls at all the same points as their opposites on the left. (Kiesow 1962, p. 3) gives the measurements of the *avelli*, starting with the one on the extreme left, as 2.34, 2.325, 2.34, 2.34, 2.33 and 2.33 meters, but I find that no two have the same width and that the difference between the widest and narrowest is 4.1% (The Florence University survey does not measure the widths of the *avelli* and its drawings represent them as being identical.) These irregularities are not found on the long wall of tombs around the corner in the arcade which runs parallel to the nave.) We have already pointed to the asymmetry of the arrangement of the attic's row of encrusted "squares", but the asymmetry of the ribbon of sail-like figures – a Rucellai device – which stretches across the width of the façade remains to be mentioned. (It is not of much matter however, being an asymmetry only because it is a continuous strip and not divided in the

justified in not mentioning *collocatio* in his analysis of the Santa
Maria Novella façade.

A careful reading of Alberti's architectural treatise
moreover raises doubts about whether Alberti intended *finitio* to
be a standard that applies to the facades of churches or indeed of
any other buildings. In chapters 5 and 6 of the Ninth Book of *De
re aedificatoria* Alberti stated that *finitio* can be used for "squares
and open areas" and for "platforms". These of course differ from
facades by being horizontal and not vertical, and by not being
integral parts of (three-dimensional) buildings. Alberti also stated
that *finitio* can be applied to three-dimensional structures, but he
specified its use only for determining the proportions of various
kinds of rooms. He did not state or imply that *finitio* could also be
used for computing the proportions of a building's exterior.[39] *In
fact, as far as we know Alberti never addressed the question of what the
proportions of a church façade should be.* He did, to be sure, declare
(VII. 14), that façades of basilicas should be one and a half times
as high as they are wide, but it is clear that he did not intend this
formula to apply to the dimensions of a church façade, for his
discussion explicitly contrasts the design of basilicas (as palaces
or halls of justice) to that of "temples", or churches.[40] In any case,

---

middle with the figures on one half reversing the direction of those on
the other.) The specific distributions of asymmetric features in the lower
portions of the structure do not appear to be guided by any aesthetic or
philosophical program. Such, seemingly unprogrammatic, asymmetries
are a commonplace in medieval churches. In Florence alone we can point
to the earlier facades of Santa Croce and Santa Maria del Fiore and the
surviving facades of Santa Maria Maggiore, San Miniato al Monte, and
San Lorenzo, among others, as examples of medieval asymmetric
facades. It scarcely needs saying that the irregular, asymmetric features
of the Santa Maria Novella façade – so different from its representation
in the line drawings of Wittkower, Borsi, Tavernor and others - preclude
the possibility that a consistent system of proportion was applied to its
design.

[39] In *De re aedificatoria* (IX.6) Alberti stated that harmonic chords can
determine the ratios of "all the three lines of any body whatsoever" but
he specified only the dimensions of "public halls, council chambers and
the like" (IX.5), while ratios "not derived from harmony" are used for
"the three relations of an apartment"(IX.6).

[40] Rykwert, Leach and Tavernor (Alberti 1988, p. 396) write: "One of the
problems of reading this passage is that no ancient basilica façade
survives, and the passage must therefore be read in reference to Alberti's
church facades". This is surely (1) a *non sequitur* and (2) mistaken on two
grounds, the first being that Alberti was very clear that his discussion of

even if we do not yet know the overall dimensions of the Santa Maria Novella façade if is obvious that its height is very much less than one-and-a-half times its width.

It is baffling that Alberti never did prescribe the proportions of a church's façade.[41] Failing that, however, one might have expected to find a rational relationship between the dimensions of the Santa Maria Novella façade and the plan of the church, yet no such relationship seems to exist.

We are thus left to assume that the proportions of the façade were determined by contingent factors that are unknown to us today but are likely to have included the requirement imposed on the 15th century architect to preserve certain portions of the original structure. The fact that the façade does not reflect any of Alberti's ideas regarding *concinnitas* could perhaps suggest that the question of his contribution to the façade's design ought to be revisited. [42]

We conclude then that Wittkower's attempt to establish that the standard by which Alberti designed the façade of Santa Maria Novella was the criterion of proportion – *finitio* - has nothing to commend it. Moreover, Wittkower's attempt to exclude *collocatio* from Alberti's *concinnitas* also has nothing to commend it and calls for no revision of our position that, for Alberti, symmetry was a requirement of the utmost importance. In fact, as we look at the three church facades – those of San Sebastiano and San Andrea in Mantua, and the unfinished facade

---

basilicas did *not* apply to churches; and the second being that none of Alberti's church facades is anywhere nearly one and a half times as tall as it is wide. (There is some dispute as to whether Alberti's *latitudo spatii* in this passage refers to the internal measurements of the nave rather than to the width of the whole façade; but Leoni, from Bartoli, gives it as the façade's width, and Orlandi accepts this.) Of course, if Rykwert, Leach and Tavernor were correct that Alberti required church facades to be 3:2 Wittkower's entire hypothesis would be disproved and we would also have good reason to conclude that the Santa Maria Novella façade we see today could not have been the work of Alberti!

[41] Lang (1965) proposed that the "all-embracing key to the whole building" in Alberti's theory is the ground plan. Ackerman (1954) made a similar point, though about 16th century architects, not 15th century ones as Lang mistakenly wrote. Yet the fact remains that Alberti said nothing on this subject, and not even the most audacious have suggested that Alberti designed or rebuilt the *plan* of Santa Maria Novella.

[42] See Lorch (1999) for differing views of Alberti's contribution to the overall design.

of San Francesco at Rimini - that we know are by Alberti we find, ironically, that while they conform to the standard of *collocatio* (which is to say that they are symmetric), they do not conform to the proportions Wittkower defined as "*eurhythmic*".[43]

---

[43] Wittkower 1971, pp. 55, 58. Wittkower's drawing of the hypothetical complete façade of S. Francesco (*ibid*, p. 46) renders its proportions as approximately 1:1.10 - not an "eurhythmic" ratio!

WORKS CITED OR CONSULTED

Ackerman, James (1951): review of Wittkower's *Architectural Principles... Art Bulletin,* v. 33.

-- (1954): "Architectural Practice in the Italian Renaissance", *J. Society of Architectural Historians* v. 13.

-- (1986): *The Architecture of Michelangelo.*

-- (1991): *Distance Points. Essays in theory and Renaissance art and architecture.*

-- (2002): *Origins, Imitation, Conventions.*

-- (1991): *Palladio.*

Adam, Leonhard (1936): "North-West American Indian Art and its Early Chinese Parallels", *Man* v. 36.

Adams, Henry (1933): *Mont-Saint-Michel and Chartres.*

Addison, Joseph (1705): *Remarks on Several Parts of Italy etc., in the years 1701, 1702, 1703.* (Kindle ed.)

-- (1712): *The Spectator* No. 411, June 21, 1712.

Adler, Borrmann, Doerpfeld and others (1966): *Baudenkmaeler von Olympia* (repr. ed.)

Ahuja, Dilip and M.B. Rajani (2016): "On the symmetry of the central dome of the Taj Mahal", *Current Science* [India], v. 110.

Alberti, Leon Battista (1965): *The Ten Books on Architecture.* Leoni translation, ed. Rykwert.

-- (1966a): eds. Orlando and Portoghesi, *Leon Battista Alberti l'Architettura [De re aedificatoria].*

-- (1966b): *On Painting (della Pittura)* (rev. ed., trans. Spencer).

-- (1972): *On Painting and On Sculpture: The Latin texts of De Pictura and De Statua,* ed. Grayson.

-- (1988): ed. and trans. Rykwert, Leach and Tavernor, *Leon Battista Alberti on the Art of Building in Ten Books.*

Albertini, Francesco (1510): *Memoriale di molte pitture e statue che sono nella inclyta ciptà di Florentia* (eds. Milanesi, Guasti and Milanesi, 1863).

Alexander, Christopher, Hansjoachim Neis and Maggie Moore Alexander (2013): *The Battle for the Life and Beauty of the Earth.*

Allen, Grant (1879): "The Origin of the Sense of Symmetry", *Mind,* v. 4.

Alsop, Joseph (1982): *The Rare Art Traditions.*

Anderson, P. W. (1972): "More is Different", *Science,* v. 177.

(Anon) (1894): "Snow Crystals", *Symons Monthly Meteorological Magazine,* v.336, pp.177-8.

Anon [John Gwynn] (1742): *The Art of Architecture.*

Arnheim, Rudolf (1966): *Toward a Psychology of Art.*

-- (1977): *The Dynamics of Architectural Form.*

Bacon, Francis (1909): "Of Gardens" in *Essays, Civil and Moral* (Harvard Classics)

Bailey, Anthony (2011): *Velazquez and the Surrender of Breda.*

Bailey, Gauvin Alexander et al eds., (2005): *Hope and Healing. Painting in Italy in a time of Plague, 1500-1800.*

Baetjer Katharine and J. G. Links (1989): *Canaletto.*

Balanos, Nicolas (1938): *Les monuments de l'Acropole. Relevement et conservation.*

Baldinucci, Filippo (1697): *Vocabolario Toscana dell' Arrte del Disegno.*

Bardeschi, Marco Dezzi (1970): *La Facciata di Santa Maria Novella.*

Barrow, John (1990): *The World Within.*

-- (n.d.): "Simplicity versus Complexity: Plato and Aristotle revisited",athensdialogues.chs.harvard.edu/cgiin/WebObjects /athensdialogueswoa/wa/dist? dis=15

Bassin, Joan (1979): "The English Landscape Garden in the Eighteenth Century", *Albion*, v. 11.

Battisti, Eugenio (1981): *Filippo Brunelleschi the complete work.*

Beazley, J.D. and Bernard Ashmole (1966): *Greek Sculpture and Paintings.* (repr. ed).

Bell, Corydon (1957): *The Wonder of Snow.*

Bell, Malcolm (1980): "The stylobate and roof in the Olympeion at Akagras", *American J. of Archaeology*, v. 48.

Benelli, Francesco (2015): "Rudolf Wittkower versus Le Corbusier: A matter of proportion". *Architectural Histories*, v. 3.

Benes and Harris, eds. (2001): *Villas and Gardens in Early Modern Italy and France.*

Bentley and Humphreys (1961): *Snow Crystals* (Dover ed.)

Berenson, Bernard (1953): *Aesthetics and History.*

-- (1960): *The Passionate Sightseer.*

Bergman, David J. and Jacob S. Ishay (2007): "Do bees and hornets use acoustic resonance to monitor and coordinate comb construction?", *Bulletin of Mathematical Biology*, v. 69.

Bernal, J. D. (1937): "Art and the scientist" in eds. Martin, Nicholson and Gabo, *Circle: International Survey of constructive art.*

-- (1955): review of Weyl's "Symmetry", *The British J. for the Philosophy of Science*, v. 5.

Bernard, Emile (1912): *Souvenirs de Paul Cezanne.*

Betts, Richard J. (1993): "Structural Innovation and Structural Design in Renaissance Architecture", *J. Society of Architectural Historians*, v. 52.

Bialostocki, Jan (1963): "Renaissance Concept of Nature and Antiquity", *Acts of the Twentieth International Congress of Art History.*

Biondo, Flavio (2005): *Italy Illuminated*, ed. White.

Bloch. Marc (1954): *The Historian's Craft* (trans. Putnam).

Blomfield, Reginald (1892): *The Formal Garden in England* (repr. ed.)

Blunt, Anthony (1940): *Artistic Theory in Italy 1450-1600.*

Boardman, John (1968): *Archaic Greek Gems. Schools and artists in 6th and early 5th centuries.*

— ed. (1993): *The Oxford History of Classical Art.*

Boas, Franz (1907): "Notes on the blanket designs" in George T. Emmons, "The Chilkat Blanket", American Museum of Natural History Memoirs v. 3, as reprinted in Jonaitis (1995).

-- (1927): Primitive Art (Dover repr. 1955).

Boase, T. S. R. (1979): Giorgio Vasari. The man and the book.

Bober and Rubinstein (2010): Renaissance Artists and Antique Sculpture: a handbook of sources (2nd. ed.)

Boccaccio, Giovanni (1972): The Decameron (trans. McWilliam).

Bonsanti, Giorgio (1997): The Basilica of St. Francis of Assisi Glory and Destruction.

Borsi, Franco: (1986): Leon Battista Alberti The Complete Works.

Boucher, Bruce (2000): "Nature and the Antique in the works of Andrea Palladio", J. Society of Architectural Historians v. 59.

Boyle, Robert (1979): Selected Philosophical Papers of Robert Boyle (ed. Stewart).

Bridbury, A.R. (1973): "The Black Death", Economic History Review, vol. 26.

Brink, Joel (1978): "Carpentry and Symmetry in Cimabue's Santa Croce Crucifix", Burlington Magazine, v. 120.

Brown, Beverly Louise and Diana E.E. Kleiner (1983): "Giuliano da Sangallo's drawings after Ciriaco d'Ancona: Transformation of Greek and Roman antiquities in Athens", J. Society of Architectural Historians v. 42.

Brown, Patricia F. (1992): "The Antiquarianism of Jacopo Bellini", Artibus et Historiae, v. 13.

Bucher François (1972): "Medieval Architectural Design Methods, 800-1560", Gesta, v.11.

Buddenseig, T. (1971): "Criticism and Praise of the Pantheon", in ed. R. R. Bolgar, Classical Influences on European Culture A.D. 500-1500.

Bunzel, Ruth L. (1972): The Pueblo Potter a study of creative imagination in primitive art.

Burckhardt, Jakob (1985): The Architecture of the Italian Renaissance (ed. Murray).

Burke, Edmund (1796): "Letter to a Noble Peer".

-- (1939): On the Sublime and Beautiful (Harvard Classics)

Burke, John G. (1966): Origins of the Science of Crystals.

Burns, H. (1971): "Quattrocento Architecture" in ed. R.R. Bolgar, Classical Influences on European Culture AD 500-1500.

Businani, Alberto and Raffaello Bencini (1993): Le Chiese di Firenze Quartiere di San Giovanni.

Caglioti, Giuseppe, (1992): The Dynamics of Ambiguity.

Cantor, Norman (2001): In the Wake of the Plague.

Cardini, Franco and Massimo Miglio (2002): Nostalgia del Paradiso: Il giardino medievale.

Carmichael, Ann G. (1986): Plague and the Poor in Renaissance Florence.

Caroti, G. and A. de Falco (2002): "Geometric Survey for the Structural Assessment of the Architectural Heritage: the Case of the Cupola of the Baptistery of S. Giovanni e Reparata in Lucca", *International Archive of the Photogrammetry … Sciences*, v. 34.

Caruth, Cathy, ed. (1995): *Trauma: Explorations in Memory.*

Castell, Robert (1728): *The Villas of the Ancients.*

Chambers William (1772): *Dissertation on Oriental Gardening.*

Chiarini, Marco and Alessandro Marabottini, eds.(1994): *Firenze e la sua immagine. Cinque secoli di vedutismo.*

Choisy, Auguste (1865): "Note sur la courbure dissymétrique des degrés qui limitent au couchant la plate-forme du Parthenon", *Academie des Inscriptions et Belles-Lettres*, NS v. I

-- (1996): *Histoire de l'Architecture* (repr.ed.)

Clark, H.F. (1944): "Lord Burlington's Bijou, or Sharawaggi at Chiswick", *Architectural Review*, v. 110.

Clark, Kenneth (1981): *The Art of Humanism.*

-- (1974): *The Gothic Revival. An essay in the history of taste* (Icon ed.)

-- (1949): *Landscape into Art.*

Close, Frank (2000): *Lucifer's Legacy. The Meaning of Asymmetry.*

Cohn, Samuel K. (2002): *The Black Death Transformed: disease and culture in early Renaissance Europe.*

Cole, Bruce (1973): "Old and New in the Early Trecento", *Mitteilungen des Kunsthistorischen Institutes in Florenz*, v. XVII.

-- (1976): *Giotto and Florentine Painting 1280-1375.*

Collyer, Mary (1749): *Felicia to Charlotte: being letters from a young lady in the country…*

Colonna, Francesco (1499): *Hypnerotomachi Poliphili ubi humana omnia non nisi somnium esse ostendit at que obiter plurima scitu sanequam digna commemorat* (London, 1904 facsimile ed.)

-- (1890): *The strife of love in a dream: being the Elizabethan version of the first book of the Hypnerotomachia of Francesco Colonna.* Andrew Lang, ed., repr. of London, 1592 English translation by "R.D." [=Robert Dallington?]

-- (1999): *Hypnerotomachia Poliphili*, trans. Godwin.

Comito, Terry (1957): *The Idea of the Garden in the Renaissance.*

Cook, R.M. (1972): *Greek Art. Its development, character and influence.*

Cooper, Frederick A. (1996): *The Temple of Apollo Bassitas.* v. 1.

Corbusier (1986): *Towards a New Architecture* (trans. Etchells).

Cotman, John Sell and Dawson Turner (1822): *Architectural Antiquities of Normandy.*

Cotton, Charles (1683): *Wonders of the Peake* (2nd. ed.)

Coulton, G.G. (1930): *The Black Death.*

Coulton, J. J. (1982): *Ancient Greek Architects at Work. Problems of structure and design*, (2nd ed.)

Crawford, Virginia (1978): "Northwest Coast Indian Art". *Bulletin of the Cleve land Museum of Art*, v. 65.

Cresti, G. et al. (1987): *L'Avventura della Facciata.*

Crick, F. (1988): *What Mad Pursuit: a Personal View of Scientific Discovery.*

Crisp, Frank (1924): *Medieval Gardens.*

Crosby, Sumner McKnight (1987): *The Royal Abbey of Saint-Denis from its beginnings to the death of Suger, 475 - 1151.*

Cudworth, Ralph (1678): *True Intellectual System of the Universe...*

D'Alembert, J-B. le R.(1757) *"Reflections on the Use and Abuse of Philosophy in Matters that are properly relative to Taste"*, lecture delivered to the French Academy March 14, 1757, trans. in Gerard 1759.

Dallington, Robert (1605): *A Survey of the Great Dukes State of Tuscany.*

Darwin, Charles (1998): *Origin of Species,* (4th ed; Modern Library)

Davis, Michael (2002): "On the Drawing Board: Plans of the Clermont Cathedral Terrace" in Wu, 2002.

Deaux, George (1960): *The Black Death 1347.*

Dee, John (1570): *The Mathematical Preface to Elements of Geometry of Euclid of Megara,* repr. ed., n. d.

Defoe, Daniel (1927): *Tour through the Whole Island of Great Britain,* (Everyman ed.)

Delaine, J. (1997): *The Baths of Caracalla: A study in the design, construction, and economics of large-scale building projects in Imperial Rome.*

Delougaz, P. (1960): "Architectural Representations on Steatite Vases", *Iraq* v. 22.

Descartes, Renee (1824): *Oeuvres de Descartes Les Météores.*

Desgodetz, Antoine (1682): *Les edifices antiques de Rome dessinés et mesurés très exactement.*

Dietterlin, Wendel (1968): *Architectura von Austheilung Symetria und Proportion der fünf Seulen.*

Dillingham, Rick (1992): *Acoma & Laguna Pottery.*

Dimacopoulos, Jordan (1985): "Anastylosis and Anasteloseis", *Icomos Information,* v.1.

Dinsmoor, W. B. (1950): *The Architecture of Ancient Greece,* (3rd ed.)

Doerpfeld, Wilhelm (1892): *Olympia.*

Doumas, Christos (1983): *Cycladic Art. Ancient Sculpture and Pottery from the N. P. Goulandris Collection.*

Downing, A.J. (1844): *Cottage Residences; or a series of designs ...* (2nd. ed.)

Dresser, William W. and Michael C. Robbins (1975): "Art Styles, Social Stratification and Cognition: an analysis of Greek vase painting", *American Ethnologist* v. 2.

Duddy, Michael C. (2008): "Roaming Point Perspective: a dynamic interpretation of the visual refinements of the Greek Doric temple", *Nexus Network Journal* v. 10.

Due Granduchi (1987): *Due Granducchi Tre Re e Una Facciata.*

Dupree, A. Hunter (1951): "Some letters from Charles Darwin to Jeffries Wyman" *Isis,* v. 42.

Durand, J-N-L: (2000): *Precis of the lectures on architecture,* (trans. Pritt).

Dutton, Ralph (1950): *The English Garden* (2nd.ed.)

Eco, Umberto (1986): *Art and Beauty in the Middle Ages.*

Ekrami, Omid, Stefan Van Dongen, and Peter Claes (2018): "Fluctuating Asymmetry, Sexual Dimorphism and Attractiveness in Humans...", Paper presented to "Symmetry 2017 – The First International Conference on Symmetry, Barcelona, Spain, 16–18 October 2017."

Eisenman, Russel and H. K. Gellens (1968): "Preferences for Complexity-Simplicity and Symmetry-Asymmetry", *Percep-tual and Motor Skills* v. 26.

Emmons, George T. 1907): "The Chilkat Blanket", *American Museum of Natural History Memoirs* v. 3, pt. 4.

Eriksen, Roy (2010): *The Building in the Text: Alberti to Shakespeare and Milton.*

Etlin, Richard A. (1987): "Le Corbusier, Choisy, and French Hellenism: The Search for a New Architecture", *Art Bulletin,* v. 69.

-- (1994): *Frank Lloyd Wright and Le Corbusier. The romantic legacy.*

Evans, Abel (1713): *Vertumnus. An epistle to Mr. Jacob Bobart, Botany Professor to the University of Oxford and Keeper of the Physick-Garden.*

Evans, Robert (1995): *The Projective Cast. Architecture and its Three Geometries.*

Evelyn, John (2017): *An Account of Architects and Architecture* published as introduction to Freart, 2017.

-- (1906): *The Diary of John Evelyn* (ed. Dobson).

Ferguson, Kitty (2008): *The Music of Pythagoras.*

Fergusson James, (1849): *An Historical Inquiry into the true principles of beauty in art.*

Festinger, Leon (1956): *When Prophecy Fails.*

Fewkes, Jesse Walter (1895-6):"Designs on Prehistoric Hopi Pottery" in *Seventeenth Annual Report of the Bureau of American Ethnology to the Secretary of the Smithsonian Institution.*

Filarete (1965): *Filarete's Treatise on Architecture* (ed. Spencer).

Fischer, J.L. (1961): "Art styles as cultural cognitive maps", *American Anthropologist* v. 63.

Fisher, Sally (1995): *The Square Halo and other mysteries of western art.*

Fitzpatrick, Simon (2013): "Simplicity in the Philosophy of Science", *Internet Encyclopedia of Philosophy.*

Foat, F.W.G. (1915): "Anthropometry of Greek Statues", *J. of Hellenic Studies* v. 35.

Foster, Philip (1981): "Lorenzo de' Medici and the Florence Cathedral façade", *Art Bulletin,* v. 63.

Frank, F.C. (1974): "Descartes' Observations on the Amsterdam Snowfalls of 4, 5, 6 and 9 February, 1634", *J. of Glaciology,* v. 13.

Frankfort, Henry (1969): *The Art and Architecture of the Ancient Orient.* (4th ed.)

Fréart. Roland, John Evelyn and L. B. Alberti (2017), *Architecture and Sculpture Three Essays*: comprising Fréart's *A Parallel of the Antient Architecture with the Modern (trans. Evelyn)*; Evelyn's *Account of Architects and Architecture;* and Alberti's *Of Statues* (trans. Evelyn), repr. of the London, 1664 edition.

Friedlaender, Max J. (1969): *Reminiscences and Reflections* (trans. Magurn).

Friedman, Alice T. (1998): "John Evelyn and English Architecture" in eds. Therese O'Malley and Joachim Wolschke-Bulmahn, *John Evelyn's "Elysium Britannicum and European Gardening.*

Frisch, Karl von (1975): *Animal Architecture* (trans. Gombrich).

Frommel and Adams (2000): *The Architectural Drawings of Antonio da Sangallo the Younger and his Circle.*

Furukawa, Yoshimoro (1997): "Faszination der Schneekristalle - wie ihre bezaubernden Formen..." *Chemie in unsererZeit*, v. 31.

-- and Wettlaufer, John S. (2007): "Snow and ice Crystals", *Physics Today.*

Gadol, Joan (1969): *Leon Battista Alberti Universal Man of the Renaissance.*

Gal, Joseph (2011): "Louis Pasteur, language, and molecular chirality. I. Background and dissymmetry", *Chirality.*

Gandy, Joseph (1805): *Designs for Cottages, Cottage Farms, and other Rural buildings, including gates and lodges.*

Gardner, Helen (2005): *Gardner's Art through the Ages.*

Gerard, Alexander (1759): *An Essay on Taste,*

Germann, Georg (1973): *Gothic Revival in Europe and Britain.*

Gibney, Elizabeth (2014): "Force of Nature gave life its Asymmetry", *Nature, International Weekly of Science.*

Gillerman, David M. (1999): "Cosmopolitanism and *Campanilisimo*: Gothic and Romanesque in the Siena Duomo Façade", *College Art Bulletin* v. 81.

Gilman, Ernest B. (2009): *Plague Writing in Early Modern England.*

Giorgio, Francesco di Martini (1967), *Tratatti di Architettura* (ed. Maltese).

Girouard (1978): *Life in the English Country House.*

Goethe, J. W. von (1963): "Kunst und Altertum", in *Aus einer Reise am Rhein, Main und Neckar.*

-- (1982): *Italian Journey* (trans Auden and Mayer).

Goldthwaite, Richard A. (1993): *Wealth and the Demand for Art in Italy 1300-1600.*

Gombrich, E.H. (1966): *Norm and Form.*

-- (1979): *The Sense of Order. A study in the psychology of decorative art.*

-- (2002): *The Preference for the Primitive. Episodes in the history of western taste and art.*

Goodyear, W.H. (1905): *Illustrated Catalogue of Photographs and Surveys of Architectural Refinements in Medieval Buildings.*

-- (1912): *Greek Refinements. Studies in temperamental architecture.*

Grafton, Anthony (2002): *Leon Battista Alberti, Master Builder of the Renaissance.*

Grew, Nehemiah (1673): "On the Nature of Snow", *Philosophical Transactions of the Royal Society*, v. 8.

Gunther, R. W. T (1928): *The Architecture of Sir Roger Pratt.*

Gwynn, John (1742): *The Art of Architecture. A poem in imitation of Horace's "Art of Poetry".*

Hall, James (2005): *Michelangelo and the Reinvention of the Human Body*

-- (2008): *The Sinister Side. How left-right symbolism shaped Western art.*

Hargittai, Istvan and Magdolna (1986): *Symmetry through the eyes of a chemist.*

-- (1994): *Symmetry a unifying concept.*

Hart, Joan Goldhammer (1981): *Heinrich Wölfflin: An Intellectual Biography.* Unpublished PhD dissertation, Dept. of Art History, U.C. Berkeley.

Haselberger, Lothar (1999): *Appearance and Essence, Refinements of Classical Architecture: Curvature.*

Hatfield, Rab (2004): "The funding of the façade of Santa Maria Novella", *J. of the Warburg and Courtauld Institutes*, v. 67.

Hattenhauer, Darryl (1984): "The Rhetoric of Architecture: A Semiotic Approach", *Communication Quarterly*, v. 32.

Hauser, Arnold (1965): *Mannerism. The Crisis of the Renaissance & the Origin of Modern Art.*

Havell, E. B. (1913): *Indian Architecture. Its Psychology, Structure and History from the first Muhammadan Invasion to the Present Day.*

Havens, Raymond (1953): "Simplicity, a changing concept", *J. of the History of Ideas*, v. 14.

Hawthorne, Nathaniel (1874): *Passages from the French and Italian Notebooks.*

Hays, David L. (2001): "'This is not a Jardin Anglais'. Carmontelle, the Jardin de Monceau, and Irregular Garden Design in Late-Eighteenth Century France", in eds. Benes and Harris, *Villas and Gardens in early modern Italy and France".*

Heely, Joseph (1775): *A description of Hagley, Envil and the Leasowes wherein all the Latin Inscriptions are translated, and every particular beauty described.*

Hellmann, Prof. Dr. (1893): *Schneekrystalle: Beobachten und Studien.*

Hemelrijk, Jaap M. (1984): *Caeretan Hydriai, Forschungen zur antiken Keramik* (2nd series, *Kerameus* v. 5).

--. (2000): "Three Caeretan Hydriai in Malibu and New York" *Greek Vases in the J. Paul Getty Museum*, v. 6.

Hemsterhuis, François (1769): *Lettre sur la Sculpture à Monsieur Théod. De Smeth.*

Herlihy, David (1997): *The Black Death and the Transformation of the West.*

Hersam, C., Nathan Guisinger, and Joseph Lyding (n.d.): "Silicon-Based Molecular nanotechnology", foresight.org/conference/MNT7/Papers/Hersam

Hersey, David (1977): review of Heydenreich (1996): *JSHA*, v. 36.

Heydenreich, L.H. (1937): "Pius II als Bauherr von Pienza", *Z. für Kunstgeschichte*, v. 6.

-- (1996): *Architecture in Italy 1400-1500* (rev. ed.)

Hiscock, Nigel (2000): *The Wise Master Builder: Platonic Geometry in Plans of Medieval Abbeys and Cathedrals*.

Hiscock, Nigel (2002): "A Schematic Plan for Norwich Cathedral" in Wu, 2002.

Hobbes, Thomas (1636): *De Mirabilibus Pecci*.

Hogarth, William (2015): *The Analysis of Beauty* (Dover ed.)

Holm, Bill (1965): *Northwest Coast Indian Art. An analysis of form*.

Holanda, Francisco de (2006): *Dialogues with Michelangelo* (ed. Hemsoll).

- (1963): Portuguese text (first two dialogs only): *Historia e antalogia da Literatura Portuguesa* v. 16.

Hommel, Hildebrandt (1987): *Symmetrie in Spiegel der Antike*.

Hon, Giora and Bernard Goldstein (2008): *From Summetria to Symmetry, the making of a revolutionary scientific concept*.

Honour, Hugh (1961): *Chinoiserie. The vision of Cathay*.

Hooke, Robert (1665): *Micrographia*.

Hopkins, Clark (1979): *The Discovery of Duro-Europos*.

Horn and Born (1979): *The Plan of St Gall. A study of the architecture and economy of, and life in a paradigmatic Carolingian monastery*.

Hubbs, Carl and Laura (1944): "Bilateral asymmetry and bilateral variation in fishes", *Papers of the Michigan Academy of Science, Arts, and Letters*, v. 30.

Huizinga, J. (1954): *The Waning of the Middle Ages* (Anchor ed.)

Humphreys, A. R. (1937): *William Shenstone. An eighteenth-century portrait*.

Hurston, Zora Neale (1983): "Characteristics of Negro Expression" (1934) repr. in *Zora Neale Hurston, The Sanctified Church*.

Hurwit, Jeffrey (1997): "Image and Frame in Greek Art", *American J. of Archaeology*, v. 81.

-- (2002): "Reading the Chigi Vase", *Hesperia* v. 71.

Hussey, Christopher (1967): *English Gardens and Landscapes 1700-1750*.

Huxley, T. H. (1888): *Physiography, an introduction to the study of nature*.

Hyams, Edward (1971): *A History of Gardens and Gardening*.

Iacopi, Irene (2008): *The House of Augustus Wall Paintings*.

Jablan, Slavik A. (1955): *Theory of Symmetry and Ornament*.

Jacobsthal, Paul (1925): "The Ornamentation of Greek Vases", *Burlington Magazine*, v. 47.

Jaeger, F.M. (1917): *Lectures on the principle of symmetry and its application in all natural sciences*.

James, John (1982): *Chartres, the masons who built a legend*.

James, M.R. (1895): *A descriptive catalogue of the manuscripts in the library of Eton College.*

James, P.D. (2005): *The Lighthouse.*

Johnson, Lee (1991): "The Art of Delacroix" in *Eugène Delacroix (1798-1863) Paintings, Drawings and Prints from North American Collections.*

Johnson, Samuel (1810): *The Lives of the Most Eminent English Poets with critical observations on their works* (new ed.)

Jonaitis, Aldona ed., (1995): *A Wealth of Thought. Franz Boas on Native American Art.*

Jourdain, Margaret (1948): *The Work of William Kent.*

Kahn, Charles H. (1960): *Anaximander and the Origins of Greek Cosmology* (repr. ed.)

Kallenberg, Mary Hunt and Anthony Berland, (1972): *The Navajo Blanket.*

Kames, Henry Home Lord (1845): *Elements of Criticism.* New Edition.

Kellogg, Rhoda (1970): *Analyzing Children's Art.*

Kemp, Martin (1977): "From 'Mimesis' to 'Fantasie', the Quattrocento Vocabulary of Creation, Inspiration and Genius in the Visual Arts", *Viator*, v. 8.

-- (1989): *Leonardo on Painting. An Anthology...*

Kepler, Johannes (2010): *The Six-Cornered Snowflake.*

Ker, J.B. (1840): *Essay on the Archaeology of our popular phrases.*

Kiesow, Gottfried (1962): "Die gotische Suedfassade von S. Maria Novella in Florenz", *Zeitschrift für Kunstgeschichte*, v. 25.

Kirk, G.S., J.E. Raven and M. Schofield (1983): *The Pre-Socratic Philosophers* (2nd. ed.)

Klarreich, Erica (2000):"Foams and Honeycombs", *American Scientist*, v. 88.

Kliger, S. (1952): *The Goths in England.*

Koch, H (1955): Studien zum Theseustempel. *Abhandlung des Saechsischen Akademie der Wissenschafft*, v. 47.

Koestler, Arthur 1963): *The Sleepwalkers* (Universal Library ed.)

Korres M. (1999): *"Refinements of Refinements"* in Haselberger 1999.

Krautheimer, Richard (1969): *Studies in early Christian, Medieval and Renaissance art.*

_ with Trude Krauthimer-Hess (1956): *Lorenzo Ghiberti.*

Kruft, H-w (1994): *A History of Architectural Theory.*

Kunst, Christiane (n.d.): "Paestum Imagery in European Architecture", dialnet. unirioja.es/servlet/articulo? Codigo =2663376

LaChapelle, Edward (1960): *Field Guide to Snow Crystals.*

Ladis, Andrew (2008): *Giotto's O: Narrative, Figuration and Pictorial Ingenuity in the Arena Chapel.*

Landauro, Inti (2013): "A Greek Goddess Gets A Makeover", *Wall Street Journal* Aug. 28, 2013, sect. D1.

Lang, S. (1965): "L. B. Alberti's Use of a Technical Term", *J. of the Warburg and Courtauld Institutes*, v. 28.

-- and N. Pevsner (1949):"Sir William Temple and Sharawaggi", *Architectural Review*, v. 106.

Langley, Batty (1728): *New Principles of Gardening ... after a more grand and rural manner than has been done before.*

Lawrence, A.W. (1967): *Greek Architecture.*

Le Clerc, Sebastien (1732): *A Treatise of Architecture.*

Lederman, Leon M. and Christopher T. Hill (2008): *Symmetry and the Beautiful Universe.*

Lefaivre, Liane (1997): *Leon Battista Alberti's Hypnerotomachia Poliphili.*

Leisinger, Hermann (1957): *Romanesque Bronzes. Church Portals in Medieval Europe.*

Lepper, Frank and Sheppard Frere (1988): *Trajan's Column. A new edition of the Cichorius Plates.*

Lewis and Short (1879): *A Latin Dictionary.*

Lewis, W. S. (1960): *Horace Walpole.*

Libbrecht, Ken (2005): "The physics of snow crystals", *Rep. Prog. Phys.* v. 68.

-- (2006): Ken Libbrecht's Field Guide to Snowflakes.

Licht, Kjeld de Fine (1966?): *The Rotonda in Rome. A Study of Hadrian's Pantheon.*

Livio, Mario (2006): *The equation that couldn't be solved. How mathematical genius discovered the language of symmetry.*

Lloyd, Joan E. Barclay (1986): "The Building History of the Medieval Church of S. Clemente in Rome", *J. Society of Architectural Historians*, v. 45.

Lloyd, Seton (1980): *Foundations in the Dust*, (rev .ed.)

Lobell, Jarrett A. (2016): "A new view of the birthplace of the Olympics", *Archaeology*, v. 69.

Lorand, Ruth (2003-4): "The Role of Symmetry in Art", *Symmetry: Culture and Science*, vols. 14-15.

Lorch, Ingomar (1999): *Die Kirchenfassade in Italien von 1450 bis 1527: Die Grundlagen durch Leon Battista Alberti und die Weiterentwicklung des Basilikalen Fassadenspiegels.*

Lorenzoni, Mario, editor (2007.): *La Facciata del Duomo di Siena. Iconografia, Stile, Indagini Storiche e Scientifiche.*

Lorris, Guillaume de and Jean de Meun (1995): *The Romance of the Rose* (3rd ed., trans. Dahlberg).

Lovejoy, Arthur O. (1932): "The First Gothic Revival and the Return to Nature", *Modern Language Notes*, v. 47.

-- (1955a): "The Chinese Origin of a Romanticism", repr. in *Essays in the History of Ideas.*

-- (1955b): "The First Gothic Revival", repr. in *ibid.*

-- (1960): *The Great Chain of Being. A study in the history of ideas.* (Harper Torchbooks edition).

Lowic, Lawrence (1983): "The Meaning and Significance of the Human Analogy in Francesco di Giorgio's Trattato", *J. Society of Architectural Historians*, v. 42.

Luecke, D. (1994): in eds. Rykwert and Engel: *Leon Battista Alberti*.

MacDonald, William L. (1982): *The Architecture of the Roman Empire An Introductory Study*.(rev. ed)

Machover, Karen (1949): *Personality Projection in the Drawing of the Human Figure*.

Mack, Charles (1974): "The Rucellai Palace: Some New Proposals", *Art Bulletin*, v. 56.

-- (1987): *Pienza: the Creation of a Renaissance City*.

Mackay, Charles (1932): *Extraordinary Popular Delusions and the Madness of Crowds*.

Magnus, Olaus (1996): *Historia de Gentibus Septentrionalibus*, (trans. Fisher and Higgins).

Magono, Choji and Chung Woo Lee (1966): "Meteorological Classification of Natural Snow Crystals", *J. of the Faculty of Science, Hokkaido University*, Ser.VII, II:4.

Mainzer, Klaus (1996): *Symmetries of Nature*.

Mallgrave, Harry Francis and Eleftherios Ikonomou (1994): *Empathy, Form, and Space. ööProblems in German Aesthetics 1987-1893*.

Malins, Edward (1966): *English Landscaping and Literature 1660-1840*.

Mancini, G. (1882):*Vita di Leon Battista Alberti*.

Manetti, Antonio (1970): *The Life of Brunelleschi*, (ed. Saalman).

Martin, Constance (1988): "William Scoresby (1789-1857) and the open polar sea", *Arctic* v. 41.

Marvell, Andrew (1927): *The Poems and Letters of Andrew Marvell* (ed. Margoliouth).

Mason,B.J. (1992):"Snow crystals, natural and man-made, *Contemporary Physics*, v. 33

Mason, William (1772): *The English Garden, A Poem*.

Masson, Georgina (1961): *Italian Gardens*.

Mateer, David (2000): *Courts, Patrons and Poets*.

Mattusch, Carol C. (1988): *Greek Bronze Statuary*.

McBeath, Michael, Diane Schiano and Barbara Tversky (1997): "Three-dimensional Bilateral symmetry bias in judgments of figural identity and orientation", *Psychological Science* v. 8.

McCann, A. M. (1978): *Roman Sarcophagi in the Metropolitan Museum of Art*.

McCarthy, Michael (1987): *Origins of the Gothic Revival*.

McManus, I.C. (2004): "Right-Left and the Scrotum in Greek Sculpture", *Laterality*, v. 2.

-- (2005): "Symmetry and Asymmetry in Aesthetics and the Arts", *European Review*, v. 13.

Meiss, Millard (1951): *Painting in Florence and Siena after the Black Death*.

Meyboom, Paul G. M. and Eric M. Moormann (2013): *Le Decorazioni Dipinte e Marmoree ella Domus Aurea di Nerone a Roma.*

Michelis, P.A. (1955): "Refinements in Architecture", *J. of Aesthetics and Art Criticism,* v. 14.

Millon, H. (1958): "The Architectural Theory of Francesco di Giorgio", *Art Bulletin.*

-- (1972): "Rudolf Wittkower's 'Architectural Principles', its influence on the development and interpretation of modern architecture", *J. Society of Architectural Historians,* v. 31.

-- (1994): "Models in Renaissance Architecture" in Millon and Lampugnani, eds., *The Renaissance. From Brunelleschi to Michelangelo. The Representation of Architecture.*

Milton, John (1958): *The Poems of John Milton,* (ed. Darbishire).

Mitten, David Gordon and Suzannah F. Doeringer (1967): *Master Bronzes From The Classical World.*

Mokhopadhyay, Swapna (2009): "The decorative impulse: ethnomathematics and Tlingit basketry", *ZDM Mathematics Education,* v. 41.

Molini, Giuseppe (1820): *La Metropolitana fiorentina illustrata.*

Montesquieu (1759): "Essay on Taste" in Gerard 1759.

-- (1777): *The Complete Works of M. de Montesquieu.*

-- (1825): *Oeuvres* (ed. De Plancy).

Morgan, Luke (2006): *Nature as Model: Salomon de Caus and early 17th century landscape design.*

Mormando, Franco, and Thomas Worcester (2007): *Piety and Plague from Byzantium to the Baroque.*

Morolli, Luchinat and Marchetti, eds. (1992): *L'Architettura di Lorenzo il Magnifico.*

Morris, William (2004): *News from Nowhere* (Dover reprint).

Morrison, Alan S, J. Kirshner and A. Molho (1985): "Epidemics in Renaissance Florence", *Am. J. Public Health,* v. 75.

Morse, Edward S. (1972): *Japanese Homes and their Surroundings.*

Muet, Pierre Le (1623): *Manière de bâtir pour toutes sortes de personnes.*

-- (1670): *The Art of Fair Building,* (trans. Pricke).

Mullett, Charles F. (1936): "The English Plague Scare of 1720-1723", *Osiris,* v. 2.

Murray, Ciaran (1998): "Sharawadgi Resolved", *Garden History,* v. 26.

Nakaya, Ukichiro, (1954): *Snow Crystals, natural and artificial.*

Naredi-Rainer, Paul (1977): "Exkurs zum Problem der Proportionen bei Alberti", *Zeitschrift fuer Kunstgeschichte,* v. 40.

Neagley, Linda Elaine (1992): "The Late Gothic Plan Design of St.-Maclou in Rouen", *Art Bulletin* v. 74.

Needham, Joseph (1963): "Poverty and Triumphs of the Chinese Scientific Tradition" in ed. Crombie, *Scientific Change. Historical Studies in the Intellectual, Social and Technical Conditions for Scientific Discovery and Technical Invention, from Antiquity to the*

Present. Symposium on the History of Science University of Oxford 9-15 July 1961

-- and Lu Gwei-Djen (1970): "The Earliest Snow Crystal Observations", in Needham et al., *Clerks and Craftsmen in China and the West.*

Negra, Ricardo Della (2004): *La Cupola di Santa Maria del Fiore. Il rilievo fotogrammatico.*

Neville, Anthony C. (1976): *Animal Asymmetry.*

Nocentinni, Carlo (1992): in eds., Morolli, Luchinat and Marchetti, *L'Architettura di Lorenzo il Magnifico.*

Norman, Diana, (1995): *Siena, Florence and Padua. Art, Society and Religion 1280-1400.*

Norton, John D. (2000): "Nature is the Realisation of the Simplest Conceivable Mathematical Ideas": Einstein and the Canon of Mathematical Simplicity *Stud. Hist. Phil. Mod. Phys.*, v. 31, No. 2.

Nyman, Michael (1980): "Against Intellectual Complexity in Music", *October*, vol. 13.

Oechslin, Werner (1985): "Symmetrie – Eurythmie: oder, 'Ist Symmetrie schön?", *Daidalos*, v. 15.

Ogden, H. V. S. (1949): "The principles of variety and contrast in seventeenth century aesthetics, and Milton's poetry", *J. of the History of Ideas*, v. 10.

Onians, John, (1992): "Architecture, Metaphor and Mind", *Architectural History*, v. 35.

Ostwald, Michael J (2000): "Under Siege: the Golden Mean in Architecture", *Nexus Network Journal*, v. 2.

Palladio, Andrea (1570): *I Quattro Libril dell' Architettura.*

-- (1997): *The Four Books on Architecture* (trans. Tavernor and Schofield).

Panofsky, Erwin (1955): *Meaning in the Visual Arts.*

-- (1957): *Gothic Architecture and Scholasticism.*

-- (1972): *Renaissance and Renascences in Western Art.*

Paret, Peter (1997): *Imagined Battles. Reflections of War in European Art.*

Pasteur, Louis (1874): *Works,* v. I. (from *Comptes Rendus de l'Académie des Sciences,* June 1, 1874).

Payne, Alina (1994): "Rudolf Wittkower and architectural principles in the age of modernism", *J. Society of Architectural Historians,* v. 53.

-- (1999): *The Architectural Treatise in the Renaissance.*

-- (2012): *From Ornament to Object. Genealogies of Architectural Modernism.*

Pelt, Robert and Carroll Westfall (1993): *Architectural Principles in the Age of Historicism.*

Pennick, Nigel (2012): *The Sacred Architecture of London.*

Penrose, Frances Cranmer (1888): *An Investigation of the Principles of Athenian Architecture or the results of a survey conducted chiefly with reference to the optical refinements exhibited in the construction of ancient buildings in Athens.* (New and enlarged ed.)

Perrault, Claude (1993): *Ordonnance for the five kinds of columns after the methods of the ancients* (trans. McEwen).

Perry, Ellen E. (2000): "Notes on *Diligentia* as a term of Roman Art Criticism", *Classical Philology*, v. 95.

Petkau, Karen (n.d.): "Baskets: Carrying a Culture. The Distinct-ive Regional Styles Of Basketmaking Nations in the Pacific Northwest", langleymuseum.org/baskets/pd fs/carrying-a-culture11.pdf downloaded Feb.4, 2010.

Pevsner, N (1969): *Ruskin and Viollet-le-Duc. Englishness and Frenchness in the appreciation of Gothic architecture.*

-- (1964): *The Englishness of English Art.*

-- (1974): *The Picturesque Garden and its influence outside the British Isles.*

-- and M. Aitchison (2010): *Visual Planning and the Picturesque.*

Pfaff, Christopher A. (2003): *The Argive Heraion, v. 1: The Architecture of the Classical Temple of Hera.*

Philipp, Hanna (1999): "Curvature: Remarks of a Classical Archaeologist" in Haselberger, *supra.*

Picard, Gilbert Charles (1970): *Roman Painting.*

Pius II (1962): eds. Gragg and Gable, *Memoirs of a Renaissance Pope.*

-- (1984): *Pii II commentarii rerum memorabilium, que temporibus suis contigerunt*, Van Heck, ed.

-- (2003): ed. Meserve, *Pius II Commentaries*, v. 1.

Plato (1966): trans. Fowler, *Plato in Twelve Volumes*, v. 1.

Pliny (1952): *Letters* (Loeb ed.)

Plommer, Hugh (1960): "The Archaic Acropolis: Some Problems", *J. Hellenic Studies*, v. 80.

Plutarch (1874): *Plutarch's Morals*, ed. Goodwin.

-- (1932): *Lives*, ed. Perrin.

Poeschke, J., and C. Syndikus, eds., (2008): *Leon Battista Alberti Humanist Architekt Kunsttheoretiker.*

Pope, Alexander (1831): *The Poetical Works of Alexander Pope* (Aldine ed.)

Pope-Hennessy, John (1980): *The Study and Criticism of Italian Sculpture.*

-- (1991): *Paradiso. The illustrations to Dante's Divine Comedy by Giovanni di Paolo.*

Portoghesi, Palo (1972): *Rome of the Renaissance* (trans. Sanders).

Powell, Melissa S. and C. Jill Grady, eds., (2010): *Huichol Art and Culture: balancing the world.*

Prak, N.L. (1966): "Measurements of Amiens Cathedral", *J. Society of Architectural Historians*, v. 25.

Prigogine, Ilya and Isabelle Stengers (1984): *Order out of Chaos. Man's New Dialogue with Nature.*

Prochaska, Frank (2013): *The Memoirs of Walter Bagehot.*

Prest, John (1981): *The Garden of Eden, the Botanic Garden and the Re-Creation of Paradise.*

Puffer, E. D. (1905): *The Psychology of Beauty.*

Ragghianti, Licia Collobi (1979): *National Archaeological Museum Athens.*

Ramberg, Walter Dodd (1960): "Some Aspects of Japanese Architecture", *Perspecta*, v. 6.

Redford, Bruce (2002): "The Measure of Ruins: Dilettanti in the Levant 1750-1770", *Harvard Library Bulletin* v. 13.

Reichard, Gladys A. (1922): "The Complexity of Rhythm in Decorative Art", *American Anthropologist* n.s., v. 24.

-- (1933): *Melanesian Design. A study of style in wood and tortoiseshell carving.*

Reinberger, Stefanie (2016): "Decoding Biological Asymmetry", *Medical Press*, July 12, 2016.

Reynolds, Joshua (1997): *Discourses on Art*, ed. Robert R. Wark.

Richards, Charles M. (n.d.): "Ralph Cudworth", *Internet Encyclopedia of Philosophy*, accessed Aug. 27, 2013.

Richter, Gisela A. M. (1946): *Attic Red-Figured Vases. A Survey.*

Robertson, D. S. (1954) *Handbook of Greek and Roman Architecture.*

Robison, Elwin C: (1998-1999): "Structural Implications in Palladio's Use of Harmonic Proportions", *Annali di architettura*, vv. 10-11.

Rochberg, George (1997): "Polarity in Music: Symmetry and Asymmetry and their Consequences", *Proceedings of the American Philosophical Society*, v. 141.

Rodin, August (1965): *Cathedrals of France*, trans. Geissbuhler from the French edition, 1914.

Romanes, G. R. (1882): *Animal Intelligence* (2nd ed.)

Rosen, Joe (1975): *Symmetry Discovered. Concepts and applications in nature and science.*

Rossini, Orietta (2007): *Ara Pacis.*

Ruskin, John (1903-1912): *Complete Works* (eds. Cook and Wedderburn).

Rykwert, Joseph (1972): *On Adam's House in Paradise. The Idea of the Primitive Hut in Architectural History.*

-- Engel, Anne, eds., (1994): *Leon Battista Alberti.*

Saalman, Howard (1959): "Early Renaissance Architectural Theory and Practice in Antonio Filarete's *Trattato di Archittetura*", *Art Bulletin*, v. 41.

Saarinen, Eliel (1985): *The Search for Form in Art and Architecture*, (Dover ed.)

Sackville-West, Vita (1929): *Andrew Marvell.*

Sakka, Niki (2013): "'A Debt to Ancient Wisdom and Beauty': The Reconstruction of the Stoa of Attalos in the Ancient Agora of Athens", *Hesperia*, v. 82.

Samonà, Giuseppe et al (1977): *Piazza San Marco l'architettura la storia e funzioni* (2nd. ed.)

Sapirstein, Philip (2016): "The Columns of the Heraion at Olympia: Dorpfeld and early Ionic Architecture", *American J. Of Archaeology*, v. 120.

Sartwell, Crispin (2016): "Beauty", *Stanford Encyclopedia of Philosophy*.

Sautoy, Marcus du (2008): *Symmetry, a journey into the patterns of Nature.*

Saxl, F. (1979): "Jacopo Bellini and Mantegna as Antiquarians" in Saxl, *A Heritage of Images.*

Schlosser, Julius von (1929): "Ein Künstlerproblem der Renaissance: L. B. Alberti", *Akademie der Wissenschaften in Wien, Phil-Hist Klasse*, v. 2.

– (1988): "On the History of Art Historiography – the 'Gothic'", in ed. Schiff, *German Essays on Art History.*

Schneider, Lambert A. (1973): *Asymmetrie griecheischer Koepfe vom 5.Jh. bis zum Hellenismus.*

Scholfield, P. H: (1958): *The Theory of Proportion in Architecture.*

Scholten, Fritz (2011): "Frans Hemsterhuis's memorial for Herman Boerhaave: a monument of wisdom and simplicity", *Simiolus: Netherlands Quarterly for the History of Art*, v. 35.

Scoresby, William (1820): *An account of the arctic regions.*

Scranton, Robert L. (1967): "The Architecture of the Sanctuary of Apollo Hylates at Kourion", *Trans. American Philosophical Society* N.S. v. 57.

-- (1974): "Vitruvius' Art of Architecture", *Hesperia*, v. 43.

Seligman, G. (1980): *Snow Structure and Ski Field* [1936] (repr. ed.)

Selzer, Michael (2021): *Byzantine Aesthetics and the Concept of Symmetry.*

Semper, Gottfried (2004): *Style in the Technical and Tectonic Arts; or, Practical Aesthetics* (trans. Mallgrave and Robinson).

Serle, J. (1745): *A Plan of Mr. Pope's Garden, as it was left at his death: with a plan and perspective view of the grotto. All taken by J. Serle, his Gardener* (Augustan Reprint Society ed).

Serlio, Sebastiano (1982): *The Five Books of Architecture*, Peake trans. (Dover ed.)

-- (1996, 2001): *Serlio on Architecture*, (ed. and trans., Hart and Hicks.)

Shahbazi, Marta, Eric Siggia, Magdalena Zernicka-Goetz (2019): "Self-organization of Stem Cells into Embryos", *Science*, v.364.

Shaftesbury, Earl of (1749): *Characteristicks of Men, Manners, Opinions, Times.*

Shaw, Joseph (1978): "Evidence from the Minoan Tripartite Shrine", *American J. of Archaeology*, v. 82.

Shelby, L. R. (1972): "The Geometrical Knowledge of Medieval Master Masons", *Speculum* v. 47.

Shepard, Anna O (1948): *The Symmetry of Abstract Design with special reference to ceramic decoration.* (Contributions to American Anthropology and History no. 47. Carnegie Institute of Washington Publication no. 574.)

Sherbo, Arthur (1972): "Paradise Lost IV. 239", *Modern Language Review,* v. 67.

Shrewsbury, J. F. D. (1970): *A History of the Bubonic Plague in the British Isles.*

Shute, John (1563): *The First and Chief Groundes of Architecture...*

Sieveking, A. F., ed. (1908): *Sir William Temple upon the Gardens of Epicurus, with other XVIIth Century Garden Essays.*

Simson, Otto von (1962): *The Gothic Cathedral. Origins of Gothic Architecture and the medieval concept of order.* (2nd ed.)

Sitte, Camillo (2006): *City Planning According to Artistic Principles,* in trans. Collins and Collins, *Camillo Sitte: The Birth of Modern City Planning,* pp. 133 – 332.

Six, J. (1885): "Some Archaic Gorgons in the British Museum". *J. Hellenic Studies* v. 6.

Smith, C. (1992): *Architecture in the Culture of Early Humanism: ethics, aesthetics, and eloquence, 1400-1470.*

Smith, Graham (2000): "Gaetano Baccani's 'Systematization' of the Piazza del Duomo in Florence", *J. Society of Architectural Historians,* v. 59.

Soles, Jeffrey T. (1991): "The Gournia Palace", *American J. of Archaeology,* v. 95.

Southern, R.W. (1970): *Medieval Humanism and other Studies.*

--   (1995): *Scholastic Humanism and the Unification of Europe* v.1.

Sparavigna, A.C (2013.): "Reflection and Refraction in Robert Grosseteste's De Lineis, Angulis et Figuris", *International Journal of Sciences,* v. 2.

Spectator (1891): *The Spectator* ed. Morley.

Spence, Joseph (1964): *Anecdotes, Observations and Characters of Books and Men* (repr. ed.)

Spenser, Edmund (1993): *Edmund Spenser's Poetry* (eds. Maclean and Prescott, 3rd. ed.)

Stendhal (1962): *Memoirs of a Tourist* (trans. Seager).

Stern, Judith (2000): "The Eichmann Trial and its Influence on Psychology and Psychiatry", *Theoretical Inquiries in Law,* v. 1.

Stevens, Gorham (1943): "The Curve of the North Stylobate of the Parthenon", *Hesperia,* v. 12.

Stewart, Dugald (1855): *Collected Works,* v. 5.

Stewart, Ian, (2001): *What Shape Is A Snowflake?*

Stewering, Roswitha (2000): "Architectural Representation in the 'Hypnerotomachia Poliphili'", *J. Society of Architectural Historians,* v. 59.

Stockstad, M., and J. Stannard (1983): *Gardens of the Middle Ages.*

Streatfield, David C., and Alistair Duckworth, eds. (1981): *Landscape in the Gardens and Literature of eighteenth-century England.*

Striker, Cecil L. (1981): *The Myrelaion (Bodrum Camii) in Istanbul.*

Strong, Roy (1998): *The Renaissance Garden in England.*

Stuart, James, and Nicholas Revett (2008): *Antiquities of Athens. (repr. ed.)*

Sturgis, Russell (1905): *A Dictionary of Architecture and Building.*

Summers, David (1981): *Michelangelo and the language of art.*

Summerson, John (1963): *Heavenly Mansions and other essays on architecture.*

Sutton, Peter (1988): *Dreamings: The Art of Aboriginal Asia.*

Sydow, Eckart von (1932): *Die Kunst der Naturvölker und der Vorzeit.*

Tavernor, Robert (1998): *On Alberti and the Art of Building.*

Taylor, Rabun (2003): *Roman Builders A Study in Architectural Process.*

Tellez, Trinidad Ruiz and Anders Pape Moller (2006): "Fluctuating Asymmetry of Leaves in *Digitalis thapsi* under Field and Common Garden Conditions", *International Journal of Plant Sciences*, v. 167.

Temko, Allan (1959): *Notre-Dame of Paris* (Viking Compass ed.)

Temple, William (1731): *Works*, (2nd ed.)

Thacker, Christopher (1979): *History of Gardens.*

Thiersch, August (2017): *Proportion in Architecture* and Heinrich Wölfflin, *A Theory of Proportion* (trans. Michael Selzer).

Thompson, D'Arcy (2005): *On Growth and Form* (repr. of abridged edition, 1961).

Thornton, Peter (1998): *Form and Decoration. Innovation in the décorative arts 1470-1870.*

Tietze-Conrat, E. (1955): *Mantegna. Paintings, Drawings, Engravings.*

Tobin, Richard (1981): "The Doric Groundplan", *American J. of Archaeology*, v. 85.

Torbrügge, Walter (1968): *Prehistoric European Art.*

Townsend, P.S. (1818): "Memoir on the Crystallization of Snow; read before the Lyceum of Natural History, New-York, April 8th, 1817", *American Monthly Magazine and Critical Review*, v. III, 1.

Trachtenberg, Marvin (1997): *Dominion of the Eye. Urbanism, Art, and Power in Early Modern Florence.*

Tuchman, Barbara (1978): *A Distant Mirror. The Calamitous 14th Century.*

Tunnard, Christopher (1978): *A World with a View.*

Turner, A. Richard (1966): *The Vision of Landscape in Renaissance Italy.*

Vasari, Giorgio (1908): *Le vite dei più celebri pittori, scultori e architetti* (ed. Salani).

-- (1996): *Lives of the Painters, Sculptors and Architects* (trans. de Vere).

Venturi, Robert (1977): *Complexity and Contradiction in Architecture* (2nd ed.)

Vermeule, Cornelius Clarkson (1977): *Greek Sculpture and Roman Taste.*

Vershbow (2013): *The Collection of Arthur and Charlotte Vershbow* Part One (Christie's NY catalog).

Vickers, Michael (1987): "Eighteenth-Century Taste and the Study of Greek Vases", *Past and Present*, v. 116.

Villard de Honnecourt (1959): *The Sketchbook of Villard de Honnecourt* (ed. Bowie).

Vinci, Leonardo da (1958): *The Notebooks of Leonardo da Vinci* (ed. MacCurdy).

Viollet-le-Duc, E-E. (1868): *Dictionnaire raisonné de l'architecture française du XIe au XVIe siècle*, v. 7 (article "Symétrie").

-- (1987): *Lectures on Architecture*.

Vitruvius (1931): ed. and trans. Frank Granger, *Vitruvius on Architecture*.

Voloshinov, Alexander (1996): "Symmetry as Superprinciple of Science and Art", *Leonardo*, v. 29.

Voltaire, F-M, A. (1757): "*Essay on Taste*" in Gerard, 1759.

Walpole, Horace (1891): *Letters of Horace Walpole* (ed. Cunningham).

-- (1995): *The History of the Modern Taste in Gardening* (repr. ed.)

Ward-Perkins, John, and Amanda Claridge (1978): *Pompeii AD 79*.

Washburn, Dorothy and Donald W. Crowe (1988): *Symmetries of Culture*.

Washburn, Dorothy K. (1995):"Symmetry Clues to the Puebloan Lifeway", in *Symmetry: Culture and Growth*, v. 6.

-- (1999) "Perceptual Anthropology: the cultural salience of symmetry", *American Anthropologist*, v. 101.

Watkins, R. N. (1972): "Petrarch and the Black Death", *Studies in the Renaissance*, v. 19

Webster, T. B. L. (1939): "Tondo composition in Archaic and Classical Greek Art" *J. of Hellenic Studies*, v. 59.

Weil-Garris, and John F. D'Amico (1980): "The Renaissance Cardinal's Ideal Palace" in ed. Henry A. Millon, *Studies in Italian Art and Architecture 15th through 18th centuries*. Memoirs of the American Academy in Rome, v. 35.

Weinberger, Martin (1941): "The First Façade of the Cathedral of Florence", *J. of the Cortauld and Warburg Institutes*, v. 4.

Weyl, Hermann (1952): *Symmetry*.

White, Arthur (2014): *Plague and Pleasure. The Renaissance World of Pius II*.

White, John (1967): *The Birth and Rebirth of Pictorial Space*.

White, John (1973): "Giotto's Use of Architecture in 'The Expulsion of Joachim' and 'The Entry into Jerusalem' at Padua", *Burlington Magazine*, v. 115.

Whittow, Mark (1996): *The Making of Orthodox Byzantium 600-1025*.

Wiencke, Martha Heath (2000): *The architecture, stratification, and pottery of Lerna*, v. III.

Weiner, Gordon M. (1970: "The Demographic Effects of the Venetian Plagues of 1575-77 and 1630-31", *Genus*, v. 26.

Wilczek, Frank (2015): *A Beautiful Question. Finding Nature's Deep Design*.

Williman, Daniel (ed.) 1982: *The Black Death. The Impact of the Fourteenth-Century Plague.* Medieval and Renaissance Texts and Studies, v. 13.

Willmott, Hugh, et al (2020): "A Black Death Mass Grave at Thornton Abbey [Lincolnshire]. The Discovery and Examination of a 14th century rural catastrophe", *Antiquity,* v.94, pp.179-96.

Wilson Jones, Mark (2001): "Doric Measure and Architectural Design 2", *A. J. Archeology,* v. 4.

-- (2003): *Principles of Roman Architecture.*

Wilson, Michael I. (1977): *The English Country House and its furnishings.*

Winckelmann, Johann Joachim (1968): *History of Ancient Art* (trans. Alexander Gode).

Wind, Edgar (1986): *Hume and the Heroic Portrait. Studies in Eighteenth-Century Imagery,* ed. Jaynie Anderson.

Wittkower, Rudolf (1940-1941): "Alberti's Approach to Antiquity in Architecture", *J. of the Warburg and Courtauld Institutes,* v. 4.

-- (1949): *Architectural Principles in the Age of Humanism,* First ed.

-- (1952): *Architectural Principles in the Age of Humanism,* Second ed.

-- (1962): *Architectural Principles in the Age of Humanism,* Third ed.

-- (1971): *Architectural Principles in the Age of Humanism,* Fourth ed.

- (1974a*): Gothic vs. Classic. Architectural projects in seventeenth-century Italy.*

-- (1974b): *Palladio and English Palladianism.*

-- (1978): "Changing Concept of Proportion", in *Idea and Image.*

Wölfflin, Heinrich (1889): "Zur Lehre von den Proportionen" in *Kleine Schriften* (ed. Gantner); translated as "A Theory of Proportion" in Thiersch, 2017.

-- (1952): *Classic Art. An Introduction to the Renaissance* (trans. Peter and Linda Murray).

-- (1966): *Renaissance and Baroque* (trans. Kathrin Simon).

-- (2017): *Prolegomena to a Psychology of Architecture* (trans. Michael Selzer).

Woodbridge, Homer (1940): *Sir William Temple. The man and his work.*

Woodford, Susan (1988): *Introduction to Greek Art.*

Wotton, Henry (1968): *Elements of Architecture* (repr. ed.)

Wu, Nancy, editor (2002): *Ad Quadratum: The practical application of geometry in medieval architecture.*

Wyman, Jeffries (1866): *Notes on the Cells of Bees.*

Yeroulanou, Marina (1998): "Metopes and Architecture: The Hephaiston and the Parthenon", *Annual of the British School of Athens,* v. 93.

Zarnecki, George (1975): *Art of the Medieval World.*

Zocchi, Giuseppe (1757): *Vedute delle ville, e d'altri luoghi della Toscana* (3rd.ed.)

Zucker, Paul (1959): *Town and Square. From the agora to the village green.*

INDEX

A

Ackerman, James: 81, 120, 127, 286, 299, 302, 302, 379, 380, 389, 392
Acropolis (Athens): 87, 88, 95, 107-109, 406
Addison, Joseph: 314, 350-353, 357-359, 373, 392
Alberti, L. B: 11, 18-19, 23-24, 55, 60, 284-287, 291, 300-306, 313, 331-339
Albertini, Francesco: 283, 392
Amiens, cathedral of: 115, 407
Anaximander: 61, 401
Anderson, P.W: 10, 392
Apollo Bassitas, temple of: 93, 104, 187, 395
Ara Pacis (Rome): 56, 75, 83, 180, 408
Aristotle: 56, 286, 330, 341, 393
Arnheim, Rudolf: 329, 392

B

Bacon, Francis: 345, 392
Badia, Fiesole: 117, 119, 199
Balanos, Nicolas: 85, 92-102, 108, 186, 393
Ball, Philip: 46
Bardeschi, Marco: 381-383, 393
Barrow, John: 47, 287, 393
Beazley, J. D: 68, 393
Bell, Corydon: 40, 393
Benelli, Francesco: 69, 371, 377, 393
Bentley, W. A: 44, 393
Bergman and Ishay: 28, 393
Bernal, J. D: 46-47, 393
Bernini, G: 284, 363
Black Death (see also Plague): 129, 250, 294-298, 394-396, 400, 404, 412
Blake, William: 9, 59, 315
Bloch, Marc: 281, 393
Boardman, John: 69, 76, 79, 393
Bobart, Jakob: 290, 397
Bober, Phyllis: 76, 334, 394
Boccaccio, Giovanni: 128, 292, 294-295, 297, 394
Borsi, Franco: 331, 371, 376, 383-4, 386, 388, 394
Botticelli: 244, 283
Bridgeman, Charles: 357
Bronte, Charlotte: 315
Brougham, Henry: 26
Brown, Capability: 342

Browne, Sir Thomas: 291, 347
Brunelleschi, Filippo: 338, 380, 386, 393, 403-404
Buffon, G-L. L: 26, 28
Burckhardt, Jakob: 326, 333, 394
Burke, Edmund: 301, 303, 315, 319-320, 353, 357-359, 394
Burke, John: 41, 394
Burlington, 3rd Earl of: 353, 360, 394-395, 401

C
Caen, cathedral of: 115
Cage, John: 287
Canaletto: 362, 393
Cassini, Giovanni: 30
Castell, Robert: 345, 395
Cezanne, Paul: 286, 393
Chambers, William: 318, 350, 395
Chambord, chateau de: 125, 220
Chartres, cathedral of: 14, 110-113, 125, 188, 392, 401
Chatsworth House: 288, 293, 342, 355
Chaucer, Geoffrey: 128, 297
Choisy, Auguste: 53, 87-89, 395
Cicero: 8, 55, 61, 300
Cimaroli, Giambattista: 130
Clark, Kenneth: 281, 292, 294, 318, 334, 339, 395
Coecke, Pieter: 309, 313
Collyer, Mary: 345-346, 395
Colonna, Francesco (*Hypnerotomachia Poliphili*): 72-73, 77, 127, 305-308, 312, 336, 395
Cook, R.M: 65-66, 69, 87-88, 395
Cooper, Frederick: 93, 104, 187, 395
Copernicus, Nicolaus: 31
Corbusier: 114., 129-131, 285-286, 365-366, 393, 395, 397
Corpus Vasorum Antiquorum: 62
Cortesi, Paolo: 339
Cosimo, Piero di: 296
Cotton, Charles: 288, 293, 342-343, 395
Coulton, G. G: 295, 395
Coulton, J. J: 373, 395
Coutances, cathedral of: 115
Crick, Francis: 287, 396
Crisp, Frank: 127, 282, 396
Crosby, Sumner: 114, 396

D

Dallington, R: 313, 395-396
Dante: 136, 282-283, 292, 294, 407
Darwin, Charles: 26-29, 44, 48, 396-397
Defoe, Daniel: 298, 343, 396
Delacroix, Eugène: 287, 401
Descartes, René: 29-30, 33, 35-6, 60, 394, 396
Desgodetz, Antoine: 76, 373, 396
Dietterlin, Wendel: 283, 321, 396
Dinsmoor, W. B: 85, 88, 92-93, 103, 396
Diocletian, baths of: 312
Doi, Toshitsura: 34
Donne, John: 296
Doumas, Christos: 67, 396
Downing, A. J: 254, 328, 396
Dura Europos: 81, 184, 400
Dürer, Albrecht: 301, 338

E

Eco, Umberto: 327, 397
Eden, Garden of: 128, 292, 346-347, 407
d'Este, Villa: 243, 291
Etlin, Richard: 88, 105, 328, 361, 397
Euclid: 23, 35, 60, 312, 334, 396
Evelyn, John: 291, 298, 314, 330, 338, 347, 362, 376, 397-398
Exekias: 65, 79, 161

F

Farnese, Palazzo: 242, 282, 342,
Fergusson, James: 90-91, 280, 397
Ferrara, cathedral of: 118, 203
Festinger, Leon: 79, 397
Filarete: 283-284, 301, 331, 350, 370, 386, 397, 408
Florence, cathedral of – Santa Maria del Fiore: 120, 208,380, 386,
    398, 405
Fontenelle, Bernard Le B: 25, 48
Fowler, Henry - *Dictionary of Modern English Usage*: 11
Fréart, Roland: 284-285, 314, 317, 330, 337-338, 376, 397-398
Friedlaender, Max: 92, 341, 396
Frisch, Karl: 28, 398
Furukawa, Yoshimoro: 43, 398

G

Gadol, Jane: 19, 332, 369,371, 398

Galen: 59
Gandy, Joseph: 51, 316, 398
Gardner, Helen: 55, 398
Ginzburg, M: 19
Giotto: 119, 128-129, 143, 222, 295, 304, 395
Goethe, J. W: 1, 280, 322-323, 398
Goldthwaite, Richard: 126, 280-281, 399
Gombrich, E. H: 260, 262, 276-277, 365, 398-399
Goodyear, William: 12, 86, 102, 106, 112, 124, 399
Gournia [Crete], palace of: 102, 409
Grew, Nathaniel: 33, 399
Guarico, Pomponio: 303

H
Hargittai, Istvan and Magdolna: 39-40, 399
Hauser, Arnold: 31, 39
Hephaisteion (Athens): 104
Heraion (Argive): 102, 404, 406
Heraion (Olympia): 103-4, 408
Heydenreich, L. H: 331, 380, 400
Hobbes, Thomas: 343, 363, 400
Hogarth, William: 52, 287, 302, 319, 359, 400
Hollando, Francesco: 283, 321
Hon and Goldstein: 60, 301-2, 309, 400
Honour, Hugh: 318, 400
Hooke, Robert: 32-33, 35, 43, 400
Hopkins, Clark: 81, 400
Huizinga, J: 295, 400
Humphreys, A.R: 344, 400
Huxley, T.H: 33, 36, 320, 400
Hypnerotomachia Poliphili: see under "*Colonna, Francesco*"

I
Iacopi, Irene: 56, 74, 401

J
Jaeger, F. M: 45-46, 401
James, P. D: 315
Johnson, Samuel: 315, 353, 14, 351, 401
Jupiter Capitolinus, temple of (Rome): 58

K
Kahn, Charles: 61, 401
Kames, Lord: 19, 50, 285, 401

Kent, William: 355, 357, 360, 401
Kepler, Johannes: 23-24, 27-32, 35, 37, 39, 41-45, 48, 61, 401
Ker, J.B: 316, 401
Kiesow, Gottfried: 376, 387, 401
Kirk, Raven and Schofield: 61, 401
Klarreich, Erica: 25, 401
Koch, H: 104, 401
Koenig, Samuel: 25
Kruft, H-w: 19, 294, 331, 492

L

LaChapelle, Edward: 40, 402
Landauro, Inti: 55, 402
Landino, Cristoforo: 303, 369
Le Clerc, Sebastian: 307 ,402
Lederman, Leon: *titlepage*, 19, 41, 48, 402
Leibnitz, Gottfried: 25
Le Muet, Pierre: 254, 316-317
Leonardo da Vinci: 14, 16, 285, 294, 301, 33, 338-339, 401, 411
Libbrecht, Ken: 37-38, 41-42, 44, 156, 402
Licht, Kjeld: 83, 402
Lorand, Ruth: 52, 402
Lovejoy, Arthur: 112, 318, 347, 359, 403

M

Mack, Charles: 304-305, 370, 403
Magono and Lee: 43, 403
Manetti, Antonio: 303, 307-308, 403
Mann, Thomas: 38, 479
Mantegna, Andrea: 82-83, 334-335, 48, 411
Maraldi, Giacomo: 24-26, 28
Marsh, Ngaio: 18
Marvell, Andrew: 288-9, 403, 408
Mason, B. J: 315, 403
Mattusch, Carol: 68, 404
McBeath, Michael: 19
McManus, I. C: 16, 68, 129, 404
Michelangelo: 282-283, 302, 321, 392, 339-340, 404, 410
Milton, John: 52, 288-289, 320, 330, 404-405
Montesquieu, C.-L. de S: 55, 288-289, 320, 330, 404
Morris, William: 299, 404

N

Nakaya, Ukichuro: 34, 38-44, 405

Naredi-Rainer, Paul: 384-385, 405
Nemi, Lake of: 334
Newton, Isaac: 25
Norwich, cathedral of: 116, 121, 194, 400
Notre Dame (Paris): 113-114, 125, 129-131, 188, 329, 410

O

Olaus Magnus: 21-23, 34-35, 38, 40
Ostwald, Michael: 371, 383, 405
Otranto, cathedral of: 118, 202

P

Padua, baptistery and cathedral: 118, 200-201
Palladio, Andrea: 18, 83, 119, 125, 241, 282, 286, 301-302, 337,
    353, 371, 377, 392, 394, 405, 413, 407,
Panofsky, Erwin: 282, 303, 338, 374, 379, 406
Pantheon (Rome): 73, 82-84, 87, 175, 306, 329, 350-351, 394, 402
Paolo, Giovanni di: 293, 300
Pappus of Alexandria: 23-24, 27-28
Parthenon (Athens): 48, 53, 84-5, 87, 89-108, 136, 326, 329, 365,
    395, 410, 413
Pascal, Blaise: 302, 304, 320
Payne, Alina: 301, 371, 406
Peak District: 342, 393, 395
Penrose, F. C: 84, 89, 92, 96-97, 105-106, 406
Pepys, Samuel: 298
Perrault, Claude: 83-84, 317, 328, 373, 406
Petrarch: 292, 294-295, 412
Pevsner, Nikolaus: 55, 318, 341, 361, 402, 406
Pfaff, Christopher: 112, 406
Photogrammetry: 93, 103, 380-381, 385-386, 395
Pico della Mirandola: 287, 320
Pius II, Pope: 52, 89, 251, 297, 303-305, 308, 310, 3122, 328, 334,
    336, 400, 406, 412
Plague (see also *Black Death*): 129, 249-250, 292-300, 342-343,
    358, 362-363, 393-395, 398, 404-405, 409, 412
Plato: 9, 31, 35, 41, 48, 56-58, 60, 127, 393, 400, 406
Pliny: 55, 58, 345, 362, 406
Plutarch: 58-59, 400
Pollaiuolo, Antonio: 248, 293
Polykleitos: 58
Pope, Alexander: 255, 344, 346, 352-3357, 403, 409
Prague, street lights: 29
Pratt, Roger: 302, 314, 349

Prest, John: 347, 363, 407
Puffer, E. D: 44, 268,407

R
Ragghianti, Licia: 5, 79-80, 407
Ramberg, Walter: 318, 360, 407
Réaumur, R.A.F: 25
Reinberger, Stefanie: 20, 407
Repton, Humphrey: 340
Reynolds, Joshua: 283, 341, 350, 360, 407
Richter, Gisela A. M: 64, 68, 87-88
Rilke, Rainer Maria: 359
Rosen, Joe: 41, 408
Rossini, Orietta: 75, 408
Rouen, cathedral: 114, 190, 380, 405
Ruskin, John: 12, 111, 124, 131-145, 282-283, 315, 363, 406, 408
Rykwert, Joseph: 36, 332, 338, 385, 388-389, 392, 403, 408

S
Saarinen, Eero & Eliel: 7, 15, 326, 408
St. Denis abbey church: 114, 191, 396
St. Lo cathedral: *frontispiece,* 14
 St. Maclou church, Rouen: 380, 405
Salute church, Venice: 298
St. Peter's basilica, Vatican: 50, 119, 129, 351
S. Clemente church, Rome: 122, 212, 402
S. Lorenzo church, Florence: 380, 388
S. Marco basilica and *piazza*, Venice: 124, 204, 363-364, 408
S. Miniato al Monte church, Florence: 119, 379, 388
S. Petronio church, Bologna: 120
S. Spirito church, Florence: 380
S. Spirito church, Siena: 298
S. Zeno church, Verona: 122, 213-214
Sangallo, Giuliano da: 82, 331,394, 398
Santa Croce church (Florence): 120-122, 206-207, 339, 388, 394
Santa Maria del Fiore *see under Florence Cathedral*
Santa Maria del Pieve church, Arezzo: 215-216
Santa Maria Maggiore church, Florence: 388
Santa Maria Novella church, Florence: 12, 14-15, 61, 119, 258,
      368-390, 393, 399, 401
Sargon's palace: 81, 85
Sautoy, Marcus: 29, 61, 408
Schlosser, Julius: 306, 339, 408
Schneider, Lambert: 87-89, 408

Scholfield, P. H: 371-372, 408
Scoresby, William: 33-36, 48, 320, 403, 408
Scranton, Robert: 103, 373, 408
Selinus, Temple C: 79, 85, 103
Semper, Gottfried: 36, 323-324, 408
Serlio, Sebastiano: 59, 81-82, 125, 252-253, 249, 307-313, 317,
    327, 333, 337, 339
Shaftesbury, third earl of: 357, 357, 409
Shakespeare, William: 341, 397
Sharawaggi or *Sharawadgi*: 318, 349, 395, 402, 405
Shute, John: 312-313, 409
Siena: 119, 121-122, 129,209-211, 295-296, 298-299, 329,398, 415
Signoria, *Piazza & Palazzo della* (Florence): 88, 125-126, 329, 363
Simson, Otto: 113, 126, 286, 379, 409
Sitte, Camillo: 56, 59, 280, 321, 323, 364-365, 409
Spenser, Edmund: 341, 344, 410
Stewart, Dugald: 361, 410
Stewart, Ian: 36-37, 410
Strong, Roy: 291, 410
Stuart & Revett: 92
Sturgis, Russell: 55, 410

T
Taj Mahal: 15, 392
Tavernor, Robert: 59, 302,332,371,385-386, 388-389, 392, 405,
    410
Temple, William: 343, 347, 402, 409, 413
Thiersch, August: 370-371, 379, 410, 413
Thompson, D'Arcy: 25-26, 28, 410
Tobin, Richard: 105, 411
Trachtenberg, Marvin: 88, 411
Tuchman, Barbara: 297, 411
Tunnard, Christopher: 319, 411

U
Uccello, Paolo: 245-246, 293

V
Vasari, Georgio: 126, 282-283, 302, 313, 336-337, 339, 342, 369-
    370, 380, 382, 394, 411
Vernon, Francis: 91
Vickers, Michael: 66, 411
Villani, Giovanni: 128, 294
Viollet-le-Duc, E-E: 55, 91, 280, 361, 406, 411

Vitruvian Man (see under "Leonardo")
Vitruvius: 9, 59-60, 128, 143-144, 301-303, 312, 326-327, 337, 339, 372-374, 376, 378-379, 408, 411

W

Walpole, Horace: 282, 290, 318-319, 348-349, 353, 355-357, 360, 402, 411
Weyl, Hermann: 11, 13, 28, 38, 41, 48, 61, 80-81, 113, 155, 328-329, 393, 412
White, John:12, 15, 342, 412
Wilczek, Frank: 10, 29, 31, 412
Wilson Jones, Mark: 70, 81-83, 86-88, 90, 392-393, 412
Winckelmann, J. J: 87-89
Wittkower, Rudolf: 61, 120, 127, 284, 32, 305, 327-328, 331-332, 368 – 390, 412
Wölfflin, Heinrich: 258, 284,302, 323-6, 330-331, 364, 370-371, 383, 399, 410, 413
Woodford, Susan: 67, 413
Wotton, Henry: 284-285, 313, 334-337, 344-345, 372, 413
Wren, Christopher: 282
Wyman, Jeffries: 20, 24, 26-28, 44, 46, 153, 397, 413

Y

Yeroulanou, Marina: 95, 413